For other titles published in this series, go to
http://www.springer.com/series/4318

Davide L. Ferrario · Renzo A. Piccinini

# Simplicial Structures
# in Topology

Dr. Davide L. Ferrario
Università di Milano-Bicocca
Dipartimento di Matematica e Applicazioni
Via R. Cozzi 53
20125 Milano
Italy
davide.ferrario@unimib.it

Dr. Renzo A. Piccinini
Department of Mathematics and Statistics
Dalhousie University
Halifax, Nova Scotia, B3H 3J5
Canada
renzo@mathstat.dal.ca

*Editors-in-Chief*
*Rédacteurs-en-chef*
K. Dilcher
K. Taylor
Department of Mathematics and Statistics
Dalhousie University
Halifax, Nova Scotia, B3H 3J5
Canada
cbs-editors@cms.math.ca

ISSN 1613-5237
ISBN 978-1-4614-2698-1          ISBN 978-1-4419-7236-1 (eBook)
DOI 10.1007/978-1-4419-7236-1
Springer New York Dordrecht Heidelberg London

Springer is part of Springer Science+Business Media (www.springer.com)

*To Albrecht Dold,*
*friend and mentor.*

# Foreword to the English Edition

Except for a few added comments, this is a faithful translation of the book *Strutture simpliciali in topologia*, published by Pitagora Editrice, Bologna 2009, as part of the collection *Quaderni* of the Italian Mathematical Union.

It should be noted that this book is neither a comprehensive text in algebraic topology nor is it a monograph on simplicial objects in the modern sense. Its focus is instead on the role of finite simplicial structures, and the algebraic topology deriving from them.

We wish to thank Maria Nair Piccinini, who carefully and expertly translated the Italian monograph. We should also thank Karl Dilcher for suggesting and working toward its inclusion in the monograph series of the Canadian Mathematical Society. Finally, we thank Marcia Bunda, Assitant Editor, and Vaishali Damle, Senior Editor of Springer for managing all the technical and bureaucratic aspects of the publication.

# Preface

On a dit, écrivais je (ou à peu près)
dans une préface, que la géometrie est
l'art de bien raisonner sur des figures
mal faites.

HENRI POINCARÉ [28]

In 1954, One hundred years after Henri Poincaré's birth, there was a special session during the International Congress of Mathematics, in the Netherlands, in honor of this great mathematician. The Russian mathematician, Pavel S. Aleksandrov, chosen to bring this about, started his speech by saying: *«To the question of what is Poincaré's relationship to topology, one can reply in a single sentence: he created it; but it is also possible to reply with a course of lectures in which Poincaré's fundamental topological results would be discussed in greater or lesser detail»* (see [3]). This is in part what we set out to do in this book.

Topology is a branch of mathematics that deals with the study of the qualitative properties of figures. Johann Benedikt Listing was among the first mathematicians who dedicated themselves to studying geometry in this sense and in 1847, he published a paper [23] in which he coined the term Topology. Bernhard Riemann's contribution to the birth of topology was also remarkable; after Riemann, Enrico Betti [5] studied manifolds through an invariant that generalizes Euler's for convex polyhedra. Betti's work provided Poincaré with the basis for his work in topology (called *Analysis Sitûs* by Poincaré).

Poincaré realized the importance of his new theory and wrote some twelve papers on Analysis Sitûs; indeed, here is what he stated in the paper he wrote at the request of the Swedish mathematician Gösta Mittag-Leffler [28]: *«Quant à moi, toutes les voies diverses où je m'étais engagé successivement me conduisaient à l'Analysis Sitûs. J'avais besoin des données de cette science pour poursuivre mes études sur les courbes définies par les équations différentielles et pour les étendre aux équations différentielles d'ordre supérieur et en particulier à celles du problème des trois corps. J'en avais besoin pour l'étude des périodes des intégrales multiples et pour l'application de cette étude au développement de la fonction pérturbatrice. Enfin j'entrevoyais dans l'Analysis Sitûs un moyen d'aborder un problème importante de la théorie des groupes, la recherche des groupes discrets ou des groupes finis contenus dans un groupe continu donné.»*

In his first paper on Analysis Sitûs, published in 1895 [27], Poincaré defined manifolds in spaces with dimension greater than three and introduced the basic concept of *homeomorphism* defined as *the relation between two manifolds with the same qualitative properties* (see [28]). From a more up-to-date point of view, we may say that topology is the branch of mathematics concerning spaces with a certain structure (topological spaces), which are invariant under homeomorphisms; in other words, functions that are injective, surjective, and bicontinuous. The concept of homeomorphism allows us to group topological spaces into equivalence classes and consequently, to know in practical terms whether two of them are equal. This is the main purpose of topology.

This book consists of six chapters. In the first one, we present the basic concepts in Topology, Group Actions and Category Theory needed for developing the remainder of the book. The part concerning Category Theory is especially important since this book has been set in categorial terminology. This chapter could also serve as a basic text for a mini-course on General Topology.

In the second chapter, we study the category of simplicial complexes **Csim**, and two important covariant functors: the *geometric realization functor* from **Csim** to the category of topological spaces, and the *homology functor* from **Csim** to the category of graded Abelian groups. The geometric realization $-|K|-$ of a simplicial complex K is a polyhedron, assumed to be compact throughout this book. This is the chapter closest to Poincaré's initial paper. The Swiss mathematician Leonhard Euler was among the first ones to study one- and two-dimensional simplicial complexes; indeed, he used one-dimensional simplicial complexes and their geometric realization (graphs) in the famous problem about the seven bridges of Königsberg; later on, he also used two-dimensional simplicial complexes when he noticed that the relation

$$v - e + f = 2,$$

– where $v$ is the number of vertices, $e$ is the number of edges, and $f$ is the number of faces – holds for every convex polyhedron in the elementary sense (namely, every edge is common to two faces and every face leaves the entire polyhedron to one of its sides).

By defining the so-called *Betti Numbers* for polyhedra of any dimension, Enrico Betti gave an initial generalization to the relation above; these are invariant under homeomorphims and, therefore, useful for classifying polyhedra. Based on these numbers, Poincaré developed a more complete characterization of polyhedra; in fact, in his 1895 [27] paper, Poincaré linked the Betti numbers to certain finitely generated Abelian groups associated with a polyhedron (integral homology groups of the polyhedron) and pointed out that the Betti numbers are ranks of homology groups. However, since homology groups are finitely generated Abelian groups, besides its free part (the one which gives its rank) they also have a torsion part; this too is an invariant by homeomorphisms. The combination of Betti numbers and torsion coefficients allows for a more complete analysis of polyhedra.

In Chap. II, homology groups are considered strictly from the simplicial point of view; in other words, the geometric structure of polyhedra is overlooked. In order to develop the theory (which is, at this point, of algebraic nature), one needs to define concepts in Homological Algebra (a branch of Algebra that sprang partly from Algebraic Topology): among other things, we prove the important Long Exact

Sequence Theorem in Sect. II.3, and in Sect. II.5 we define homology groups with coefficients in an arbritary group. The Long Exact Sequence Theorem appears in simplicial homology in the case of a pair $(K,L)$ of (oriented) simplicial complexes where $L$ is a subcomplex of $K$; in fact, we may define an exact sequence of Abelian groups

$$\ldots \to H_n(L) \xrightarrow{H_n(i)} H_n(K) \xrightarrow{q_*(n)} H_n(K,L) \xrightarrow{\lambda_n} H_{n-1}(L) \to \ldots$$

where $H_n(K,L)$ are the relative homology groups.

The homology of polyhedra is defined in Chap. III which is, therefore, more geometric in nature than the previous one. The Simplicial Approximation Theorem (first proved by the American mathematician James Wadell Alexander [1]) provides the means by which one can pass from a simplicial approach to a geometric one. In practice, this theorem states that every map between two polyhedra may be "approached" by the geometric realization of a simplicial function between two triangulations of the polyhedra (reminding that any continuous curve on a plane can be approached by a polygonal line). The Long Exact Sequence Theorem appears also in this chapter, but here, the relative homology groups have a clearer meaning, since pairs of polyhedra $(|K|,|L|)$ have the Homotopy Extension Property, which allows us to prove that the group $H_n(|K|,|L|)$ is the homology group of the "quotient" polyhedron $|K|/|L|$. Samuel Eilenberg and Norman Steenrod [13] provided the formal method needed for constructing the long exact sequence of a pair of simplicial complexes or of polyhedra.

This chapter also contains an important application of the homology of polyhedra to the Theory of Fixed Points, namely, the Lefschetz Fixed Point Theorem, as well as several corollaries such as the Brouwer Fixed Point Theorem and the Fundamental Theorem of Algebra.

Computing homology groups of a polyhedron may pose serious difficulties, as in the case of the projective real spaces $\mathbb{R}P^n$. This is why we define *block homology*, based on ideas found in two classical books: Seifert–Threlfall [30] and Hilton–Wylie [17]. In closing this chapter, we dedicated Sect. III.6 to the proof of Eilenberg–MacLane's Acyclic Models Theorem [12], which allows us to compute the homology groups of the product of two polyhedra in terms of the homology groups of its factors. Somehow, this problem was already solved by H. Künneth [21] in 1924, when he established a relation between the Betti numbers and torsion coefficients of the product with the ones of each factor; the main difficulty resided precisely in strengthening Künneth's result by describing it in terms of homology groups. The paper [12] is one of the first written in terms of category theory, created in 1945 by Samuel Eilenberg and Saunders MacLane [11].

In Chap. IV we study cohomology: the homology groups

$$H_n(K;\mathbb{Q})$$

of an oriented complex $K$ with rational coefficients have the structure of vector spaces on the rational field and may, therefore, be dualized. The possibility of dualizing such vector spaces led several mathematicians to consider "dualizing" homology groups, also when the coefficients are in an arbitrary Abelian group $G$. One of the first to study this possibility was James W. Alexander [2]; Solomon Lefschetz [22] wrote a detailed account of this paper. The new homology theory was soon called *cohomology* (it seems that it was Hassler Whitney [35] who coined this new term). As we might expect, the cohomology groups $H^n(|K|;\mathbb{Z})$ of a polyhedron $|K|$ are contravariant functors. The cohomology groups of a polyhedron are related to its homology groups by the Universal Coefficient Theorem; its proof (in terms of homologycal algebra) is given in this chapter. The cohomology of a polyhedron is an invariant stronger than the homology, since the cohomology with coefficients in a commutative ring with identity element (for instance, the ring of integers $\mathbb{Z}$) is also a ring. The product in such a ring is called *cup product*. In this way, we may obtain more precise information on the nature of the polyhedron. This chapter also introduces the *cap product* which is a bilinear relation of the type

$$\cap: H^p(|K|;\mathbb{Z}) \times H_{p+q}(|K|;\mathbb{Z}) \longrightarrow H_q(|K|;\mathbb{Z}) ;$$

the cap product will be used in Chap. V for proving Poincaré's Duality Theorem.

Chapter V is divided into three sections: Manifolds, Closed Surfaces, and Poincaré Duality. In the first one, we introduce $n$-dimensional manifolds (without boundary) and triangulable $n$-manifolds. Then, we study closed surfaces, namely, path-connected, compact 2-manifolds: by a theorem due to Tibor Radó [29], these surfaces are triangulable. We prove that these manifolds can be classified into three types: the sphere $S^2$, the connected sums of two-dimensional tori (the torus $T^2$ and spheres with $g$ handles), and the connected sums of real projective planes. Subsequently, by using block homology with coefficients in $\mathbb{Z}$, we prove that these three kinds of spaces are not homeomorphic. Finally, we prove Poincaré's Duality Theorem for connected, triangulable, and orientable $n$-manifolds $V$, that is to say, for triangulable $n$-manifods $V$ such that $H_n(V;\mathbb{Z}) \cong \mathbb{Z}$. This very important theorem states that for every $0 \leq p \leq n$, $H^{n-p}(V;\mathbb{Z}) \cong H_p(V;\mathbb{Z})$ holds true.

In the last chapter (Chap. VI), we introduce another very important functor, from the category of polyhedra to that of groups (not necessarily Abelian), namely, *the fundamental group*, defined by Poincaré (see [27]). Next, we study a family of functors from the category of polyhedra to that of Abelian groups; we are talking about the *(higher) homotopy groups* $\pi_n(|K|,x_0)$ with $n \geq 2$. Only after Heinz Hopf wrote his 1931 paper [18], did mathematicians show interest in higher homotopy groups. In this paper, Hopf proved the existence of infinitely many different homotopy classes of maps from $S^3$ to $S^2$ (Satz 1); indeed, the isomorphism $\pi_3(S^2,\mathbf{e}_0) \cong \mathbb{Z}$ (see [26]) is deduced from the exact sequence of homotopy groups associated to the fibration $S^3 \to S^2$ with fiber $S^1$.

It is interesting to note that higher homotopy groups had already been introduced by Eduard Čech [6] during the Zürich International Mathematics Congress, in 1932; after that, Witold Hurewicz [19] studied these groups in depth. We approach homotopy groups by considering the set $[S^n,|K|]_*$ of all based homotopy classes of all maps $S^n \to |K|$ of a polyhedron $|K|$ and providing this set with a group operation, by means of a natural comultiplication of $S^n$ (this is a map from $S^n$ to the wedge product $S^n \vee S^n$). In this way, we bring about two endofunctors $\Sigma$ and $\Omega$ of the category **Top**$_*$ of based spaces, namely, the *suspension functor* and the *loop space functor*; this idea is based on the work of Beno Eckmann and Peter Hilton ([9], for instance). We close this chapter (and the book) with a small section on Obstruction Theory which combines cohomology and homotopy groups; Samuel Eilenberg [10] created this theory when researching the possibility of extending maps from a subpolyhedron of a polyhedron to the polyhedron itself.

Sections II.3, II.5, III.6, and IV.1 could be the basis for a course on Homological Algebra, a theory developed from the study of the homology of polyhedra; for further reading on this subject, we suggest either the book by Karl Gruenberg [15] or the book by Peter J. Hilton and Urs Stammbach [16].

In its first century of existence, Algebraic Topology has made remarkable progress, giving rise to new theories in mathematics, forging its way into various other mathematical branches, and solving seemingly unrelated yet important problems. The reader could, therefore, wish to go farther into this subject and so we give here some suggestions for further reading: we recommend the books by Peter Hilton and Shaun Wylie [17] (the reader may, at first, have some difficulty with its notation), Albrecht Dold [7] (a classic), and Edwin Spanier [32].

For the readers who wish to learn Homotopy Theory in more depth, we suggest the book by George Whitehead [33]. Finally, for reading on topological spaces and cellular structures in topology (CW-complexes), we recommend the book by Rudolf Fritsch and Renzo Piccinini [14].

The chapters of this book were developed from the material taught in the undergraduate courses on higher geometry given by the authors at the University of Milano and the University of Milano–Bicocca. A first version was prepared by R. Piccinini some years ago, when he was a professor at the University of Milano–Bicocca. This volume is essentially the revision and completion of that material. Our many thanks to several colleagues and friends who have read the rough copy and made worthwhile suggestions: Keith Johnson, Sandro Levi, Augusto Minatta, Claudio Pacati, Robert Paré, Petar Pavešić, Nair Piccinini, Dorette Pronk, Alessandro Russo, Mauro Spreafico, and Richard Wood. Our colleague Delfina Roux must be thanked for having read with great attention the first draft of this volume, correcting the errors of language in the first Italian draft.

We conclude with our sincere thanks to the referees for their excellent work and remarkable patience in pointing out the many flaws in the manuscript; their suggestions have greatly improved the final text.

Milano 2008                                                     Davide L. Ferrario, Renzo A. Piccinini

Knowledge is of two kinds. We know a subject ourselves,
or we know where we can find information upon it.

SAMUEL JOHNSON
*(Letter to Lord Chesterton, February 1755)*

# Contents

# Chapter I
# Fundamental Concepts

## I.1 Topology

### I.1.1 Topological Spaces

Let $X$ be a given set. A *topology* on $X$ is a set $\mathfrak{U}$ of subsets of $X$ satisfying the following properties:

**A1** $\emptyset, X \in \mathfrak{U}$;

**A2** if $\{U_\alpha \mid \alpha \in J\}$ is a set of elements of $\mathfrak{U}$, then

$$\bigcup_{\alpha \in J} U_\alpha \in \mathfrak{U};$$

**A3** if $\{U_\alpha \mid \alpha = 1, \ldots, n\}$ is a finite set of elements of $\mathfrak{U}$, then

$$\bigcap_{\alpha=1}^{n} U_\alpha \in \mathfrak{U}.$$

A *topological space* or, simply, a *space* is a set $X$ with a topology. The elements of $\mathfrak{U}$ are the *open sets* of $X$; axioms A1, A2, and A3 above state that $\emptyset, X$ are open sets, that the union of any number of open sets is open, and that the intersection of any finite number of open sets is open.

The complement of an open set is a *closed set* and, therefore, $\emptyset$ and $X$ are both open and closed sets. A topology $\mathfrak{U}$ may also be studied (and characterized) through the set of all closed subsets of the topological space. Let $\mathfrak{C}$ be such a set; by definition, a set $C$ is closed in $X$ if and only if its complement $X \smallsetminus C$ is open in $X$, and we write

$$C \in \mathfrak{C} \iff X \smallsetminus C \in \mathfrak{U}.$$

Moreover, the set $\mathfrak{C}$ of all closed subsets satisfies the following properties which characterize any set of closed sets of $X$:

D.L. Ferrario and R.A. Piccinini, *Simplicial Structures in Topology*,
CMS Books in Mathematics, DOI 10.1007/978-1-4419-7236-1_I,
© Springer Science+Business Media, LLC 2011

**C1** $\emptyset, X \in \mathfrak{C}$;

**C2** if $\{U_\alpha \mid \alpha = 1, \ldots, n\}$ is any finite set of elements of $\mathfrak{C}$, then

$$\bigcup_{\alpha=1}^{n} U_\alpha \in \mathfrak{C}$$

(in other words, any finite union of closed sets is a closed set);

**C3** if $\{U_\alpha \mid \alpha \in J\}$ is any set of elements of $\mathfrak{C}$, then

$$\bigcap_{\alpha \in J} U_\alpha \in \mathfrak{C}$$

(in other words, any intersection of closed sets is a closed set).

For any space $X$, the *closure* $\overline{Y}$ of a subset $Y \subset X$ is the intersection of all closed subsets of $X$ which contain $Y$; it is, obviously, the smallest closed subset containing $Y$, and $\overline{Y} \in \mathfrak{C}$. The *interior* $\mathring{Y}$ of $Y$ is the union of all open subsets of $X$ contained in $Y$; it is the largest open set contained in $Y$, and $\mathring{Y} \in \mathfrak{U}$. The following lemma is a useful characterization of the closure of a set.

**(I.1.1) Lemma.** *Let* $Y \subset X$*; then* $y \in \overline{Y}$ *if and only if every open set* $U$ *containing* $y$ *intersects* $Y$.

*Proof.* In fact, if $U \cap Y = \emptyset$, then $X \smallsetminus U$ is a closed subset of $X$ which contains $Y$; since $y$ belongs to the closure of $Y$, it follows that $y \in X \smallsetminus U$, a contradiction.

Conversely, suppose that there exists a closed set $C \subset X$ containing $Y$ and which does not contain $y$. Then $X \smallsetminus C$ is an open set, $y \in X \smallsetminus C$, and $(X \smallsetminus C) \cap Y = \emptyset$, a contradiction. ∎

For any set $X$, there are two topologies that come immediately to mind:

1. The *discrete topology* whose open sets are all the subsets of $X$, that is to say, $\mathfrak{U}^d = \mathfrak{P}(X) = 2^X$.
2. The *trivial topology* with $\mathfrak{U}^b = \{\emptyset, X\}$.

Clearly, $\mathfrak{U}^d \supset \mathfrak{U}^b$ and this fact brings to mind the idea of comparing two topologies on the same set: we shall say that topology $\mathfrak{U}'$ is *finer* than topology $\mathfrak{U}$ if $\mathfrak{U}' \supset \mathfrak{U}$; hence, the discrete topology of $X$ is finer than the trivial one.

A set $\mathfrak{B}$ of subsets of $X$ is a *basis* for a topology on $X$ or a *basis of open sets* for $X$ if:

**B1** For each $x \in X$, there is a $B \in \mathfrak{B}$ such that $x \in B$ (that is to say, $X = \bigcup_{B \in \mathfrak{B}} B$).

**B2** If $B_1, B_2 \in \mathfrak{B}$ and $B_1 \cap B_2 \neq \emptyset$, then for each $x \in B_1 \cap B_2$ there exists $B_3 \in \mathfrak{B}$ such that $x \in B_3 \subset B_1 \cap B_2$.

A basis $\mathfrak{B}$ generates a topology $\mathfrak{U}$ on $X$ automatically, by requiring that $U \subset X$ be open in the topology $\mathfrak{U}$ if, and only if, for each $x \in U$, there exists $B \in \mathfrak{U}$ such that $x \in B \subset U$; formally,

$$U \in \mathfrak{U} \iff (\forall x \in U)(\exists B \in \mathfrak{B}) \, x \in B \subset U.$$

We need to show that, with this definition, the set $\mathfrak{U}$ of open sets satisfies properties A1, A2, and A3 for open sets. Property A1 is easily verified. Let us prove now that A2 is true: let $\{U_\alpha \mid \alpha \in J\}$ be a set of elements of $\mathfrak{U}$; we want to show that $U = \bigcup_{\alpha \in J} U_\alpha \in \mathfrak{U}$. In fact, for each $x \in U$ there is $\alpha \in J$ such that $x \in U_\alpha$; hence, there is a $B \in \mathfrak{B}$ such that $x \in B \subset U_\alpha \subset U$. We prove A3 by induction. Let us consider two elements $U_1, U_2 \in \mathfrak{U}$. If $U_1 \cap U_2 = \emptyset$, then $U_1 \cap U_2 \in \mathfrak{U}$. Otherwise, for each $x \in U_1 \cap U_2$, we may find two sets $B_1, B_2 \in \mathfrak{B}$ such that $x \in B_1 \subset U_1$ and $x \in B_2 \subset U_2$; since $\mathfrak{B}$ is a basis, there exists an element $B_3 \in \mathfrak{B}$ such that $x \in B_3 \subset B_1 \cap B_2$. We conclude that $x \in B_3 \subset U_1 \cap U_2$ and so, $U_1 \cap U_2 \in \mathfrak{U}$. Suppose now that A3 holds for any intersection of $n-1$ elements of $\mathfrak{U}$. Let $U_1, U_2, \ldots, U_n$ be elements of $\mathfrak{U}$; we write

$$\bigcap_{i=1}^{n} U_i = \left( \bigcap_{i=1}^{n-1} U_i \right) \cap U_n$$

and note that $\bigcap_{i=1}^{n-1} U_i \in \mathfrak{U}$ by the induction hypothesis; hence, the intersection of all the given elements $U_i$ belongs to $\mathfrak{U}$. We also note that all elements of a basis are elements of the topology it generates (that is to say, all elements of a basis are automatically open).

We give now an important example of a topological space which illustrates well the concepts given so far. Let $\mathbb{R}^n$ be the *Euclidean n-dimensional space*, namely, the set of all $n$-tuples $x = (x_1, \ldots, x_n)$ of real numbers $x_i$. Let $d: \mathbb{R}^n \times \mathbb{R}^n \to \mathbb{R}^n$ be the function defined in the following manner: for each $x = (x_1, \ldots, x_n)$ and $y = (y_1, \ldots, y_n)$ in $\mathbb{R}^n$,

$$d(x,y) = \sqrt{\sum_{i=1}^{n} (x_i - y_i)^2}.$$

This function, called *Euclidean distance* (or *metric*) has the following properties:

$(\forall x, y \in \mathbb{R}^n)\ d(x,y) = d(y,x),$
$d(x,y) = 0 \iff x = y,$
$(\forall x, y, z \in \mathbb{R}^n)\ d(x,z) \leq d(x,y) + d(y,z).$

For every $x \in \mathbb{R}^n$ and every real number $\varepsilon > 0$, we define the *n-dimensional open disk* (or *open n-disk*) with centre $x$ and radius $\varepsilon$ as the set

$$\mathring{D}_{\varepsilon}^{n}(x) = \{y \in \mathbb{R}^n \mid d(x,y) < \varepsilon\}.$$

The set

$$\mathfrak{B} = \{\mathring{D}_{\varepsilon}^{n}(x) \mid x \in \mathbb{R}^n,\ \varepsilon > 0\}$$

is a basis for the Euclidean topology of $\mathbb{R}^n$. In fact, it is evident that condition B1 above holds; let us prove condition B2: let $\mathring{D}_{\varepsilon}^{n}(x)$ and $\mathring{D}_{\delta}^{n}(y)$ be two elements of $\mathfrak{B}$ with non-empty intersection; for each $z$ in this intersection, consider the real numbers $\gamma_1 = \varepsilon - d(x,z)$ and $\gamma_2 = \delta - d(y,z)$; let $\mu$ be the minimum of $\gamma_1$ and $\gamma_2$. Then, as we can see from Fig. I.1,

$$z \in \mathring{D}_{\mu}^{n}(z) \subset \mathring{D}_{\varepsilon}^{n}(x) \cap \mathring{D}_{\delta}^{n}(y).$$

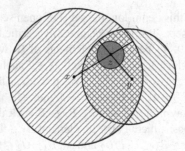

Fig. I.1

The topology $\mathfrak{U}$ generated by a basis $\mathfrak{B}$ may be described in another way:

**(I.1.2) Lemma.** *Let $X$ be a set, let $\mathfrak{B}$ be a basis of open sets of $X$, and let $\mathfrak{U}$ be the topology generated by $\mathfrak{B}$. Then, $\mathfrak{U}$ coincides with the set of all unions of elements of $\mathfrak{B}$.*

*Proof.* Given any set $\{U_\alpha \mid \alpha \in J,\ U_\alpha \in \mathfrak{B}\}$ of open sets of the basis $\mathfrak{B}$, since all elements $\mathfrak{B}$ in are open sets and $\mathfrak{U}$ is a topology, the union $\bigcup_{\alpha \in J} U_\alpha$ is also open. Conversely, given $U \in \mathfrak{U}$, for each $x \in U$, there exists $B_x \in \mathfrak{B}$ such that $x \in B_x \subset U$. Therefore, $U = \bigcup_{x \in U} B_x$, that is to say, $U$ is a union of elements of $\mathfrak{B}$. $\blacksquare$

We now look into the concept of sub-basis. A set of subsets $\mathfrak{S}$ of X is a *sub-basis* for $X$ if the union of all elements of $\mathfrak{S}$ coincides with $X$ (in other words, if property B1, given above, holds). The next result provides a good reason to work with sub-bases.

**(I.1.3) Theorem.** *Let $\mathfrak{S}$ be a sub-basis of a set $X$. Then, the set $\mathfrak{U}$ of all unions of finite intersections of elements of $\mathfrak{S}$ is a topology.*

*Proof.* It is enough to prove that the set $\mathfrak{B}$ of all finite intersections of elements of $\mathfrak{S}$ is a basis and then apply Lemma (I.1.2).

B1: For each $x \in X$ there exists $B \in \mathfrak{S}$ such that $x \in B$; however, $B \in \mathfrak{B}$, since all elements of $\mathfrak{S}$ are elements of $\mathfrak{B}$.

B2: Given $B_1 = \bigcap_{i=1}^{n} C_i^1$ and $B_2 = \bigcap_{j=1}^{m} C_j^2$ in $\mathfrak{B}$, the property holds because

$$B_1 \cap B_2 = \left( \bigcap_{i=1}^{n} C_i^1 \right) \cap \left( \bigcap_{j=1}^{m} C_j^2 \right) \in \mathfrak{B}. \qquad \blacksquare$$

Clearly, any basis is a sub-basis. For instance, the set $\{\{x\}' \mid x \in X\}$ is a sub-basis for the discrete topology on $X$ as well as a basis for that same topology. On the contrary, the family of all open intervals of $\mathbb{R}$ with length 1 is an example of a sub-basis that is *not* a basis. Property B1 holds but not property B2. The basis it generates consists of all open intervals of $\mathbb{R}$ and we have, therefore, the Euclidean topology.

Let $X$ and $Y$ be two topological spaces with respective topologies $\mathfrak{U}$ and $\mathfrak{V}$. The *product topology* of the set $X \times Y$ is generated by the basis

$$\mathfrak{B} = \{U \times V \mid U \in \mathfrak{U},\ V \in \mathfrak{U}\}.$$

Let $X$ be a topological space with topology $\mathfrak{U}$. A subset $A \subset X$ canonically inherits a topology from $\mathfrak{U}$, namely, the *induced topology* in $A$ whose open sets are the intersections of the open sets of $X$ with $A$:

$$\mathfrak{B} = \{U \cap A \mid U \in \mathfrak{U}\}.$$

**(I.1.4) Lemma.** *Let $X$ be a topological space with topology $\mathfrak{U}$, $Y$ be a set, and $q: X \to Y$ be a function. The set of subsets with anti-images open in $X$*

$$\mathfrak{V} = \{U \subset Y \mid q^{-1}(U) \in \mathfrak{U}\}$$

*is a topology on $Y$.*

*Proof.* We need to verify axioms A1, A2, and A3.
A1: $q^{-1}(\emptyset) = \emptyset \in \mathfrak{U}$, $q^{-1}(Y) = X \in \mathfrak{U}$.
A2: It holds because

$$q^{-1}\left(\bigcup_{\alpha \in J} U_\alpha\right) = \bigcup_{\alpha \in J} q^{-1}(U_\alpha).$$

A3: Likewise,

$$q^{-1}\left(\bigcap_{i=1}^{n} U_i\right) = \bigcap_{i=1}^{n} q^{-1}(U_i). \qquad \blacksquare$$

When the function $q$ is a surjection, the topology $\mathfrak{V}$ of $Y$, defined above, is called the *quotient topology* on $Y$ induced by $q$. We note that the set $Y$ could be given by a partition of $X$ in disjoint classes whose union is precisely $X$. To make this fact clear, we now give three examples.

**(I.1.5) Example.** Let $D^2 = \{(x,y) \in \mathbb{R}^2 \mid x^2 + y^2 \le 1\}$ be the two-dimensional unit disk with boundary $S^1 = \{(x,y) \in \mathbb{R}^2 \mid x^2 + y^2 = 1\}$ (one-dimensional sphere). Let $D^2_{\equiv}$ be the set whose elements are

1. $\{(x,y)\}$ for the points $(x,y) \in D^2$ such that $x^2 + y^2 < 1$
2. $\{(x,y),(-x,-y)\}$ for the boundary points $(x,y) \in S^1$

in this case, we say that the boundary points $(x,y)$ and $(-x,-y)$ are being *identified*, as in Fig. I.2. In this way, we obtain the topological space $D^2_{\equiv}$, with the quotient topology induced by the epimorphism (that is to say, by the surjective function) $q: D^2 \to D^2_{\equiv}$; this is the *real projective plane* $\mathbb{RP}^2$. We shall return to this meaningful example in Sect. I.3.

**(I.1.6) Example.** Consider $I^2 = I \times I$ ($I$ is the unit interval $[0,1]$) with the product topology. Let us now take the set $I^2_{\equiv}$ whose elements are the following sets (see also Fig. I.3):

1. $\{(x,y)\}$, if $0 < x < 1$, $0 < y < 1$
2. $\{(x,0),(x,1)\}$, if $0 < x < 1$

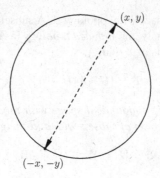

**Fig. I.2**

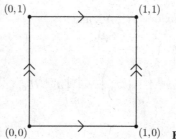

**Fig. I.3**

3. $\{(0,y),(1,y)\}$, if $0 < y < 1$
4. $\{(0,0),(1,0),(1,1),(0,1)\}$

The set $I^2_{\equiv}$ is a partition of $I^2$; let $I^2_{\equiv}$ be the space with the quotient topology given by the surjection $q\colon I^2 \to I^2_{\equiv}$ which takes every $(x,y) \in I^2$ in its own class. Later on in this section, we shall prove that $I^2_{\equiv}$ may be viewed as the *two-dimensional torus* $T^2 = S^1 \times S^1$.

**(I.1.7) Example.** Let us take the unit disk $D^2$ once more and let $D^2_{\equiv}$ be the set with the following elements:

1. $\{(x,y)\}$ for the points $(x,y) \in D^2$ such that $x^2 + y^2 < 1$
2. $\{(x,y),(x,-y)\}$, for the points $(x,y) \in D^2$ such that $x^2 + y^2 = 1$

(see also Fig. I.4). We shall prove later on that the space $D^2_{\equiv}$ with the quotient topology induced by the surjection $q\colon D^2 \to D^2_{\equiv}$ may be viewed as the two-dimensional sphere $S^2 = \{(x,y,z) \in \mathbb{R}^3 \,|\, x^2 + y^2 + z^2 = 1\}$.

The reader must have realized that we may describe the previous situation by means of an equivalence relation: let $\equiv$ be an equivalence relation in a space $X$; this relation defines a surjection

$$q\colon X \to X_{/\equiv}$$

on the set of equivalence classes (a partition of $X$) which determines the quotient topology on $X_{/\equiv}$.

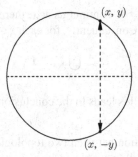

$(x, y)$

$(x, -y)$

**Fig. I.4**

Between topological spaces, it is natural to consider the functions which are continuous. Let $X$, $Y$ be topological spaces with the topologies $\mathfrak{U}$ and $\mathfrak{V}$, respectively. A function $f \colon X \to Y$ is said to be *continuous* (or *a map*) if for every open set $V$ of $Y$, $f^{-1}(V)$ is an open set of $X$; formally, we write

$$f \text{ continuous} \iff (\forall V \in \mathfrak{V}) \, f^{-1}(V) \in \mathfrak{U}.$$

It is easily proved that the constant function $f \colon X \to Y$ at a point $y_0 \in Y$ (in other words, $f(x) = y_0$ for all $x \in X$) is continuous: In fact, for each open set $V$ of $Y$, $f^{-1}(V)$ is either $X$ or $\emptyset$, which are open. It is also easy to see that the identity function $1_X \colon X \to X$ (namely, $1_X(x) = x$ for each $x \in X$) and the inclusion function $i_A \colon A \to X$ of a subspace $A$ of $X$ are continuous.

We add another example to this list: let $q \colon X \to Y$ be a surjection from a topological space $X$ on a set $Y$; let us now give $Y$ the quotient topology; by the definition of quotient topology and Lemma (I.1.4), we conclude that the function $q$ is continuous. We may also describe the concept of continuity locally; in a more precise manner, we say that $f \colon X \to Y$ is *continuous at the point* $x \in X$ if for each open set $V$ of $Y$ containing $f(x)$ (in other words, such that $f(x) \in V$) there is an open set $U$ of $X$ containing $x$ such that $f(U) \subset V$.[1]

**(I.1.8) Theorem.** *A function $f \colon X \to Y$ is continuous if and only if it is continuous at every point $x \in X$.*

*Proof.* Suppose that $f$ is continuous. Let $\mathfrak{U}$ and $\mathfrak{V}$ be the topologies of $X$ and $Y$, respectively; let $x \in X$ and let $V$ be a neighbourhood of $f(x)$. Then,

$$x \in U = f^{-1}(V) \in \mathfrak{U}, \, f(U) = f(f^{-1}(V)) \subset V$$

and, therefore, $f$ is continuous at $x$.

Conversely, suppose that $f$ is continuous at every $x \in X$, and let $V$ be any open set of $Y$; we want to prove that $U = f^{-1}(V)$ is an open set of $X$. Indeed, for every

---

[1] Open sets containing a point $x$ are called *neighbourhoods* of $x$.

$x \in U$, $f(x) \in V$ and, since $f$ is continuous at $x$, there is an open set $U_x$ of $X$ such that $x \in U_x$ and $f(U_x) \subset V$. Consequently, for each $x \in X$, $x \in U_x \subset U$ and so

$$U \subset \bigcup_{x \in X} U_x \subset U,$$

that is to say, $U = \bigcup_{x \in X} U_x$; this leads to the conclusion that $U$ is open, as a union of open sets of $X$.                                                                           ∎

The continuity of a function between two topological spaces may be characterized in two other ways, as follows.

**(I.1.9) Theorem.** *Let $X$ and $Y$ be topological spaces and $f : X \to Y$ be a function. The following conditions are equivalent:*

1. *The function $f$ is continuous.*
2. *For every subset $U \subset X$, $f(\overline{U}) \subset \overline{f(U)}$.*
3. *For every closed subset $C$ of $Y$, $f^{-1}(C)$ is closed in $X$.*

*Proof.* $1 \Rightarrow 2$. Assume that $f$ is continuous; we want to prove that $f(x) \in \overline{f(U)}$ for every $x \in \overline{U}$. Let $V$ be a neighbourhood of $f(x)$; since $f$ is continuous, $f^{-1}(V)$ is an open set of $X$ which contains $x \in \overline{U}$ and thus, $f^{-1}(V) \cap U \neq \emptyset$ (see Lemma (I.1.1)).
$2 \Rightarrow 3$. Assuming $C \subset Y$ to be closed, we want to prove that the anti-image $F = f^{-1}(C)$ is closed in $X$. Since $F \subset \overline{F}$, we have to prove that, for every $x \in \overline{F}$, $x \in F$. In fact,

$$f(x) \in f(\overline{F}) \subset \overline{f(F)} \subset \overline{C} = C$$

(the last inclusion is due to the fact that $f(F) \subset C$) and so,

$$x \in f^{-1}(C) = F.$$

$3 \Rightarrow 1$. Let $V$ be any open set of $Y$. It follows from condition 3 that $f^{-1}(Y \smallsetminus V)$ is closed in $X$. We now note that

$$f^{-1}(V) = f^{-1}(Y \smallsetminus (Y \smallsetminus V)) = f^{-1}(Y) \smallsetminus f^{-1}(Y \smallsetminus V) = X \smallsetminus f^{-1}(Y \smallsetminus V)$$

and this last set is open in $X$.                                                          ∎

**(I.1.10) Corollary.** *Given two topological spaces $X$ and $Y$, where $X = C_1 \cup C_2$ is the union of two subspaces $C_1$ and $C_2$ closed in $X$, let $f : X \to Y$ be a function whose restrictions to the closed sets $C_1$ and $C_2$ are continuous. Then, $f$ is continuous.*

*Proof.* Let us write $f_1 = f|C_1$ and $f_2 = f|C_2$. For each closed set $V \subset Y$,

$$f^{-1}(V) = f_1^{-1}(V) \cup f_2^{-1}(V).$$

By the previous theorem, $f_1^{-1}(V)$ and $f_2^{-1}(V)$ are closed in $X$; therefore, $f^{-1}(V)$ is closed in $X$ and, consequently, $f$ is continuous.                                          ∎

If the function $f : X \to Y$ is bijective (injective and surjective), then there exists an inverse function $f^{-1} : Y \to X$ such that $f^{-1}f = 1_X$ (the identity function from $X$ onto itself) and $ff^{-1} = 1_Y$; in this case, if also $f^{-1}$ is continuous, we say that $f$ is a *homeomorphism* between $X$ and $Y$. A homeomorphism is really a 1-1 correspondence between points of $X$ and $Y$ which induces an 1-1 correspondence between their respective topologies (that is to say, between the open sets of $X$ and the open sets of $Y$). In practice we make no distinction between two homeomorphic spaces. Let $q\colon X \to Y$ be a surjection from a space $X$ onto a space $Y$ with the quotient topology induced by $q$. Then the function $q$ is continuous; the function $q\colon X \to Y$ is called *quotient map*.

The next result provides a link between homeomorphisms and quotient spaces.

**(I.1.11) Lemma.** *Let $f\colon X \to Y$ be a homeomorphism and let $\equiv_X$, $\equiv_Y$ be equivalence relations in $X$ and $Y$, respectively. Then $X/\equiv_X$ and $Y/\equiv_Y$ are homeomorphic, provided that*

$$x \equiv_X x' \iff f(x) \equiv_Y f(x').$$

*Proof.* Consider the following commutative diagram (that is to say, such that $F q_X = q_Y f$)

$$
\begin{array}{ccc}
X & \xrightarrow{\ f\ } & Y \\
\left\downarrow{\scriptstyle q_X}\vphantom{\int}\right. & & \left\downarrow{\scriptstyle q_Y}\vphantom{\int}\right. \\
X/_{\equiv_X} & \cdots\!\!\xrightarrow{\ F\ }\!\!\cdots & Y/_{\equiv_Y}
\end{array}
$$

where $F$ is defined as follows: for each $[x] \in X/_{\equiv_X}$, $F([x]) := [f(x)]$. $F$ is a function: if $x \equiv_X x'$, then $f(x) \equiv_Y f(x')$ and therefore the entire class $[x]$ is transformed univocally into class $[f(x)]$. Since the composite function $q_Y f$ is continuous, so is $F q_X$; but the space $X/\equiv_X$ has the quotient topology and therefore (see *and do* Exercise 7 on p. 27) $F$ is continuous.

At this point, let us consider the inverse function $f^{-1}\colon Y \to X$ and, as in the case of $f$, let us construct the function

$$F'\colon Y/_{\equiv_Y} \to X/_{\equiv_X},\ F'([y]) := [f^{-1}(y)]$$

for $[y] \in Y/\equiv_Y$. Also $F'$ is a function, for the same reason given for $F$.

The function $F'$ defined above is also continuous and

$$F'F = 1_{Y/\equiv_Y},\ FF' = 1_{X/\equiv_X}$$

in other words, $F$ is a homeomorphism. ∎

## I.1.2 Connectedness

**(I.1.12) Theorem.** *Let $X$ be a topological space. The following statements are equivalent:*

*(i) The empty set $\emptyset$ and the set $X$ itself are the only two subsets of $X$ that are both open and closed.*

(ii) $X$ is not the union of two non-empty subsets $U$ and $V$, which are open and disjoint.

(iii) Let $\{0,1\}$ be the discrete topological space with exactly two points; then any continuous function $f: X \longrightarrow \{0,1\}$ (also called a two-valued map) is constant.

*Proof.* $(i) \Rightarrow (ii)$: Suppose that $X = U \cup V$ with

$$U \neq \emptyset, \; V \neq \emptyset \text{ and } U \cap V = \emptyset.$$

Then, the subset $V = X \smallsetminus U$ is not empty and is both open and closed, in contradiction to $(i)$.

$(ii) \Rightarrow (iii)$: Let $f: X \longrightarrow \{0,1\}$ be a continuous function that we assume not to be constant. Since $\{0\}$ and $\{1\}$ are open in $\{0,1\}$ and $f$ is continuous, $U = f^{-1}(\{0\})$ and $V = f^{-1}(\{1\})$ are open in $X$; moreover, $U \cap V = \emptyset$ and $U \cup V = X$, which contradicts $(ii)$.

$(iii) \Rightarrow (i)$: Let $U$ be a non-empty, proper subset of $X$, both open and closed. Then

$$X = U \cup (X \smallsetminus U)$$

with $X \smallsetminus U \neq \emptyset$ and $U \cap (X \smallsetminus U) = \emptyset$. We now define the function $f: X \longrightarrow \{0,1\}$ by $f(U) = 0$ and $f(X \smallsetminus U) = 1$. This function is continuous but not constant, contrary to $(iii)$. ∎

A topological space that satisfies one of the equivalent conditions of Theorem (I.1.12) is said to be *connected*. Condition $(iii)$ is probably the most useful; here are some of its applications, joined in a single theorem.

**(I.1.13) Theorem.** *The following results are true:*

1. *Let $f: X \to Y$ be a continuous function, where $X$ is connected; then the image $f(X)$ is a connected space.*

2. *Let $\{X_j \mid j \in J\}$ be a set of connected subspaces of a space $X$ with $\bigcap_j X_j \neq \emptyset$; then $X = \bigcup_j X_j$ is a connected space.*

3. *If $X$ and $Y$ are connected, then $X \times Y$ is connected.*

*Proof.* 1. Let $g: f(X) \to \{0,1\}$ be any two-valued map; the map $gf: X \to \{0,1\}$ is two-valued and, because $X$ is connected, $gf$ is constant; it follows that $g$ is constant.

This result shows in particular that if $X$ is connected and $Y$ is homeomorphic to $X$, also $Y$ is connected.

2. Let $f$ be a two-valued map of $X$, and let $x_j \in X_j$ and $x_{j'} \in X_{j'}$ be given arbitrarily; furthermore, let us choose any point $y \in X_j \cap X_{j'}$. Since $X_j$ and $X_{j'}$ are connected, $f(x_j) = f(y)$ and $f(x_{j'}) = f(y)$; therefore, $f(x_j) = f(x_{j'})$ and we conclude that the map $f$ is constant.

3. Let $f: X \times Y \to \{0,1\}$ be a map; in order to prove that $f$ is constant, we choose $(x,y), (x',y') \in X \times Y$ arbitrarily; the space $\{x\} \times Y$ is homeomorphic to $Y$ and is, therefore, connected; it follows that $f(x,y) = f(x,y')$. Similarly, we prove that $f(x,y') = f(x',y')$. Then, $f(x,y) = f(x',y')$. ∎

We recall that an *interval* of $\mathbb{R}$ (or more generally, of an ordered set) is a subset $A \subset \mathbb{R}$ such that, for each $a, b, x \in \mathbb{R}$ with $a < x < b$, if $a, b \in A$, we have $x \in A$. Real line intervals are important examples of connected spaces, as shown by the next theorem. It is not difficult to prove that every (non-empty) interval of $\mathbb{R}$ must be of one of the following types.[2] The set

$$[a,b] = \{x \in \mathbb{R} \mid a \le x \le b\}$$

is the *closed interval* with end-points $a, b$; the set

$$(a,b) = \{x \in \mathbb{R} \mid a < x < b\}$$

is the *open interval* with end-points $a, b$; the sets

$$(a,b] = \{x \in \mathbb{R} \mid a < x \le b\} \text{ and } [a,b) = \{x \in \mathbb{R} \mid a \le x < b\}$$

are *semi-open intervals*. The sets

$$(a,+\infty) = \{x \in \mathbb{R} \mid a < x\}, \ [a,+\infty) = \{x \in \mathbb{R} \mid a \le x\},$$
$$(-\infty,b) = \{x \in \mathbb{R} \mid x < b\} \text{ e } (-\infty,b] = \{x \in \mathbb{R} \mid x \le b\}$$

are *infinite intervals* (and naturally $\mathbb{R} = (-\infty, \infty)$ is the maximal interval).

**(I.1.14) Theorem.** *Any interval of $\mathbb{R}$ is a connected space.*

*Proof.* Let $X \subset \mathbb{R}$ be a closed interval, say $X = [a,b]$. Let $U \ne \emptyset$ be a subset of $X$, both open and closed; we wish to prove that $U = X$. Since $U$ is open in $X$, we may choose $u \in U$ such that $u \in \mathring{X}$. Let

$$s = \sup\{x \in X \mid [u,x) \subset U\};$$

clearly, $u < s$. We prove that $[u,s) \subset U$. Indeed, for each $v \in [u,s)$, there exists $x \in X$ such that $v < x$ and $[u,x) \subset U$; hence, $v \in U$. We now prove that $s = b$. In fact, if $s \ne b$, then $s < b$; since $U$ is closed in $X$, we conclude that $s \in U$ and so, $[u,s] \subset U$. However, $U$ is also open and consequently there is $\varepsilon > 0$ such that $[u,s+\varepsilon) \subset U$; but this contradicts the definition of least upper bound. It follows that $s = b$ and $[u,b) \subset U$. Similarly, $(a,u] \subset U$ which implies that $X = (a,b) \subset U \subset X$, and we conclude that $U = X$.

The reader may verify that this proof applies to the other types of interval (open, semi-open or infinite); alternatively, once every finite or infinite, open or closed interval is the telescopic union of a sequence of closed intervals, for instance,

$$(0,1) = \bigcup_{n \ge 2} \left[\frac{1}{n}, 1 - \frac{1}{n}\right],$$

the remainder of the proof follows from part 2 of Theorem (I.1.13). ∎

---

[2] It is enough to consider the extrema of the interval $a = \inf A$ and $b = \sup A$, if they exist. If $\inf A$ does not exist, set $a = -\infty$; if $\sup A$ does not exist, set $b = +\infty$.

In particular, the real line $\mathbb{R}$ is connected.

The previous theorem has a converse:

**(I.1.15) Theorem.** *Any connected subspace of $\mathbb{R}$ is an interval.*

*Proof.* If $X \subset \mathbb{R}$ is not an interval, we can find real numbers $a, b \in X$ and $c \notin X$ such that $a < c < b$. In this case,

$$U = (-\infty, c) \cap X \text{ and } V = X \cap (c, +\infty)$$

are non-empty ($a \in U$ and $b \in V$), open in $X$, disjoint, and such that $X = U \cup V$. Then $X$ is not connected, contradicting the hypothesis. ∎

The last two theorems have an immediate and interesting consequence:

**(I.1.16) Corollary.** *Let $X$ be a connected space and $f \colon X \to \mathbb{R}$ be a map. Then $f(X)$ is an interval.*

**(I.1.17) Remark.** The concept of connectedness provides a simple method for establishing when two spaces are homeomorphic. The criterion goes as follows: suppose that $f \colon X \to Y$ is a homeomorphism; then, for every $x \in X$, the restriction of $f$ to $X \smallsetminus \{x\}$ is a homeomorphism from $X \smallsetminus \{x\}$ onto $Y \smallsetminus \{f(x)\}$; therefore, $X \smallsetminus \{x\}$ is connected if and only if $Y \smallsetminus \{f(x)\}$ is connected.

So, how does this method work? We wish to prove, for instance, that a semi-open interval $(a, b]$ cannot be homeomorphic to an open interval $(c, d)$. Suppose that a homeomorphism $f \colon (a, b] \to (c, d)$ could exist; then, we would have a homeomorphism

$$(a, b) = (a, b] \smallsetminus \{b\} \cong (c, d) \smallsetminus \{f(b)\}$$

which is impossible, for the first space is connected (it is an interval), but the second is not!

Here is an useful result.

**(I.1.18) Theorem.** *Let $X$ be a topological space and $A$ one of its subspaces such that its closure in $X$ coincides with $X$. Then, if $A$ is connected, so is $X$.*

*Proof.* Let $f \colon X \to \{0, 1\}$ be a two-valued map of $X$. Its restriction $f|A$ is a two-valued map of $A$ and, since $A$ is connected, $f|A$ is constant; we may suppose that $f|A = 0$ with no loss in generality. On the other hand, since $\overline{A} = X$ and $f$ is continuous, we have

$$f(\overline{A}) \subset \overline{f(A)} = \overline{\{0\}} = \{0\}$$

and so $f$ is constant in $X$. Therefore, $X$ is connected. ∎

**(I.1.19) Remark.** Let $X$ be a topological space and $x$ be one of its points; let $\{X_j, j \in J\}$ be the set of all connected subspaces of $X$ which contain $x$; then, by Theorem (I.1.13) (part 2), the union

$$C_x = \bigcup_{j \in J} X_j$$

is a connected space that contains $x$, known as a *connected component* of $x$. The space $C_x$ is maximal in the sense that if a subset $M \subset X$ is connected and contains $C_x$, then $C_x = M$ (in other words, the connected component $C_x$ is the largest connected subspace that contains $x$). Moreover, $C_x$ is a closed subspace of $X$; in fact, by Theorem (I.1.18), $\overline{C_x}$ is connected and since $C_x \subset \overline{C_x}$, then $C_x = \overline{C_x}$. The connected components of two points $x, y \in X$ are either disjoint or coincide: in fact, if $z \in C_x \cap C_y$, the subspace $C_x \cup C_y$ is a connected subspace of $X$ that contains both spaces $C_x$ and $C_y$; but the connected components are maximal and so $C_x = C_x \cup C_y = C_y$. From this fact, we conclude that the relation "$x, y \in X$ are in the same connected component" is an equivalence relation. Therefore, a topological space $X$ is a disjoint union of maximal, closed, connected subspaces. A space is connected if and only if it has only one connected component.

Finally, we note that two topological spaces with different numbers of connected components cannot be homeomorphic. This is another criterion for verifying whether two spaces are homeomorphic.

There is another type of connectedness, called *path-connectedness*. A *path* in a topological space $X$ is a map $f \colon [0,1] \longrightarrow X$; two points $x_0, x_1 \in X$ are *joined by a path* if there is path $f$ of $X$ such that $f(0) = x_0$ and $f(1) = x_1$. We say that a space $X$ is *path-connected* if and only if any two points $x_0, x_1 \in X$ may be joined by a path in $X$.

The results of Theorem (I.1.13) hold true for path-connectedness; as a matter of completion (and to follow the preceding model), we present these results as a single theorem.

**(I.1.20) Theorem.** *The following statements are true:*

1. *Let $f \colon X \to Y$ be a continuous function, where $X$ is path-connected; then the space $f(X)$ is path-connected.*
2. *Let $\{X_j \mid j \in J\}$ be a set of path-connected subspaces of a space $Y$, with $\bigcap_j X_j \neq \emptyset$; then $X = \bigcup_j X_j$ is a path-connected space.*
3. *If $X$ and $Y$ are path-connected, then $X \times Y$ is path-connected.*

*Proof.* 1. Given any two points $y_0, y_1$ of $f(X)$, we choose $x_0, x_1 \in X$ such that $y_0 = f(x_0)$ and $y_1 = f(x_1)$. Because $X$ is path-connected, there is a path $g \colon [0,1] \to X$ such that $g(0) = x_0$ and $g(1) = x_1$. So, the path $fg \colon [0,1] \to f(X)$ links $y_0$ to $y_1$ and $f(X)$ is, therefore, path-connected.

In particular, if a space $X$ is path-connected, any space $Y$ homeomorphic to $X$ is path-connected.

2. Given any two points $x_0, x_1 \in X$, suppose that $x_0 \in X_{i_0}$ and $x_1 \in X_{i_1}$; let $a \in X_{i_0} \cap X_{i_1}$. By the hypothesis, there are two continuous functions

$$f_0 \colon [0,1] \to X_{i_0} \,,\; f_0(0) = x_0 \,,\; f_0(1) = a,$$
$$f_1 \colon [0,1] \to X_{i_1} \,,\; f_1(0) = x_1 \,,\; f_1(1) = a.$$

We now define the function $f\colon [0,1] \to X$

$$(\forall t \in [0,1])\ f(t) = \begin{cases} f_0(2t) & \text{if } 0 \leq t \leq \tfrac{1}{2} \\ f_1(2-2t) & \text{if } \tfrac{1}{2} \leq t \leq 1. \end{cases}$$

Since $f(\tfrac{1}{2}) = f_0(1) = f_1(0) = a$, it follows from Corollary (I.1.10) that the function $f$ is continuous. However, $f(0) = x_0$, $f(1) = x_1$ and so $f$ is a path that links $x_0$ to $x_1$.

3. Given $(x_1,y_1),(x_2,y_2) \in X \times Y$, choose two paths

$$f_X\colon [0,1] \to X \text{ such that } f_X(0) = x_1 \, , \ f_X(1) = x_2$$
$$f_Y\colon [0,1] \to Y \text{ such that } f_Y(0) = y_1 \, , \ f_Y(1) = y_2.$$

The path $(f_X,f_Y)\colon [0,1] \to X \times Y$ $t \mapsto (f_X(t),f_Y(t))$ links $(x_1,y_1)$ to $(x_2,y_2)$ and $X \times Y$ is, therefore, path-connected. ∎

The Euclidean space $\mathbb{R}^n$ is path-connected for every $n > 0$. Indeed, given $x_0,x_1 \in \mathbb{R}^n$, we define

$$f\colon [0,1] \to \mathbb{R}^n \, , \ (\forall t \in [0,1])\ f(t) = tx_1 + (1-t)x_0.$$

In particular, every interval of $\mathbb{R}$ is path-connected (as well as any convex subspace of $\mathbb{R}^n$).

**(I.1.21) Theorem.** *Any path-connected space is connected.*

*Proof.* Suppose $X$ to be the union of two non-empty, disjoint subspaces $U, V$ which are simultaneously open and closed. Take two points $x_0,x_1 \in X$ where $x_0 \in U$ and $x_1 \in V$. Since $X$ is path-connected, there exists a map

$$f\colon [0,1] \to X \, , \ f(0) = x_0 \, , \ f(1) = x_1.$$

Then,

$$(U \cap f([0,1])) \cap (V \cap f([0,1])) \neq \emptyset \, ,$$

contradicting the fact that $f([0,1])$ is connected (see Theorem (I.1.13)). ∎

In general, it is not true that a connected space is path-connected; here is an example.

Consider the following sets of points from the Euclidean plane:

$$A = \left\{ \left( 0, \frac{1}{2} \right) \right\},$$

$$B = \left\{ \left( \frac{1}{n}, t \right) \, \big| \, n \in \mathbb{N} \text{ and } t \in [0,1] \right\},$$

$$C = \{ (t,0) \, | \, t \in (0,1] \} = (0,1] \times \{0\}$$

and endow $X = A \cup B \cup C$ with the topology induced by the Euclidean topology of $\mathbb{R}^2$, as shown in Fig. I.5. It is immediate to verify that $B \cup C$ is a path-connected

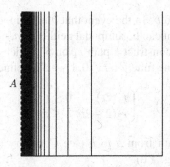

$A$

**Fig. I.5** Example of a space that is connected but not path-connected

space, which implies that $B \cup C$ is connected; moreover, the closure of $B \cup C$ in $X$ coincides with $X$ and therefore, by Theorem (I.1.18), $X$ is a connected space.

We now prove that $X$ is not path-connected. In order to come to this conclusion, we shall prove that if $f : [0,1] \to X$ is a path such that $f(0) = (0, \frac{1}{2})$, then $f$ is the constant path at the point $(0, \frac{1}{2})$. Because $X \subset \mathbb{R}^2$, we may write $f(t) = (x(t), y(t))$ for every $t \in [0,1]$; it is not difficult to prove that $x$ and $y$ are continuous functions (see Exercise 5 on p. 27). We first prove that $x$ is the constant function at 0, that is to say, $x(t) = 0$ for every $t \in [0,1]$. Indeed, suppose there exists $t' \in [0,1]$ such that $x(t') > 0$ and let $t_0 = \sup\{t \in [0,1] \,|\, x(t) = 0\}$. Since the function $x$ is continuous, we must have $x(t_0) = 0$ and $t_0 < 1$ so that we do not contradict the fact that there exists a $t'$ with $x(t') > 0$. Since $(0, \frac{1}{2})$ is the only point with zero for its first coordinate, we have $y(t_0) = \frac{1}{2}$. Since $y$ is continuous, there exists an $\varepsilon > 0$ with $t_0 + \varepsilon < 1$ and such that, for every $s \in [t_0, t_0 + \varepsilon)$, $y(s) \geq \frac{1}{4}$. Since $t_0$ is an upper bound, we can find $t_1 \in (t_0, t_0 + \varepsilon)$ such that $x(t_1) > 0$; by Corollary (I.1.16) and because $x(t_0) = 0$, we have $x([t_0, t_1]) \supseteq [0, x(t_1)]$. Consequently, we are able to find $s \in (t_0, t_1)$ such that $x(s) \neq \frac{1}{n}$ for every integer $n > 0$ (and $y(s) \geq \frac{1}{4}$). Finally, as $f(s) = (x(s), y(s)) \in X$, it follows that $y(s) = 0$ because the only points of $X$ for which $x > 0$ and $x \neq \frac{1}{n}$ are of the type $(x, 0)$, in contradiction to the fact that $y(s) \geq \frac{1}{4}$.

We conclude that for every $t \in [0,1]$, we have $x(t) = 0$ and so $f(t) = \{(0, \frac{1}{2})\}$; in other words, $f$ is constant.

**(I.1.22) Example.** For every $n \geq 1$ the unit sphere

$$S^n = \{(x_0, \ldots, x_n) \in \mathbb{R}^{n+1} \,|\, \Sigma_{i=0}^n x_i^2 = 1\}$$

is path connected and consequently, connected. Let $a = (x_0, \ldots, x_n)$ and $b = (y_0, \ldots, y_n)$ be any two points of $S^n$; we say that $b$ is *antipodal* to $a$ if $y_i = -x_i$ for every $i = 0, \ldots, n$. If $b$ is antipodal to $a$, the interval $\mathbb{R}^{n+1}$ with end points $a$ and $b$ goes through the centre $(0, \ldots, 0)$. Suppose that $b$ is not antipodal to $a$; then, for every $t \in [0,1]$, we have $(1-t)a + tb \neq 0$. It follows that

$$f : [0,1] \to S^n , \quad t \mapsto \frac{(1-t)a + tb}{|(1-t)a + tb|}$$

is a path on $S^n$ joining $a$ and $b$. In the event that $b$ is antipodal to $a$, we choose any other point $c$ of $S^n$; this point can be antipodal neither to $a$, nor to $b$. Therefore, with the preceding method, we construct a path $r_1$ on $S^n$ which links $a$ to $c$ and then a path $r_2$ which links $c$ to $b$; the function $r \colon [0,1] \to S^n$, defined by

$$r(t) = \begin{cases} r_1(2t) & 0 \le t \le \frac{1}{2} \\ r_2(2-2t) & \frac{1}{2} \le t \le 1 \end{cases}$$

for every $t \in [0,1]$, is a path from $a$ to $b$ (see Corollary (I.1.10)). Hence, $S^n$ is path-connected (see also Fig. I.6).

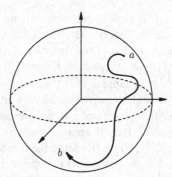

**Fig. I.6**

## I.1.3 Compactness

Let $Y$ be a topological space. A *covering* of $Y$ is a family $\mathfrak{U} = \{U_j \mid j \in J\}$ of subsets of $Y$ such that $Y = \bigcup_{j \in J} U_j$. If $Y$ is a subspace of $X$, a covering of $Y$ by subsets of $X$ is a family $\mathfrak{U}$ of subsets of $X$ whose union contains $Y$. A covering $\mathfrak{U}$ is *finite* if the set $J$ of the indexes is finite; $\mathfrak{U}$ is an *open* covering if all its elements are open in $Y$ (or, for subsets of $X$, if they are open in $X$). A *subcovering* of $\mathfrak{U}$ is a subset $\mathfrak{U}' = \{U_{j'} \mid j' \in J'\}$ of $\mathfrak{U}$ where $J' \subset J$. A covering $\mathfrak{U}$ of $Y$ is a *refinement* of a covering $\mathfrak{V}$ of $Y$ if for every $V \in \mathfrak{V}$ there exists $U \in \mathfrak{U}$ such that $U \subset V$.

A topological space $X$ is said to be *compact* if every open covering of $X$ has a finite subcovering; in other words, given any set $\mathfrak{U} = \{U_j \mid j \in J\}$ with $U_j \subset X$ open in $X$, for every $j \in J$ such that $\bigcup_j U_j = X$, there is a finite number of open sets $U_j$, for instance, $U_1, U_2, \ldots, U_n$ such that $X = U_1 \cup U_2 \cup \cdots \cup U_n$. A subspace $Y \subset X$ is compact in the induced topology on $Y$ if and only if every covering of $Y$ by open sets of $X$ has a finite subcovering (because any open set $U$ of $Y$ is of the type $U = V \cap Y$, with $V$ open in $X$).

The reader may easily prove that the space

$$X = \{0\} \cup \{1/n \mid n \in \mathbb{N}\} \,,$$

with the topology induced by the Euclidean topology of $\mathbb{R}$, is compact; on the other hand, the space $\mathbb{R}$ is not compact: indeed, the open covering

$$\mathfrak{U} = \{(n, n+2) \mid n \in \mathbb{Z}\}$$

of $\mathbb{R}$ has no finite subcovering.

**(I.1.23) Theorem.** *Any closed subspace of a compact space is compact.*

*Proof.* Let $Y$ be a closed subspace of a compact space $X$. Let $\mathfrak{C} = \{U_j \mid j \in J\}$ be a covering of $Y$ such that, for every $j \in J$, $U_j$ is an open set of $X$; then the set $\mathfrak{C} \cup \{X \smallsetminus Y\}$ is an open covering of $X$. Since $X$ is compact, it has a finite subcovering

$$U_1 \cup \ldots \cup U_n \cup (X \smallsetminus Y) = X.$$

As no $x \in Y$ is in $X \smallsetminus Y$, $\{U_1, \ldots, U_n\}$ is a covering of $Y$, and so, $Y$ is compact. ∎

We now introduce a special type of topological space which is needed for the next result. We say that a topological space $X$ is *Hausdorff*, if for every two distinct points $x, y$ of $X$, we can find two disjoint open sets $U_x$ and $U_y$ of $X$ such that $x \in U_x$, $y \in U_y$.

It is easily proved that a subspace of a Hausdorff space is also Hausdorff; and it is easily seen that the Euclidean space $\mathbb{R}^n$ is Hausdorff. Here is an example of a space which is not Hausdorff.

**(I.1.24) Example.** Let $Y = \mathbb{R} \cup \{*\}$ be the set given by the union of the real line $\mathbb{R}$ and an external point $*$; for the set of open subsets of $Y$, we choose the set consisting of the empty set and all subsets of $Y$ which are the union of an open set of $\mathbb{R}$ and the point $*$ (it is left to the reader to verify that these sets define a topology on $Y$). The space $Y$ here defined is not Hausdorff because the intersection of any two non-empty, open sets of $Y$ is not empty!

It follows from the definition that in any Hausdorff space every point is closed. More generally, all compact subspaces of a Hausdorff space are closed, as stated in the next theorem.

**(I.1.25) Theorem.** *Any compact subspace of a Hausdorff space is closed.*

*Proof.* Let $Y$ be a compact subspace of a Hausdorff space $X$. We must prove that $X \smallsetminus Y$ is open. Let us take any $x \in X \smallsetminus Y$; since $X$ is Hausdorff, for every $y \in Y$, let us take two open sets $U_y, V_y$ of $X$ such that

$$y \in U_y \ , \ x \in V_y \ , \ U_y \cap V_y = \emptyset.$$

The set $\mathfrak{U} = \{U_y \mid y \in Y\}$ is a covering of $Y$ by open sets of $X$. Since $Y$ is compact, $Y$ is covered by a finite number of these open sets, for instance, $Y \subset \bigcup_{i=1}^{n} U_i$. Let us take the set of open sets $\{V_1, \ldots, V_n\}$ corresponding to this finite covering of $Y$ and consider the open set $V = \bigcap_{j=1}^{n} V_j$. It is not difficult to prove that $V$ is an open set such that $x \in V \subset X \smallsetminus Y$. ∎

We now prove the following:

**(I.1.26) Theorem.** *Let $f: X \to Y$ be a continuous function; if $X$ is compact, $f(X)$ is compact.*

*Proof.* Let $\{U_j \mid j \in J\}$ be a covering of $f(X)$ by open sets of $Y$. The set $\{f^{-1}(U_j) \mid j \in J\}$ is an open covering of $X$ and so it has a finite subcovering, for instance, $\{f^{-1}(U_1), \ldots, f^{-1}(U_n)\}$. Then $\{U_1, \ldots, U_n\}$ is a covering of $f(X)$.  ∎

In particular, if $X$ is compact, every space $Y$ homeomorphic to $X$ is compact.

**(I.1.27) Theorem.** *Let $f$ be a continuous bijection from a compact space $X$ to the Hausdorff space $Y$. Then, $f$ is a homeomorphism.*

*Proof.* We wish to prove that the inverse function $f^{-1}: Y \to X$ is continuous, in other words, that for every open $U \subset X$, $f(U)$ is open in $Y$. In fact, by Theorem (I.1.23), $X \smallsetminus U$ is a compact subspace of $X$; it follows that $f(X \smallsetminus U)$ is a compact subspace of $Y$ (see Theorem (I.1.26)) and so, accordingly to Theorem (I.1.25), $f(X \smallsetminus U)$ is closed in $Y$; from this we conclude that $f(U) = Y \smallsetminus f(X \smallsetminus U)$ is open in $Y$.  ∎

We note that continuous functions preserve the compactness property of a space but, generally, not its Hausdorffness; here is an example. Let $X$ be a Hausdorff topological space and $Y$ be a space with at least two points and the trivial topology: then every function $f: X \to Y$ is continuous, but $f(X)$ will never be Hausdorff. Let us look into another example.

**(I.1.28) Example.** Let $Y = \mathbb{R} \cup \{*\}$ be the non-Hausdorff space from Example (I.1.24). Consider the space $X = \{(x,y) \in \mathbb{R}^2 \mid y \geq 0\}$, with the topology induced by the Euclidean topology of $\mathbb{R}^2$, and the function $f: X \to Y$ defined as follows:

$$f(x,y) = \begin{cases} x \in \mathbb{R} & \text{if } y = 0 \\ * & \text{if } y > 0 \end{cases}$$

Since every non-empty open set of $Y$ is the union of an open set of $\mathbb{R}$ and the point $*$, the counter image of any open set $U \cup \{*\}$ of $Y$ is exactly the union $f^{-1}(U) \cup f^{-1}(*)$, namely, the set $\{(x,0) \mid x \in U\} \cup \{(x,y) \mid y > 0\}$ which is the complement of the closed set $\{(x,0) \mid x \notin U\}$ of $X$. The function $f$ is, therefore, continuous; $X$ is Hausdorff but $f(X) = Y$ is not.

In the next theorem there is an instance in which the image of a Hausdorff space is Hausdorff. Before stating it, we need a definition: a map $f: X \to Y$ is *closed* if, for every closed set $C \subset X$, $f(C)$ is closed in $Y$.

**(I.1.29) Theorem.** *Let $f: X \to Y$ be a surjection where $X$ is compact and Hausdorff, and suppose that $f$ is a closed map. Then the space $Y$ is Hausdorff.*

*Proof.* Let $y_1$ and $y_2$ be two distinct points of $Y$. Since $f$ is surjective, there are two distinct points $x_1, x_2 \in X$ such that $f(x_1) = y_1$ and $f(x_2) = y_2$; since $X$ is Hausdorff, $\{x_1\}$ and $\{x_2\}$ are closed in $X$, and because $f$ is closed, then $\{y_1\}$ and $\{y_2\}$

are closed in $Y$. Therefore, $f^{-1}(y_1)$ and $f^{-1}(y_2)$ are two disjoint closed subsets of $X$ (see Theorem (I.1.9)). For every element $(x,a) \in f^{-1}(y_1) \times f^{-1}(y_2)$ choose two disjoint open sets $U_{x,a}$ and $V_{x,a}$ in $X$ such that $x \in U_{x,a}$ and $a \in V_{x,a}$. The set $\{V_{x,a} \mid a \in f^{-1}(y_2)\}$ is an open covering of $f^{-1}(y_2)$; since this space is compact (see Theorem (I.1.23)), there exists a finite subcovering $\{V_{x,a_1}, \dots, V_{x,a_n}\}$ of $f^{-1}(y_2)$. And so we have two disjoint open sets

$$U_x = \bigcap_{j=1}^{n} U_{x,a_j} \text{ and } V_x = \bigcup_{j=1}^{n} V_{x,a_j}$$

containing $x$ and $f^{-1}(y_2)$, respectively. But $\{U_x \mid x \in f^{-1}(y_1)\}$ is an open covering of $f^{-1}(y_1)$ and, since this is a compact space, there is a subcovering $\{U_{x_1}, \dots, U_{x_m}\}$ of $f^{-1}(y_1)$. We now note that the sets

$$U = \bigcup_{k=1}^{m} U_{x_k} \text{ and } V = \bigcap_{k=1}^{m} V_{x_k}$$

are disjoint open sets of $X$ which contain $f^{-1}(y_1)$ and $f^{-1}(y_2)$, respectively.

The sets $W_1 = Y \smallsetminus f(X \smallsetminus U)$ and $W_2 = Y \smallsetminus f(X \smallsetminus V)$ are open sets of $Y$ ($f$ is a closed map) containing $y_1$ and $y_2$, respectively. We wish to prove that $W_1 \cap W_2 = \emptyset$.

Suppose there exists a point $y \in W_1 \cap W_2$. Then, $y \notin f(X \smallsetminus U)$ and $y \notin f(X \smallsetminus V)$, in other words,

$$f^{-1}(y) \cap (X \smallsetminus U) = \emptyset \text{ and } f^{-1}(y) \cap (X \smallsetminus V) = \emptyset \ ;$$

hence, $f^{-1}(y) \subset U \cap V = \emptyset$, which is not possible. ∎

A first concrete example of a compact space is given by the next theorem.

**(I.1.30) Theorem.** *The unit interval $I = [0,1]$ is compact.*

*Proof.* Let $\mathfrak{U}$ be a covering of $I$ by open sets of $\mathbb{R}$. The properties of the Euclidean topology of $\mathbb{R}$ ensure that for every $x \in I$ we can find a $\delta(x) > 0$ and an element $U(x) \in \mathfrak{U}$ such that the open interval $(x - \delta(x), x + \delta(x))$ is contained in $U(x)$. The set
$$\mathfrak{I} = \{I(x) = (x - \delta(x), x + \delta(x)) \mid x \in I\}$$

is an open covering of $I$ and every interval of $\mathfrak{I}$ is contained in an open set of $\mathfrak{U}$. If there exists a finite subcovering in $\mathfrak{I}$, there is a corresponding finite subcovering in $\mathfrak{U}$; we may, therefore, assume that each open set of $\mathfrak{U}$ is an open interval of $\mathbb{R}$.

We now consider the set $\{0,1\}$ with the discrete topology and the function

$$f \colon I \to \{0,1\}$$

defined by the following conditions:

1. $f(x) = 0$ if the closed interval $[0,x]$ is covered by a finite number of elements of $\mathfrak{U}$.
2. Otherwise, $f(x) = 1$.

We wish to prove that $f$ is constant on every open set of $\mathfrak{U}$. Indeed, let $U \in \mathfrak{U}$ and $x \in U$ be given and suppose that $f(x) = 0$; then $[0,x]$ is covered by a finite subcovering $\mathfrak{V} \subset \mathfrak{U}$. But for every $y \in U$, the interval $[0,y]$ is covered by a finite family of open intervals of $\mathfrak{U}$ (that is to say, by $\mathfrak{V} \cup \{U\}$) meaning thereby that $f(y) = 0$. This shows that $f$ equals either $0$ or $1$ on the entire open interval $U$. It follows that $f$ is continuous; moreover, since $I = [0,1]$ is connected, by Theorem (I.1.12), $f$ is constant; since $f(0) = 0$, we have $f(1) = 0$, in other words, $I$ is covered by a finite subcovering of $\mathfrak{U}$.                                                    ∎

We note that, since every closed interval $[a,b]$ where $a < b$ (bounded) is homeomorphic to the unit interval $[0,1]$, every bounded and closed interval of $\mathbb{R}$ is compact. We wish to prove that a finite product of compact spaces is a compact space; this result is an immediate consequence of the following theorem:

**(I.1.31) Theorem.** *Let $B$ and $C$ be compact subspaces of the topological spaces $X$ and $Y$, respectively, and let $\mathfrak{U} = \{A_j \mid j \in J\}$ be an open covering of $B \times C$. Then there are two open sets $U \subset X$ and $V \subset Y$ such that*

$$B \times C \subset U \times V \subset X \times Y$$

*and $U \times V$ is covered by a finite number of elements of $\mathfrak{U}$.*

*Proof.* The method for proving this theorem is the one used in Theorem (I.1.29) to show the existence of two open sets $U$ and $V$ which contain $f^{-1}(y_1)$ and $f^{-1}(y_2)$, respectively. We leave to the reader the task of supplying the details needed for this proof.                                                                                  ∎

**(I.1.32) Corollary.** *Let $X_1, \ldots, X_n$ be compact spaces; then $X_1 \times \ldots \times X_n$ is compact.*

*Proof.* If $n = 2$ we have the previous theorem where $B = X = X_1$ and $C = Y = X_2$; thereafter, we proceed by induction.                                                    ∎

**(I.1.33) Corollary.** *Let $B$ and $C$ be compact subspaces of the topological spaces $X$ and $Y$ respectively. Let $W$ be open in $X \times Y$ such that $B \times C \subset W$. Then there are open sets $U \subset X$ and $V \subset Y$ such that*

$$B \times C \subset U \times V \subset W.$$

*Proof.* It is enough to set $\mathfrak{U} = \{W\}$ in Theorem (I.1.31).                          ∎

**(I.1.34) Corollary.** *Let $X$ be a Hausdorff space and let $B$ and $C$ be disjoint compact subspaces of $X$. Then there are two disjoint open sets $U$ and $V$ of $X$ with $B \subset U$ and $C \subset V$.*

*Proof.* We begin by noting that if

$$\Delta X = \{(x,x) \in X \times X\}$$

is the diagonal of $X \times X$, then $X$ is Hausdorff if and only if the diagonal $\Delta X$ is closed in $X \times X$. We now apply the previous corollary with $X = Y$ and $W = (X \times X) \smallsetminus \Delta X$. It follows from the hypothesis $B \cap C = \emptyset$ that $B \times C \subset W$. ∎

**(I.1.35) Corollary.** *Let $U$ be an open set of a compact Hausdorff space $X$. For every $x \in U$ there is an open set $V \subset X$ such that*

$$x \in V \subset \overline{V} \subset U$$

*where $\overline{V}$ is compact.*

*Proof.* The spaces $\{x\}$ and $X \smallsetminus U$ are disjoint and closed in $X$; hence by Theorem (I.1.23), they are compact. By the previous corollary we can find two disjoint open sets $V$ and $W$ of $X$ where $x \in V$, $X \smallsetminus U \subset W$. And so, we may conclude that $V \subset U$. On the other hand, $X \smallsetminus W$ is a closed set that contains $V$ and is contained in $U$; then, as we wished to prove, $\overline{V} \subset U$. The compactness of $\overline{V}$ follows from Theorem (I.1.23). ∎

The compact spaces of $\mathbb{R}^n$ are special. Before the next theorem, we give this definition: $X \subset \mathbb{R}^n$ is *bounded* if there is an $R > 0$ and an $n$-disk $D_R^n$ such that $D_R^n \supset X$.

**(I.1.36) Theorem.** *A subset $X \subset \mathbb{R}^n$ is compact if and only if $X$ is closed and bounded.*

*Proof.* $\Rightarrow$: By Theorem (I.1.25), $X$ is closed. In order to prove that $X$ is bounded, let us take a real number $\varepsilon > 0$ and the covering

$$\mathfrak{A} = \{D_\varepsilon^n(x) \,|\, x \in X\}$$

of $X$. Since $X$ is compact, there exists a finite subcovering of $\mathfrak{A}$ that contains $X$; suppose, for instance, that

$$X \subset D_\varepsilon^n(x_1) \cup \cdots \cup D_\varepsilon^n(x_r).$$

Then given any $x_0 \in X$, the disk $D_{2r\varepsilon}^n(x_0)$ contains $X$.

$\Leftarrow$: The space $X$ is contained in an $n$-disk $D_R^n$ of $\mathbb{R}^n$; but $D_R^n$ is contained in a hypercube $C^n$ of $\mathbb{R}^n$ whose edges are homeomorphic to $I$ and so $C^n$ is compact (see Theorems (I.1.30) and (I.1.31)). Therefore $X$ is a closed subspace of a compact space and, by Theorem (I.1.25), it is compact. ∎

We now are able to identify the quotient spaces defined in Examples (I.1.6) and (I.1.7) with more familiar topological spaces. Remember that the torus $T^2$ is the space $S^1 \times S^1$ with the product topology. Let $f \colon I_{\equiv}^2 \to T^2$ be the function that takes each equivalence class $[s,t] \in I_{\equiv}^2$ to $(e^{2\pi i s}, e^{2\pi i t}) \in T^2$. This is a bijective, continuous function. On the other hand, $I_{\equiv}^2$ is compact (it is the image of the compact space $I^2$ by the quotient map) and $T^2$ is Hausdorff; therefore, by Theorem (I.1.27), $f$ is a homeomorphism.

As for the Example (I.1.7), we define the function $f\colon D_{\equiv}^2 \to S^2$ such that

$$F([x,y]) = \left( x, 2\left( |y| - \frac{1}{2}\sqrt{1-x^2} \right), s(y)\sqrt{1 - \left( x^2 + 4\left( |y| - \frac{1}{2}\sqrt{1-x^2} \right)^2 \right)} \right)$$

for every equivalence class $[x,y] \in D_{\equiv}^2$, where $s(y)$ equals $+1$ if $y > 0$ and $-1$ if $y < 0$. Once more, $f$ is a bijective map from a compact space to a Hausdorff space and is, therefore, a homeomorphism.

## I.1.4 Function Spaces

Let $X$ and $Y$ be two topological spaces and let $M(X,Y)$ be the set of all maps from $X$ to $Y$. For every compact space $K \subset X$ and every open set $U \subset Y$, we define the set $W_{K,U}$ of all maps $f \in M(X,Y)$ for which $f(K) \subset U$. The set

$$\mathfrak{C} = \{ W_{K,U} \mid K \subset X \text{ compact and } U \subset Y \text{ open} \}$$

is a sub-basis for a topology of $M(X,Y)$ known as *compact-open topology*.

Let the topological spaces $X, Y, Z$ and a map $f\colon X \times Z \to Y$ be given. For every $z_0 \in Z$, the function $f(-,z_0)\colon X \to Y$ is continuous (in fact, it is the composite of the map $f$ and the inclusion $X \times \{z_0\} \to X \times Z$). Therefore, we may define a function $\hat{f}\colon Z \to M(X,Y)$ by requiring that $\hat{f}(z)(x) = f(x,z)$ for every $(z,x) \in Z \times X$. The function $\hat{f}$ is the *adjoint* of $f$.

**(I.1.37) Theorem.** *Let $M(X,Y)$ be the function space of all maps from $X$ to $Y$ with the compact-open topology. For every map $f\colon X \times Z \to Y$, its adjoint function $\hat{f}\colon Z \to M(X,Y)$ is continuous.*

*Proof.* Let $z_0 \in Z$ and let $W_{K,U}$ be an element of the sub-basis for the compact-open topology on $M(X,Y)$ such that $\hat{f}(z_0) \in W_{K,U}$, that is to say, $f(x,z_0) \in U$, for every $x \in K$. Then

$$K \times \{z_0\} \subset f^{-1}(U) \subset X \times Z$$

where $f^{-1}(U)$ is open; hence, $f^{-1}(U) \cap (K \times Z)$ is open and contains $K \times \{z_0\}$. Since $K$ is compact, there exists an open set $W \subset Z$ such that $z_0 \in W$ and $K \times W \subset f^{-1}(U)$. Therefore,

$$\hat{f}(W) \subset W_{K,U}$$

and so $\hat{f}$ is continuous. ∎

Conversely, given a function $\hat{f}\colon Z \to M(X,Y)$, we define a function $f\colon X \times Z \to Y$ by the following condition: $f(x,z) = \hat{f}(z)(x)$ for every $(x,z) \in X \times Z$; in this case too, we say that $f$ is adjoint of $\hat{f}$. The fact that $\hat{f}\colon Z \to M(M,Y)$ is continuous does not necessarily imply that $f\colon X \times Z \to Y$ is continuous; for that effect, another condition on the space $X$ is required. But first, let us prove the following:

**(I.1.38) Lemma.** *Let $X$ and $Y$ be topological spaces where $X$ is compact[3] Hausdorff. Then the function*

$$\varepsilon \colon X \times M(X,Y) \to Y$$

*(evaluation function), defined by $\varepsilon(x,f) = f(x)$ for every $x \in X$ and every $f \in M(X,Y)$, is continuous.*

*Proof.* Take arbitrarily $(x,f) \in X \times M(X,Y)$ and an open set $U \subset Y$ with $f(x) \in U$. Since $X$ is compact Hausdorff and $f^{-1}(U)$ is open in $X$, there exists an open set $V$ of $X$ such that

$$x \in V \subset \overline{V} \subset f^{-1}(U)$$

where $\overline{V}$ is compact (see Corollary (I.1.35)). We end the proof by noting that $(x,f) \in V \times W_{\overline{V},U}$ and $\varepsilon(V \times W_{\overline{V},U}) \subset U$. ∎

**(I.1.39) Theorem.** *Let $X$ be a compact Hausdorff space; then, if $\hat{f} \colon Z \to M(X,Y)$ is continuous, so is $f \colon X \times Z \to Y$.*

*Proof.* It is enough to note that $f$ is the composite of

$$1_X \times \hat{f} \colon X \times Z \to X \times M(X,Y) \quad \text{and} \quad \varepsilon \colon X \times M(X,Y) \to Y$$

and apply Lemma (I.1.38). ∎

**(I.1.40) Corollary.** *Let $q \colon X \longrightarrow Y$ be a quotient map. If $Z$ is a compact Hausdorff space, then also*

$$q \times 1_Z \colon X \times Z \longrightarrow Y \times Z$$

*is a quotient map.*

*Proof.* The quotient map $q \colon X \to Y$ has the following property which character- izes the quotient topology on $Y$: a function $g \colon Y \to W$ is continuous if and only if $gq \colon X \to W$ is continuous (see Exercise 7 at the end of this section). We must therefore prove that, for every topological space $W$ and every $g \colon Y \times Z \to W$, $g$ is continuous if and only if $h = g(q \times 1_Z)$ is continuous: If $g$ is continuous, $h$ is undoubtedly continuous. Now, the following commutative diagram

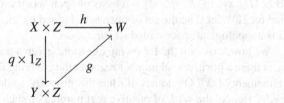

induces the commutative diagram

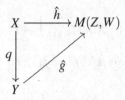

Suppose that $h$ is continuous; then, by Theorem (I.1.37), $\hat{h}$ is continuous. Since $q$ is a quotient map, $\hat{g}$ is continuous. Hence, by Theorem (I.1.39), the map $g$ is continuous. ∎

## I.1.5 Lebesgue Number

We begin this section by describing an important class of topological spaces, the class of *metric spaces*. Let $X$ be a given set; a *metric* on $X$ is a function $d$ defined from $X \times X$ to the set of the non-negative real numbers $\mathbb{R}_{\geq 0}$, with the following properties:

$$(\forall x, y \in X) \quad d(x,y) = d(y,x)$$
$$d(x,y) = 0 \quad \Longleftrightarrow \quad x = y$$
$$(\forall x, y, z \in X) \ d(x,z) \leq d(x,y) + d(y,z).$$

A first example is the Euclidean metric on $\mathbb{R}^n$; it may be generalized as follows: Let $V$ be a vector space with a norm $\| \ \|$ and let $X$ be a subset of $V$; the function $d : X \times X \to \mathbb{R}_{\geq 0}$ defined by

$$(\forall x, y \in X) \ d(x,y) = \|x - y\|$$

is a metric on $X$. In Sect. II.2 we shall give an important example of metric.

We already have seen that the Euclidean metric defines a topology on $\mathbb{R}^n$. A metric $d$ on a set $X$ defines a topology on $X$ in a similar way to the one described for $\mathbb{R}^n$. In fact, for every $x \in X$ and for every real number $\varepsilon > 0$, let $\mathring{D}_\varepsilon(x) = \{y \in X \mid d(x,y) < \varepsilon\}$ be the open disk of centre $x$ and radius $\varepsilon$; the set $\mathfrak{B} = \{\mathring{D}_\varepsilon(x) \mid x \in X, \ \varepsilon > 0\}$ is a basis of open sets for $X$ (this proof is similar to the one for $\mathbb{R}^n$); let $\mathfrak{U}$ be the set of open sets defined by $\mathfrak{B}$. The set $X$ with the topology $\mathfrak{U}$ is a topological space called *metric space*.

We now look into the following question: given a metric space $X$ and a covering $\mathfrak{A}$, is there a positive real number $r$ so that the covering $\mathfrak{B} = \{\mathring{D}_r(x) \mid x \in X\}$ of $X$ is a refinement of $\mathfrak{A}$? Obviously, if $r$ has this property, so does any positive real number $s < r$. Hence, the set $L$ of positive real numbers $r$ such that $\mathfrak{B}$ is a refinement of $\mathfrak{A}$ is either the empty set or an open interval $(0, t)$ (including the case $t = \infty$). If $L \neq \emptyset$, the real number $\ell = \sup L$ is the *Lebesgue number* of the covering $\mathfrak{A}$.

Here is an example. Consider the space $X = \mathbb{R}$ with the Euclidean topology and the open covering of $\mathbb{R}$

$$\mathfrak{A} = \{(n, n+2) \mid n \in \mathbb{Z}\}.$$

We ask reader to prove that its Lebesgue number is $\ell = \frac{1}{2}$.

The next result is of fundamental importance for proving one of the key theorems needed to define the homology of polyhedra, namely, the Simplicial Approximation Theorem (Theorem (III.2.4)).

**(I.1.41) Theorem.** *Any open covering of a compact metric space has a Lebesgue number.*

The proof of this theorem is not difficult but requires several preliminary considerations. We start by defining the *distance* $d(x, A)$ of a point $x$ of a metric space $X$ to a non-empty subspace $A \subset X$ as

$$d(x, A) = \inf\{d(x, a) \mid a \in A\}.$$

**(I.1.42) Lemma.** *The function*

$$d(-, A) \colon X \to \mathbb{R}_{\geq 0} , \ x \mapsto d(x, A)$$

*is continuous.*

*Proof.* We wish to prove that for every given $x$ and for every $\varepsilon > 0$ there exists a $\delta > 0$ such that, if $d(x, y) < \delta$, then $|d(x, A) - d(y, A)| \leq \varepsilon$. Indeed, for any $a \in A$ the inequalities:

$$d(x, a) \leq d(x, y) + d(y, a) \ \text{ and } \ d(y, a) \leq d(y, x) + d(x, a)$$

hold true and then, by applying $\inf_a$ to them, we obtain

$$d(x, A) \leq d(x, y) + d(y, A) \ \text{ and } \ d(y, A) \leq d(x, y) + d(x, A).$$

It follows that

$$|d(x, A) - d(y, A)| \leq d(x, y)$$

and the proof is completed by setting $\delta = \varepsilon$. ∎

**(I.1.43) Corollary.** $x \in \overline{A} \iff d(x, A) = 0.$

*Proof.* $\Rightarrow$: $d(-, A)^{-1}(0)$ is an open set of $X$ that contains $A$ and so it contains $\overline{A}$; then, if $x \in \overline{A}$, it is obvious that $d(x, A) = 0$.
$\Leftarrow$: If $x \notin \overline{A}$, there exists an $\varepsilon > 0$ such that $\mathring{D}_\varepsilon(x) \cap A = \emptyset$ and so $d(x, A) \geq \varepsilon > 0$. ∎

*Proof of Theorem (I.1.41).* Let $\mathfrak{A}$ be an open covering of a compact metric space $X$. Since $X$ is compact, it is possible to obtain a finite subcovering of $\mathfrak{A}$. Besides, every refinement of this subcovering is also a refinement of $\mathfrak{A}$; this allows us always to assume $\mathfrak{A}$ to be finite. Let $\mathfrak{A} = \{A_1, A_2, \ldots, A_n\}$. By Lemma (I.1.42), for every $i = 1,\ldots,n$, the metric $d(-,X \setminus A_i)\colon X \to \mathbb{R}_{\geq 0}$ is continuous. Hence, the function

$$f\colon X \to \mathbb{R}_{\geq 0}\,,\ f = \max\{d(-,X \setminus A_i) \mid i = 1,\ldots,n\}$$

is continuous. By Theorem (I.1.26), $f(X)$ is a compact subspace of $\mathbb{R}$ and then, by Theorem (I.1.25), $f(X)$ is closed in $\mathbb{R}$. On the other hand, we note that for any $x \in X$ there exists a certain $A_i$ such that $x \in A_i$ and, since $X \setminus A_i$ is closed, $d(x, X \setminus A_i) > 0$ (see Corollary (I.1.43)); and so, for every $x \in X$, $f(x) > 0$. Therefore, $\ell = \inf\{f(x) \mid x \in X\} > 0$. It follows that for every $x \in X$, $f(x) \geq \ell$ and then $d(x, A_i) \geq \ell$ for some $A_i \in \mathfrak{A}$; we conclude that $\mathring{D}_\ell(x) \subset A_i$. ∎

## Exercises

**1.** Let $X$ be a given set; prove that

$$\mathfrak{B} = \{\{x\} \mid x \in X\}$$

is a basis for the discrete topology on $X$.

**2.** Let $\mathfrak{B}$ and $\mathfrak{B}'$ be bases for the topologies $\mathfrak{A}$ and $\mathfrak{A}'$ on the set $X$. Prove that

$$\mathfrak{A}' \supset \mathfrak{A} \iff (\forall x \in X)(\forall B \in \mathfrak{B})x \in B, (\exists B' \in \mathfrak{B}')\, x \in B' \subset B,$$

that is to say, prove that $\mathfrak{A}'$ is finer than $\mathfrak{A}$ if and only if, for every $x \in X$ and every $B \in \mathfrak{B}$ containing $x$, there exists $B' \in \mathfrak{B}'$ containing $x$ and contained in $B$.

**3.** Take the following segments of the real line $\mathbb{R}$
$(a,b) = \{x \in \mathbb{R} \mid a < x < b\}$,
$[a,b) = \{x \in \mathbb{R} \mid a \leq x < b\}$,
$(a,b] = \{x \in \mathbb{R} \mid a < x \leq b\}$.
Now take the following sets of subsets of $\mathbb{R}$:
$\mathfrak{B}_1 = \{(a,b) \mid a < b\}$,
$\mathfrak{B}_2 = \{[a,b) \mid a < b\}$,
$\mathfrak{B}_3 = \{(a,b] \mid a < b\}$,
$\mathfrak{B}_4 = \mathfrak{B}_1 \cup \{B \setminus K \mid B \in \mathfrak{B}_1\}$, with $K = \{1/n \mid n \in \mathbb{N}\}$.
Prove that $\mathfrak{B}_i$, $i = 1,2,3,4$, are bases for topologies on $\mathbb{R}$ and compare these topologies.

**4.** Let $X$ and $Y$ be topological spaces with topologies $\mathfrak{A}$ and $\mathfrak{A}_Y$, respectively. Let

$$\pi_1\colon X \times Y \to X$$

$$\pi_2\colon X \times Y \to Y$$

be the $X$ and $Y$ projection.
Prove that the set

$$\mathfrak{S} = \{\pi_1^{-1}(U) \mid U \in \mathfrak{A}\} \cup \{\pi_2^{-1}(V) \mid V \in \mathfrak{A}_Y\}$$

is a sub-basis for the product topology on $X \times Y$.

**5.** Prove that a function $f\colon Z \to X \times Y$ is continuous if and only if its components $f_1 = \pi_1 f$ and $f_2 = \pi_2 f$ are continuous.

**6.** A map $f\colon X \to Y$ is *open* if and only if, for every $U \in \mathfrak{A})$, $f(U)$ is open in $Y$. Show that the projection maps

$$\pi_1 \colon X \times Y \to X$$
$$\pi_2 \colon X \times Y \to Y$$

are open.

**7.** Let $X$ and $Y$ be topological spaces and let $q\colon X \to Y$ be a surjection; give $Y$ the quotient topology relative to $q$. Prove that, for any topological space $Z$, any function $g\colon Y \to Z$ is continuous if and only if $gq\colon X \to Z$ is continuous.

**8.** Endow $\mathbb{R}$ with the Euclidean topology and let $f\colon \mathbb{R} \to \mathbb{R}$ be a given function. Prove that the following statements are equivalent.
$(\forall U \subset \mathbb{R} \,|\, U \text{ open })(f^{-1}(U) \text{ open})$;
$(\forall x \in \mathbb{R})(\forall \varepsilon > 0)(\exists \delta > 0)|x - y| < \delta \Rightarrow |f(x) - f(y)| < \varepsilon$.

**9.** Prove that if $A$ is closed in $Y$ and $Y$ is closed in $X$, then $A$ is closed in $X$.

**10.** Prove that if $U$ is open in $X$ and $A$ is closed in $X$, then $U \smallsetminus A$ is open in $X$ and $A \smallsetminus U$ is closed in $X$.

**11.** Show that a discrete space is compact if and only if it is finite.

**12.** Let $A$ and $B$ be two non-empty subsets of $\mathbb{R}^n$. The *distance* between $A$ and $B$ is defined by
$$d(A,B) = \inf \{d(a,b) \,|\, a \in A, b \in B\}.$$

Prove that if $A \cap B = \emptyset$, $A$ and $B$ are closed, and $A$ is bounded, then there exists $a \in A$ such that
$$d(A,B) = d(a,B) > 0.$$

# I.2 Categories

## I.2.1 General Ideas on Categories

A *category* $\mathbf{C}$ is a class of *objects* together with two functions, Hom and Composition, satisfying the conditions:

**Hom:** it assigns to each pair of objects $(A,B)$ of $\mathbf{C}$ a set $\mathbf{C}(A,B)$; an element $f \in \mathbf{C}(A,B)$ is a *morphism* with *domain* $A$ and *codomain* $B$;

**Composition:** it assigns to each triple $(A, B, C)$ of objects of **C** an operation

$$\mathbf{C}(A, B) \times \mathbf{C}(B, C) \to \mathbf{C}(A, C),$$

which is the *composition law* for morphisms; this law is indicated by

$$(f, g) \mapsto gf$$

for $f \in \mathbf{C}(A, B)$ and $g \in \mathbf{C}(B, C)$.

Besides, the following axioms must hold true:

**Associativity:** If $f \in \mathbf{C}(A, B)$, $g \in \mathbf{C}(B, C)$, and $h \in \mathbf{C}(C, D))$, then

$$h(gf) = (hg)f;$$

**Identity:** for every object $A \in \mathbf{C}$ there is a morphism $1_A \in \mathbf{C}(A, A)$ such that

$$f1_A = f, \ 1_A g = g$$

for every morphism $f$ with domain $A$ and every morphism $g$ with codomain $A$.

Here are some examples of categories.

1. **Set** : sets and functions between sets.
2. **Set**$_*$: based sets and based functions (namely, base preserving functions).
3. **Top**: topological spaces and *maps*, namely, continuous functions.
4. **Top**$_*$: based topological spaces and based maps (namely, base preserving continuous functions). We sometimes denote a based space $X$ with base point $x_0$ with $(X, x_0)$.
5. **CTop**: pair of spaces $(X, A)$ where $A \subseteq X$ is closed in $X$ and morphisms $f \in$ **Top**$(X, Y)$, $f|A \in$ **Top**$(A, B)$.
6. **Gr**: groups and group homomorphisms.
7. **Ab**: Abelian groups and Abelian group homomorphisms.
8. **Ab**$^{\mathbb{Z}}$: graded Abelian groups and related morphisms. We recall that a *graded Abelian group* is a succession of Abelian groups $\{C_n \mid n \in \mathbb{Z}\}$; we say that an element $x \in C_n$ has *degree n*. A *morphism (homomorphism) of degree d* between two graded Abelian groups $\{C_n\}$ and $\{C'_n\}$ is a set of homomorphisms $\{f_n \colon C_n \to C'_{n+d} \mid n \in \mathbb{Z}\}$.
9. Given two categories **C** and **D**, the *product category* of **C** and **D** is indicated with $\mathbf{C} \times \mathbf{D}$; its objects are pairs $(C, D)$, where $C \in \mathbf{C}$ and $D \in \mathbf{D}$. A morphism from $(C, D)$ to $(C', D')$ is a pair of morphisms

$$(f, g) \colon (C, D) \to (C', D'), \ f \in \mathbf{C}(C, C') \text{ and } g \in \mathbf{D}(D, D').$$

A category $\mathbf{C}'$ is a *subcategory* of a category **C** if:

**1.** $\mathbf{C}'$ is a subclass of **C**.
**2.** For every pair of objects $(X', Y')$ of $\mathbf{C}'$, the set $\mathbf{C}'(X', Y')$ is a subset of $\mathbf{C}(X', Y')$.

**3.** For every pair of morphisms $(f,g)$ with $f \in \mathbf{C}'(X',Y')$ and $g \in \mathbf{C}'(Y',Z')$, the morphism obtained through composition in $\mathbf{C}'$, namely, $gf \in \mathbf{C}'(X',Z')$, coincides with the morphism $gf \in \mathbf{C}(X',Z')$ obtained through composition in $\mathbf{C}$.

A subcategory $\mathbf{C}'$ of $\mathbf{C}$ is called *full* if, for every pair of objects $(X',Y')$ of $\mathbf{C}'$, the sets $\mathbf{C}'(X',Y')$ and $\mathbf{C}(X',Y')$ coincide. The category $C\mathbf{Top}$ is a subcategory of $\mathbf{Top} \times \mathbf{Top}$.

The set $\mathbf{Top}(X,Y)$ of all maps from $X$ to $Y$ has a very important relation called *homotopy*. Let $I$ be the closed interval $[0,1]$. Two maps $f,g\colon X \to Y$ are *homotopic* when there is a map

$$H\colon X \times I \to Y$$

such that $H(-,0) = f$ and $H(-,1) = g$. If $f$ is homotopic to $g$, we write $f \sim g$.

For instance, the maps $f,g\colon I \to I$ given by $f = 1_I$ (that is to say, $\forall x, f(x) = x$) and the constant map $g$ from $I$ to the point $0 \in I$ (in other words, $\forall x, f(x) = 0$) are homotopic; in fact, by constructing the map

$$H\colon I \times I \to I ,\ (\forall s,t \in I)\ H(s,t) = (1-t)s,$$

it is clear that $H(-,0) = f$ and $H(-,1) = g$. A homotopy may be indicated also by the notation $f_t\colon X \to Y$, where $t$ is the parameter $t \in I$ and, if $H(x,t)$ is the homotopy function, then $f_t(x) := H(x,t)$. Consequently, $f_0 \sim f_1$.

By Theorem (I.1.37), a homotopy $H\colon X \times I \to Y$ from $f$ to $g$ determines a map $\hat{H}\colon I \to M(X,Y)$, that is to say, a *path* in $M(X,Y)$ that links $f$ to $g$. Conversely, if $X$ is compact Hausdorff, a path $\hat{H}\colon I \to M(X,Y)$ determines a map $H\colon X \times I \to Y$, with $H(-,0) = f$ and $H(-,1) = g$ (Theorem (I.1.39)), in other words, a homotopy from $f$ to $g$.

**(I.2.1) Lemma.** *The homotopy relation is an equivalence relation.*

*Proof.* Clearly, any map $f$ is homotopic to itself. Suppose now that $f \sim g$; this means that there is a map

$$H\colon X \times I \to Y ,\ H(-,0) = f ,\ H(-,1) = g;$$

consider the map

$$H'\colon X \times I \to Y ,\ (\forall x \in X)(\forall t \in I)\ H'(x,t) = H(x,1-t).$$

It is immediate to show that $H'(-,0) = g$ and $H'(-,1) = f$ and so, $g \sim f$.

We finally prove that, if $f \sim g$ and $g \sim h$, then $f \sim h$. Consider the functions

$$H\colon X \times I \to Y ,\ H(-,0) = f ,\ H(-,1) = g,$$
$$G\colon X \times I \to Y ,\ G(-,0) = g ,\ G(-,1) = h.$$

Let us define a map $K\colon X \times I \to Y$ by the conditions

$$(\forall x \in x)(\forall t \in I)\ K(x,t) = \begin{cases} H(x,2t) , & 0 \leq t \leq \frac{1}{2} \\ G(x,2t-1) , & \frac{1}{2} \leq t \leq 1. \end{cases}$$

This map $K$ has the required properties. ∎

The quotient set obtained from $\mathbf{Top}(X,Y)$ and the relation $\sim$ is denoted by $[X,Y]$; its elements are equivalence classes for homotopy of maps, or *homotopy classes* of maps from $X$ to $Y$.

The category $H\mathbf{Top}$, the *homotopy category associated* with $\mathbf{Top}$, has topological spaces for objects and its morphisms are homotopy classes of maps: the set of morphisms is, therefore,

$$H\mathbf{Top}(X,Y) = [X,Y].$$

When dealing with based maps $f,g \in \mathbf{Top}_*((X,x_0),(Y,y_0))$, we say that $f \sim g$ if there is a map $H: X \times I \to Y$ such that

$$(\forall t \in I) \; H(x_0,t) = y_0 \; , \; H(-,0) = f \; , \; \text{and} \; H(-,1) = g.$$

In this case, we use the notation $[X,Y]_* = \mathbf{Top}_*(X,Y)/\sim$.

*Relative homotopy* in $C\mathbf{Top}$ is a useful concept: two maps of pairs $f,g: (X,A) \to (Y,B)$ are *homotopic relative to* $X' \subset X$ if $f|X' = g|X'$ and there exists a map

$$H: (X \times I, A \times I) \longrightarrow (Y,B)$$

such that

$$(\forall x \in X) \; H(x,0) = f(x) \; H(x,1) = g(x)$$
$$(\forall x \in X' \, , \, \forall t \in I) \; H(x,t) = f(x) = g(x).$$

This being the case, we denote the *relative* homotopy from $f$ to $g$ with

$$f \sim_{\mathrm{rel}X'} g$$

and we say that $f$ is homotopic to $g$ rel $X'$. If $X' = \emptyset$, we have a *free* homotopy in $C\mathbf{Top}$. In the category $\mathbf{Top}_*$ the based homotopy coincides with the homotopy relative to the base point.

We say that two spaces $X,Y \in \mathbf{Top}$ are of the same *homotopy type* (or simply, *type*) if there are maps $f: X \to Y$ and $g: Y \to X$ such that $gf \sim 1_X$ and $fg \sim 1_Y$; the map $f$ is called a *homotopy equivalence*.

The "functions" between categories are called *functors*, which take objects to objects and morphisms to morphisms. Specifically, a *covariant functor* or simply *functor*

$$F: \mathbf{C} \to \mathbf{C}'$$

is a relation between these two categories such that

1. $(\forall X \in \mathbf{C}) \; F(X) \in \mathbf{C}'$
2. $(\forall f \in \mathbf{C}(A,B)) \; F(f) \in \mathbf{C}'(F(A),F(B))$
3. $(\forall f \in \mathbf{C}(A,B))(\forall g \in \mathbf{C}(B,C)) \; F(gf) = F(g)F(f)$
4. $(\forall A \in \mathbf{C}) \; F(1_A) = 1_{FA}$

If conditions 2. and 3. are replaced by

> **2'.** $(\forall f \in \mathbf{C}(A,B))\, F(f) \in \mathbf{C}'(F(B),F(A))$
> **3'.** $(\forall f \in \mathbf{C}(A,B))(\forall g \in \mathbf{C}(B,C))\, F(gf) = F(f)F(g)$

we have a *contravariant* functor.

A very simple example of a (covariant) functor is the *forgetful functor* $D\colon \mathbf{Top} \to$ **Set** that merely "forgets" the topological space structure. Here is another example: Let $(X,x_0) \in \mathbf{Top}_*$ be a based space and let $F\colon \mathbf{Top}_* \to \mathbf{Set}_*$ be the function that takes each based space $(Y,y_0)$ into the set $[X,Y]_*$ of all homotopy classes of based maps $g\colon X \to Y$; for each morphism $f\colon Y \to Z$, we define $F(f)\colon [X,Y]_* \to [X,Z]_*$ as the function that takes any homotopy class $[g] \in [X,Y]_*$ into $[fg]$.

We now give an example of contravariant functor. Given a based space $(Y,y_0)$, we define $F\colon \mathbf{Top}_* \to \mathbf{Set}_*$ with the condition $F(X,x_0) = [X,Y]_*$ for every $(X,x_0)$; here, for every $f \in \mathbf{Top}_*((X,x_0),(Z,z_0))$, the function $F(f)$ may only be defined as $F(f)([g]) = [gf]$ for every based map $g\colon X \to Z$; notice that, if $f \in \mathbf{Top}_*((X,x_0),(Z,z_0))$, the arrow $F(f)$ has the opposite direction to that of $f$.

We now look into some less simple examples.

The *suspension* functor

$$\Sigma\colon \mathbf{Top}_* \to \mathbf{Top}_*$$

is defined on based spaces $(X,x_0)$ as the quotient

$$\Sigma X = \frac{I \times X}{I \times \{x_0\} \cup \partial I \times X}$$

where $\partial I$ is the set $\{0,1\}$. We shall write either $[t,x]$ or $t \wedge x$ when indicating a generic element of $\Sigma X$; then,

$$(\forall f \in \mathbf{Top}_*(X,Y))(\forall t \wedge x \in \Sigma X)\ \Sigma(f)(t \wedge x) = t \wedge f(x).$$

The base point of $\Sigma X$ is $t \wedge x_0 = 0 \wedge x = 1 \wedge x$.

The behavior of the suspension functor is particularly interesting on spheres; in fact, the suspension of an $n$-dimensional sphere is an $(n+1)$-dimensional sphere. This fact will be better understood after studying some maps, which will be useful also later on. For every $n \geq 0$, let $S^n$ be the unit $n$-sphere (it is the boundary $\partial D^{n+1}$ of the unit $(n+1)$-disk $D^{n+1} \subset \mathbb{R}^{n+1}$). Let us take the point $\mathbf{e}_0 = (1,0,\dots,0)$ as the base point for both $S^n$ and $D^{n+1}$. Let us now define the maps

$$c_n\colon I \times S^n \to D^{n+1}\,,\ (t,x) \mapsto (1-t)\mathbf{e}_0 + tx,$$

$$i_+\colon D^{n+1} \to S^{n+1}\,,\ x \mapsto \left(x,\sqrt{1-\|x\|^2}\right),$$

$$i_-\colon D^{n+1} \to S^{n+1}\,,\ x \mapsto \left(x,-\sqrt{1-\|x\|^2}\right) \text{ and}$$

$$k_{n+1}\colon I \times S^n \to S^{n+1}$$

$$k_{n+1}(t,x) = \begin{cases} i_+c_n(2t,x)\,, & 0 \leq t \leq \frac{1}{2} \\ i_-c_n(2-2t,x)\,, & \frac{1}{2} \leq t \leq 1. \end{cases}$$

The map $\dot{k}_{n+1}$ has the property

$$\dot{k}_{n+1}(0,x) = \dot{k}_{n+1}(1,x) = \dot{k}_{n+1}(t,\mathbf{e}_0) = \mathbf{e}_0$$

for every $t \in I$ and every $x \in S^n$; therefore, it gives rise to a homeomorphism

$$\widehat{k}_{n+1}: \Sigma S^n \cong S^{n+1},$$

as can be seen in Fig. I.7. We may then say that for every $y \in S^{n+1} \smallsetminus \{\mathbf{e}_0\}$ there are

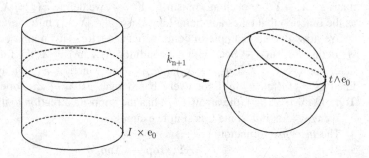

**Fig. I.7**

a unique element $x \in S^n$ and a unique $t \in I \smallsetminus \partial I$ such that $y = t \wedge x$.

The functor

$$\Omega: \mathbf{Top}_* \to \mathbf{Top}_*$$

is defined, on a given object $(X,x_0) \in \mathbf{Top}_*$, as the space

$$\Omega X = \{f \in M(I,X) \mid f(0) = \dot{f}(1) = x_0\}$$

with the topology induced by the compact-open topology of $M(I,X)$. The morphism

$$\Omega(f): \Omega X \to \Omega Y$$

corresponding to the morphism $f \in \mathbf{Top}_*(X,Y)$ is defined through composition of maps:

$$(\forall g \in \Omega X)\Omega(f)(g) = fg: I \to Y.$$

The space $\Omega X$ is called *loop space* (with base at $x_0$). The base point of $\Omega X$ is the constant path on $x_0$.

There is a special relation between the functors $\Omega$ and $\Sigma$ as follows. Let $f: I \times X \to Y$ be a map such that $f(I \times \{x_0\} \cup \partial I \times X) = y_0$; its adjoint $\hat{f}: X \to M(I,Y)$, being continuous (see Theorem (I.1.37)), is such that

$$(\forall x \in X)\ \hat{f}(x)(\partial I) = y_0$$

and so we are able to construct a function

$$\Phi: M_*(\Sigma X,Y) \to M_*(X,\Omega Y).$$

Conversely, given $\hat{f} \in M_*(X, \Omega Y)$, its adjoint $f\colon I \times X \to Y$ is continuous because $I$ is compact Hausdorff (see Lemma (I.1.39)) and $f \in M_*(\Sigma X, Y)$; this allows us to construct a function

$$\Phi'\colon M_*(X, \Omega Y) \to M_*(\Sigma X, Y)$$

such that $\Phi\Phi'$ and $\Phi'\Phi$ be equal to the respective identity functions. In other words,

$$\Phi\colon M_*(\Sigma X, Y) \to M_*(X, \Omega Y)$$

is a bijection (injective and surjective). For this reason, we say that $\Sigma$ is *left adjoint* to $\Omega$.

Given two functors $F, G\colon \mathbf{C} \to \mathbf{C}'$, a *natural transformation*

$$\eta\colon F \to G$$

is a correspondence that takes each object $A \in \mathbf{C}$ to a morphism

$$\eta(A)\colon FA \to GA$$

and such that, for every $f \in \mathbf{C}(A, B)$,

$$G(f)\eta(A) = \eta(B)F(f),$$

in other words, such that the following diagram is commutative:

$$
\begin{array}{ccc}
FA & \xrightarrow{\ \eta(A)\ } & GA \\
{\scriptstyle F(f)}\big\downarrow & & \big\downarrow{\scriptstyle G(f)} \\
FB & \xrightarrow[\ \eta(B)\ ]{} & GB
\end{array}
$$

Two functors $F, G\colon \mathbf{C} \to \mathbf{C}'$ are *equivalent* (and we write $F \doteq G$) if there are two natural transformations $\eta\colon F \to G$ and $\tau\colon G \to F$ such that

$$\tau\eta = 1_F \text{ and } \eta\tau = 1_G$$

where $1_F$ and $1_G$ equal the natural transformations given by the identity.

## I.2.2 Pushouts

Given two morphisms $f\colon A \to B$ and $g\colon A \to C$ of a category $\mathbf{C}$, a *pushout* of $(f, g)$ is a pair of morphisms $\overline{f} \in \mathbf{C}(C, D)$ and $\overline{g} \in \mathbf{C}(B, D)$ such that $\overline{g}f = g\overline{f}$ and satisfying the following *universal property*:  given $h \in \mathbf{C}(B,X)$ and $k \in \mathbf{C}(C,X)$ such that $hf = kg$, there exists a *unique* morphism $\ell \in \mathbf{C}(D,X)$ such that $\ell \bar{f} = k$ and $\ell \bar{g} = h$.

We depict this situation with the commutative diagram

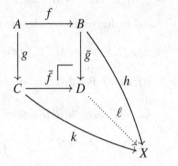

The symbol at the right lower angle of the commutative square indicates that we have a *pushout* diagram.

As a consequence of the universal property of pushouts, a pushout of $(f,g)$ is unique up to isomorphism. In fact, suppose that $(f',g')$ with $f': C \to D'$ and $g': B \to D'$ is a pushout; then by the universal property, there is a unique $\ell': D' \to D$ such that $\ell' g' = \bar{g}$ and $\ell' f' = \bar{f}$; hence, $\ell' \ell = 1_D$. Similarly, we conclude that $\ell \ell' = 1_C$. The morphism $\ell$ is an *isomorphism* and its inverse is $\ell'$.

We say that a category $\mathbf{C}$ is *closed by pushouts* or *closed regarding pushouts* if every pair of morphisms $f: A \to B$ and $g: A \to C$ of $\mathbf{C}$ has a pushout. Not every category is closed by pushouts; we now prove that the category **Top** of topological spaces and the category **Gr** of groups are closed by pushouts.

Let us start with **Top**. We define the *disjoint union* of two topological spaces $B$ and $C$ by taking a set of two points, say, $\{i,j\}$ and constructing the spaces $B \times \{i\}$ and $C \times \{j\}$, homeomorphic to $B$ and $C$, respectively. We then define the union $B \sqcup C = B \times \{i\} \cup C \times \{j\}$ with the inclusions

$$\iota_B: B \to B \sqcup C\,,\ b \mapsto (b,i)\,,$$
$$\iota_C: C \to B \sqcup C\,,\ c \mapsto (c,j)$$

and give $B \sqcup C$ the topology defined by the open sets

$$\iota_B(U) \cup \iota_C(V) = (U \times \{i\}) \cup (V \times \{j\})$$

with $U$ open in $B$ and $V$ open in $C$; in this way, we obtain the topological space called *disjoint union* of $B$ and $C$. By constructing a quotient of the disjoint union, it is possible to prove that **Top** is closed regarding pushouts.

**(I.2.2) Theorem.** *The category* **Top** *is closed by pushouts.*

*Proof.* Let $f: A \to B$ and $g: A \to C$ be any two maps in **Top**. In the disjoint union $B \sqcup C$, identify $f(a)$ with $g(a)$ for every $a \in A$; take the quotient set $B \sqcup_{f,g} C$ obtained by the identifications $f(a) = g(a)$ and then the canonic surjection

$q: B \sqcup C \to B \sqcup_{f,g} C$; finally, give $B \sqcup_{f,g} C$ the quotient topology relative to the surjection $q$ and define the maps

$$\bar{f} = q\iota_C: C \to B \sqcup_{f,g} C$$
$$\bar{g} = q\iota_B: B \to B \sqcup_{f,g} C.$$

The diagram

is commutative. Let $h: B \to X$ and $k: C \to X$ be two continuous functions such that $hf = kg$. We define the function $\ell: B \sqcup_{f,g} C \to X$, by requiring that $\ell(x) = h(x)$ if $x \in B \setminus f(A)$, $\ell(x) = k(x)$ if $x \in C \setminus g(A)$, and $\ell(x) = hf(a) = kg(a)$ for every $x \in f(A)$ or $x \in g(A)$. Since the restrictions of $\ell q$ to $B$ and $C$ are continuous, the composite function $\ell q$ is continuous; by the definition of quotient topology, we conclude that $\ell$ is continuous. It is easily proved that the map $\ell$ is unique. ∎

By the universal property, the space $B \sqcup_{f,g} C$ obtained in the pushout of $(f, g)$ is unique up to homeomorphism.

We have an important case when $A$ is closed in $C$ and $g$ is the inclusion $\iota: A \to C$. We call the space $B \sqcup_{f,\iota} C$ in the pushout of $(f, \iota)$ the *adjunction space* of $C$ to $B$ via $f$.

Let us now consider the category of groups. We first focus on some fundamental results in group theory. Let $S$ be a given set. A *word* defined by the elements of $S$ is a symbol

$$\omega = s_1^{\varepsilon_1} s_2^{\varepsilon_2} \ldots s_n^{\varepsilon_n},$$

where $s_i \in S$ and $\varepsilon_i = \pm 1$; without excluding the case where two consecutive elements $s_i$ are equal, we also request that the *length n* be finite; if $n = 0$, we say that $\omega$ is the *empty word*. Let $W_S$ be the set of all words defined by the elements of $S$. More specifically, given a set $S$, consider the set $S \sqcup S^{-1}$ whose elements are all elements of $S$ and all elements of a copy of $S$, denoted $S^{-1}$. The *words* of $W_S$ of length $n$ are precisely all $n$-tuples of elements of $S \sqcup S^{-1}$.

We now define an equivalence relation $E$ in $W_S$: $w_1 E w_2$ if $w_2$ is obtained from $w_1$ through a finite sequence of operations as follows:

1. Replacing the word $s_1^{\varepsilon_1} s_2^{\varepsilon_2} \ldots s_n^{\varepsilon_n}$ by the word $s_1^{\varepsilon_1} \ldots s_k^{\varepsilon_k} aa^{-1} \ldots s_n^{\varepsilon_n}$, or the word $s_1^{\varepsilon_1} \ldots s_k^{\varepsilon_k} a^{-1} a \ldots s_n^{\varepsilon_n}$;
2. Replacing the words $s_1^{\varepsilon_1} \ldots s_k^{\varepsilon_k} aa^{-1} \ldots s_n^{\varepsilon_n}$ or $s_1^{\varepsilon_1} \ldots s_k^{\varepsilon_k} a^{-1} a \ldots s_n^{\varepsilon_n}$ by the word $s_1^{\varepsilon_1} s_2^{\varepsilon_2} \ldots s_n^{\varepsilon_n}$,

where $0 \leq k \leq n$ and $a$ stands for any element of $S$. Let $F(S) = W_S/E$ be the set of equivalence classes defined by $E$ in $W_S$; we denote the equivalence class of a word $w$ with $[w]$. We give here the definition of a product in $F(S)$ by juxtaposition:

$$[s_1^{\varepsilon_1} \ldots s_n^{\varepsilon_n}][s_{n+1}^{\varepsilon_{n+1}} \ldots s_{n+m}^{\varepsilon_{n+m}}] = [s_1^{\varepsilon_1} \ldots s_{n+m}^{\varepsilon_{n+m}}].$$

**(I.2.3) Lemma.** *The set $F(S)$ with the operation juxtaposition is a group.*

*Proof.* Clearly, the juxtaposition is associative; the identity is the class of the empty word and is denoted with 1; finally, the inverse of $[\omega] = [s_1^{\varepsilon_1} s_2^{\varepsilon_2} \ldots s_n^{\varepsilon_n}]$ is $[\omega]^{-1} = [s_n^{-\varepsilon_1} s_{n-1}^{-\varepsilon_{n-1}} \ldots s_1^{-\varepsilon_1}]$. ∎

The group $F(S)$ is the *free group* generated by $S$; the elements of $S$ are the *generators* of $F(S)$. When $S$ has a finite number of elements, we may also write $F(\{s_1, s_2, \ldots, s_l\}) = \langle s_1, s_2, \ldots, s_l \rangle$.

It is useful to write the elements of $F(S)$ *without* the square brackets. In practice, it is usual to simplify the notation by means of integral exponents, not necessarily $\pm 1$, such as in $s_1^1 s_1^1 s_1^1 s_1^1 = s_1^4$ or $s_2^{-1} s_2^{-1} = s_2^{-2}$. It is also customary to write $s$ instead of the word $s^1$ defined by $s \in S$; all this allows us to write equalities such as $s^2 s^{-1} = s$, $ss^{-1} = 1$, and so on.

**(I.2.4) Remark.** A word $\omega = s_1^{\varepsilon_1} s_2^{\varepsilon_2} \ldots s_n^{\varepsilon_n}$ is *reduced* if, for every element $a$ of $S$, the "subword" $a^\varepsilon a^{-\varepsilon}$ does not appear in $\omega$ (that is to say, if it is not possible to do any "cancellation"). In each class $[\omega] = [s_1^{\varepsilon_1} s_2^{\varepsilon_2} \ldots s_n^{\varepsilon_n}]$, there is one and only one reduced word. We may therefore define $F(S)$ by considering only the reduced words defined by $S$.

Given a nonempty subset $R \subset F(S)$, let $\overline{R}$ be the intersection of all normal subgroups of $F(S)$ that contain $R$; the quotient group

$$F(S)/\overline{R} = F(S;R)$$

is the group *generated* by the set $S$ with the *relations* $R$ in $F(S)$. The elements of $F(S;R)$ are the lateral classes (modulo $R$) of elements of $F(S)$; for each element $\omega \in F(S)$, $\omega R$ denotes its class modulo $R$. In practice, to construct $F(S;R)$ we only take reduced words of $S$ and free them from all the subwords of $R$. If $S = \{s_1, s_2, \ldots, s_l\}$ and $R = \{w_1, w_2, \ldots, w_r\}$, we normally write

$$\langle s_1, s_2, \ldots, s_l \mid w_1 = w_2 = \cdots = w_r = 1 \rangle = F(S;R).$$

*Examples*: 1. $S = \{s\}$, $F(S) = \langle s \rangle \simeq \mathbb{Z}$; this group could also be described as $S = \{s, t\}$, $R = \{t\}$, $\langle s, t \mid t = 1 \rangle = F(S;R) \simeq \mathbb{Z}$.
2. $S = \{s\}$, $R = \{s^2\}$, $\langle s \mid s^2 = 1 \rangle \simeq \mathbb{Z}_2$.
3. $S = \{s, t\}$, $R = \{sts^{-1}t^{-1}\}$, $\langle s, t \mid sts^{-1}t^{-1} = 1 \rangle \simeq \mathbb{Z} \times \mathbb{Z}$, the *free* Abelian group generated by the elements $s$ and $t$.
4. Any group $G$ may be viewed as a group generated by elements and relations: take $S = G$ and $R_G = \{(st)^1 t^{-1} s^{-1} \mid s, t \in G\}$; then, $G \simeq F(G; R_G)$.

**(I.2.5) Lemma.** *Given a group $G$ with unity $1_G$, a group $F(S;R)$, and a function of sets $\theta: S \to G$ such that, for every $\omega \in R$, $\theta(\omega) = 1_G$, there exists a unique group homomorphism*

$$\overline{\theta}: F(S;R) \to G$$

*such that $\overline{\theta}(s) = \theta(s)$ for every $s \in S$.*[4]

*Proof.* Given any word $s_1^{\varepsilon_1} \ldots s_n^{\varepsilon_n} \in W_S$, define

$$\overline{\theta}([s_1^{\varepsilon_1} \ldots s_n^{\varepsilon_n}]) := \theta(s_1)^{\varepsilon_1} \ldots \theta(s_n)^{\varepsilon_n}. \qquad \blacksquare$$

**(I.2.6) Theorem.** *The category **Gr** is closed by pushouts.*

*Proof.* Given any pair of homomorphisms $f: G \to G_1$ and $g: G \to G_2$, we view the groups $G_1$ and $G_2$ as

$$G_i = F(G_i; R_{G_i}) \,,\ R_{G_i} = \{(xy)^1 y^{-1} x^{-1} \,|\, x, y \in G_i\} \,,\ i = 1,2$$

and consider the set

$$R_{f,g} = \{f(x)g(x)^{-1} \,|\, x \in G\}.$$

We now define the group

$$\overline{G} := F(G_1 \cup G_2; R_{G_1} \cup R_{G_2} \cup R_{f,g})$$

and the canonic homomorphisms

$$\overline{f}: G_2 \to \overline{G} \,,\ \overline{g}: G_1 \to \overline{G}.$$

Since $f(x)g(x)^{-1}$ is a relation in $F(G_1 \cup G_2)$ for every $x \in G$, the following diagram commutes:

$$
\begin{array}{ccc}
G & \xrightarrow{\ f\ } & G_1 \\
{\scriptstyle g}\downarrow & & \downarrow{\scriptstyle \overline{g}} \\
G_2 & \xrightarrow[\ \overline{f}\ ]{} & \overline{G}
\end{array}
$$

Let us prove the universal property. Given two group homomorphisms

$$h_i: G_i \to H \,,\ i = 1,2$$

such that $h_1 f = h_2 g$, the function

$$\theta: G_1 \cup G_2 \to H \,,\ (\forall x \in G_i)\ \theta(x) = h_i(x) \,,\ i = 1,2$$

---

[4] Warning: Here $s$ has two meanings. As an element of the domain of $\overline{\theta}$, it is the class $[s]$, but as an element of the domain of $\theta$, it is just the element $s$.

satisfies the equality
$$(\forall x \in G)\ \theta(f(x)g(x)^{-1}) = 1_H\ ;$$
by Lemma (I.2.5), there is a unique homomorphism
$$\overline{\theta}\colon \overline{G} \to H$$
that extends the function $\theta$ and such that
$$\overline{\theta}f = h_2\ ,\ \overline{\theta}\overline{g} = h_1.$$

The group $\overline{G}$, also denoted with $G_1 *_{f,g} G_2$, is the *amalgamated product* of the groups $G_1$ and $G_2$ with respect to the homomorphisms $f$ and $g$.

## *Exercises*

**1.** Prove that the relation of based homotopy in $\mathbf{Top}_*(X,Y)$ is an equivalence relation.

**2.** Let $A$ be a subspace of $X$ and $i\colon A \longrightarrow X$ be the inclusion map; then $A$ is a *deformation retract* of $X$ if there exists a continuous function $r\colon X \to A$ such that $ri = 1_A\colon A \to A$ and $ir \sim 1_X$. In particular, if $A \subset X$ is a deformation retract and $A = \{x_0\}$, we say that $X$ is *contractible* to $x_0$. In this case, the identity $1_X\colon X \to X$ is homotopic to the constant map $c\colon X \to \{x_0\}$. The space $A$ is called *strong deformation retract* of $X$ if $ri = 1_A$ and $ir \sim_A 1_X$. Intuitively, a subspace $A$ of $X$ is a strong deformation retract of $X$ if $X$ can be deformed over $A$ with continuity, keeping $A$ fixed during the deformation. Clearly, a strong deformation retract is a deformation retract. It follows from the definitions that, if $A$ is a (strong or not) deformation retract of $X$, then $A$ and $X$ are of the same homotopy type.

(i) Prove that the circle
$$S^1 = \{(x,y) \in \mathbb{R}^2 \,|\, x^2 + y^2 = 1\}$$
is a strong deformation retract of the cylinder
$$C = \{(x,y,z) \in \mathbb{R}^3 \,|\, x^2 + y^2 = 1\,,\, 0 \le z \le 1\}.$$

(ii) Prove that the disk
$$D^2 = \{(x,y) \in \mathbb{R}^2 \,|\, x^2 + y^2 \le 1\}$$
is contractible to $(0,0)$.

**3.** Prove that for every $X,Y \in \mathbf{Top}_*$, the function
$$[\Phi]\colon [\Sigma X, Y]_* \cong [X, \Omega Y]_*\,,\ [f] \mapsto [\Phi(f)]$$
is a natural bijection.

## I.3 Group Actions

A *topological group* is a group together with a topology such that the functions defined by the multiplication and by inverting the elements of the group are continuous; in more precise terms, let $G$ be a group with the multiplication

$$\times: G \times G \to G, \ (\forall (g,g') \in G \times G) \ (g,g') \mapsto gg';$$

let us consider $G \times G$ with the product topology; we then request that the two maps $(g,g') \mapsto gg'$ and $g \mapsto g^{-1}$ be continuous functions.

Here are some examples of topological groups; the details of the proofs are left to the reader.

1. All groups with discrete topology.
2. The additive group $\mathbb{R}$ with the Euclidean topology (given by the *distance* $(\forall x,y \in \mathbb{R}) \ d(x,y) = |x-y|$).
3. The multiplicative group $\mathbb{R}^* = \mathbb{R} \smallsetminus \{0\}$ with the topology given in the previous example.
4. The additive group $\mathbb{C}$ of the complex numbers with the topology given by the distance
$$(\forall x,y \in \mathbb{C}) \ d(x,y) = |x-y|.$$
5. Let $GL(n,\mathbb{R})$ be the multiplicative group of all real, invertible square matrices of rank $n$ (the *general linear* group). We define a topology on $GL(n,\mathbb{R})$ as follows: we note that the function

$$(a_{ij})_{i,j=1,\dots,n} \in M(n,\mathbb{R}) \mapsto (a_{11},a_{12},\dots,a_{21},a_{22},\dots,a_{nn}) \in \mathbb{R}^{n^2}$$

from the set $M(n,\mathbb{R})$ of all real matrices $n \times n$ to the set $\mathbb{R}^{n^2}$ is a bijection. This function defines a topology on $M(n,\mathbb{R})$, which derives from the Euclidean topology on $\mathbb{R}^{n^2}$ and induces a topology on $GL(n,\mathbb{R})$. With this topology, the group $GL(n,\mathbb{R})$ becomes a topological group.
6. Let $SO(n)$ be the subgroup of $GL(n,\mathbb{R})$ of the matrices $M \in GL(n,\mathbb{R})$ such that $M^{-1}$ is the transpose of $M$ and $\det M = 1$; the previously defined topology on $GL(n,\mathbb{R})$ induces a topological group structure on $SO(n)$.

Let $G$ be a topological group with neutral element $1_G$ and $X$ be a topological space. An *action* (on the right) of $G$ on $X$ is a continuous function

$$\phi: X \times G \to X$$

such that

(a) $(\forall x \in X) \ \phi(x,1_G) = x$
(b) $(\forall x \in X)(\forall g,g' \in G) \ \phi(\phi(x,g),g') = \phi(x,gg')$.

We say that $G$ acts on the right of $X$ (through the action $\phi$). To make it simple, we write $\phi(x,g) = xg$. Similarly, it is possible to define an action on the left.

**(I.3.1) Lemma.** *Let* $\phi\colon G \times X \to X$ *be an action of a topological group* $G$ *on a topological space* $X$. *For every* $g \in G$, *the function*

$$\phi(g)\colon X \to X,\ x \mapsto xg$$

*is a homeomorphism.*

*Proof.* The function $\phi(g^{-1})$ is the inverse of $\phi(g)$. The continuity of $\phi(g)$ is immediate; in fact, $\phi(g)$ is the composite of the action $\phi$ and the inclusion $X \times \{g\} \hookrightarrow X \times G$. ∎

An action $\phi\colon X \times G \to X$ gives rise to an equivalence relation $\equiv_\phi$ in $X$ (a partition of $X$ into $G$-orbits):

$$x \equiv_\phi x' \iff (\exists g \in G)x' = xg.$$

The equivalence class $[x]$ of the element $x \in X$ is the *orbit* of $x$, also denoted by $xG$ (or $Gx$, if the action is on the left); and we write $X/_G$ to indicate the set $X/\equiv_\phi$ of the orbits of $X$. The set $X/_G$ with the quotient topology given by the canonical epimorphism

$$q\colon X \to X/_G,\ x \mapsto [x]$$

is called *orbit space* of $X$ under the action of $G$.

**(I.3.2) Lemma.** *The quotient map* $q\colon X \to X/_G$ *is open. If* $G$ *is a finite group, then* $q$ *is also closed.*[5]

*Proof.* Let $U$ be any open set in $X$. Then

$$\begin{aligned}
q^{-1}(q(U)) &= \{x \in X \mid q(x) \in q(U)\} \\
&= \{x \in X \mid (\exists\,g \in G)(\exists\,y \in U)\,x = yg\} \\
&= \{x \in X \mid (\exists\,g \in G)x \in Ug\} \\
&= \bigcup_{g \in G} \phi(g)(U).
\end{aligned}$$

Since, according to Lemma (I.3.1), the functions $\phi(g)$ are homeomorphisms for every $g \in G$, $q^{-1}(q(U))$ is a union of open sets in $X$; hence $q(U)$ is open in $X/_G$.

If $G$ is finite and $K \subset X$ is closed,

$$q^{-1}(q(K)) = \cup_{g \in G}\phi(g)(K)$$

is closed as a finite union of closed sets. ∎

We recall that the real projective plane $\mathbb{RP}^2$ in Example (I.1.5) on p. 5 was constructed as the quotient space $D^2_\equiv$, obtained from the unit disk $D^2$ by identifying $(x,y) = (-x,-y)$ for every $(x,y) \in D^2$ such that $x^2 + y^2 = 1$. Then, we note that the discrete topological group $\mathbb{Z}_2 = \{1,-1\}$ acts on the unit sphere

---

[5] That is to say, it takes closed sets into closed sets.

$S^2 = \{(x,y,z) \in \mathbb{R}^3 \mid x^2 + y^2 + z^2 = 1\}$ by $\phi\colon S^2 \times \mathbb{Z}_2 \to S^2$, with $\phi((x,y,z),1) = (x,y,z)$ and $\phi((x,y,z),-1) = (-x,-y,-z)$. Therefore, we obtain the orbit space $S^2/\mathbb{Z}_2$ of this action by identifying antipodal points of the sphere $S^2$. On the other hand, let $E \cong S^1$ be the equator of $S^2$ and let the antipodal points $a$ and $-a$ of $S^2 \setminus E$ be identified. Let $S^2_+$ be the north hemisphere of $S^2$; we note that $S^2_+ \setminus E$ alone represents the classes $[a] \in S^2/\mathbb{Z}_2$ where $a \in S^2 \setminus E$. In order to obtain $S^2/\mathbb{Z}_2$, we must still identify the antipodal points of $E$. Since $S^2_+$ is homeomorphic to the unit disk $D^2$, $\mathbb{RP}^2$ is also homeomorphic to the space $S^2_+/\mathbb{Z}_2$.

In general, $\mathbb{Z}_2$ acts on the $n$-dimensional unit sphere ($n \geq 1$)

$$S^n = \left\{ (x_0, x_1, \ldots, x_n) \in \mathbb{R}^{n+1} \mid \sum_{i=0}^{n} x_i = 1 \right\}$$

by the antipodal action $\phi\colon S^n \times \mathbb{Z}_2 \to S^n$, such that

$$\phi((x_0, x_1, \ldots, x_n), 1) = (x_0, x_1, \ldots, x_n)$$

and

$$\phi((x_0, x_1, \ldots, x_n), -1) = (-x_0, -x_1, \ldots, -x_n).$$

The orbit space $S^n/\mathbb{Z}_2$ is the *n-dimensional real projective space*. When $n = 1$, we have the *real projective line*, which is homeomorphic to the circle $S^1$.

We say that a group $G$ acts *freely* on a space $X$ if

$$(\forall x \in X)(\forall g \in G,\, g \neq 1_G)\, xg \neq x.$$

## Exercises

**1.** Prove that the projective line $\mathbb{RP}^1$ is homeomorphic to $S^1$.

**2.** Prove that the action of a subgroup $H \subset G$ of a topological group $G$ given by the product $(g,h) \mapsto gh$ is an action (on the right) of $H$ on $G$. Find the quotient $G/H$ when $G = SO(2)$ and $H \subset G$ is the group generated by a rotation angle $\theta$.

**3.** Prove that the topological group $GL(n, \mathbb{R})$ is connected.

**4.** Prove that all discrete subgroups of $\mathbb{R}$ are cyclic and infinite.

**5.** Find a (non-Abelian) subgroup of $SO(3)$ that is free on two generators.

**6.** Consider the action of $\mathbb{Q}$ on $\mathbb{R}$ given by $(x,q) \mapsto x + q$. Is the quotient connected? Hausdorff? Compact?

**7.** Let $G$ be a topological group that acts on the left on two spaces $X$ and $Y$. Prove that the action of $G$

$$(g, [x \mapsto f(x)]) \in G \times [X,Y] \mapsto [x \mapsto gf(g^{-1}x)] \in [X,Y]$$

on the homotopy classes of maps is well defined.

# Chapter II
# The Category of Simplicial Complexes

## II.1 Euclidean Simplicial Complexes

Let us recall that a subset $C \subset \mathbb{R}^n$ is *convex* if $x, y \in C, t \in [0,1] \implies tx + (1-t)y \in C$. The *convex hull* of a subset $X \subset \mathbb{R}^n$ is the smallest convex subset of $\mathbb{R}^n$, which contains $X$. We say that $d + 1$ points $x_0, x_1, \ldots, x_d$ belonging to the Euclidean space $\mathbb{R}^n$ are *linearly independent* (from the affine point of view) if the vectors $x_1 - x_0$, $x_2 - x_0$, ..., $x_d - x_0$ are linearly independent. A vector $x - x_0$ of the vector space generated by these vectors can be written as a sum $x - x_0 = \sum_{i=1}^{d} r_i(x_i - x_0)$ with real coefficients $r_i$; notice that if we write $x$ as $x = \sum_{i=0}^{d} \alpha_i x_i$, then $\sum_{i=0}^{d} \alpha_i = 1$. If $\{x_0, \ldots, x_d\} \in X$ are affinely independent, the convex hull of $X$ is said to be an (Euclidean) *simplex* of dimension $d$ contained in $\mathbb{R}^n$; its points $x$ can be written in a unique fashion as linear combinations

$$x = \sum_{i=1}^{d} \lambda_i x_i,$$

with real coefficients $\lambda_i$. The coefficients $\lambda_i$ are called *barycentric coordinates* of $x$; they are nonnegative real numbers and satisfy the equality $\sum_{i=0}^{d} \lambda_i = 1$. The points $x_i$ are the *vertices* of the simplex. The *standard n-simplex* is the simplex obtained by taking the convex hull of the $n + 1$ points of the standard basis of $\mathbb{R}^{n+1}$ (see Figs. II.1 and II.2 for dimensions $n = 1$ and $n = 2$, respectively).

The *faces* of a simplex $s \subset \mathbb{R}^n$ are the convex hulls of the subsets of its vertices; the faces which do not coincide with $s$ are the *proper faces*. We can define the *interior* of a simplex $s$ as the set of all points of $s$ with positive barycentric coordinates $\lambda_i > 0$. We indicate the interior of $s$ with $\mathring{s}$. If the dimension of $s$ is at least 1, $\mathring{s}$ coincides with the topological interior. At any rate, it is not hard to prove that we obtain the interior of a simplex by removing all of its proper faces.

An Euclidean *simplicial complex* is a finite family of simplexes of an Euclidean space $\mathbb{R}^n$, which satisfies the following properties:

D.L. Ferrario and R.A. Piccinini, *Simplicial Structures in Topology*,
CMS Books in Mathematics, DOI 10.1007/978-1-4419-7236-1_II,
© Springer Science+Business Media, LLC 2011

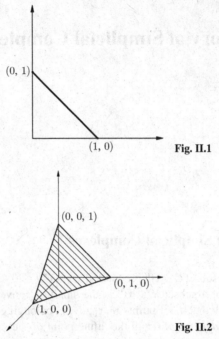

Fig. II.1

Fig. II.2

**S1** If $s \in K$, then every face of $s$ is in $K$.

**S2** If $s_1$ and $s_2$ are simplexes of $K$ with nondisjoint interiors $\mathring{s}_1 \cap \mathring{s}_2 \neq \emptyset$, then $s_1 = s_2$.

The *dimension* of $K$ is the maximal dimension of its simplexes.[1] Figure II.3 repre-

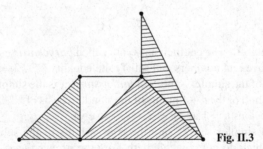

Fig. II.3

sents a two-dimensional simplicial complex of $\mathbb{R}^2$; Fig. II.4 is a set of simplexes, which is *not* a simplicial complex.

---

[1] It is possible to define Euclidean complexes with infinitely many simplexes, provided we add the *local finiteness* property that is to say, we ask that each point of a simplex has a neighborhood, which intersects only finitely many simplexes of $K$. We do this so that the topology of the (infinite) Euclidean complex $K$ coincides with the topology of the geometric realization $|\widehat{K}|$ (we are referring to the topology defined by Remark (II.2.13)) of the abstract simplicial complex $|\widehat{K}|$ associated in a natural fashion to $K$ (we shall give the definition of abstract simplicial complex in a short while).

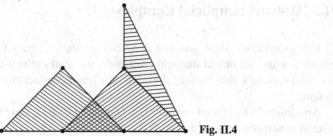

**Fig. II.4**

**(II.1.1) Example** (Euclidean polyhedra). 1. Every simplex of $\mathbb{R}^n$ together with all its faces is a simplicial complex.

2. The set of all proper faces of a $d$-dimensional simplex in $\mathbb{R}^n$ is a $(d-1)$-dimensional simplicial complex.

3. The set of all closed intervals $[1/n, 1/(n+1)]$, with $n \in \mathbb{N}$, is a simplicial complex (with infinitely many simplexes) of $\mathbb{R}$.

4. Let $P_m$ be the regular polygonal line contained in $\mathbb{C} \cong \mathbb{R}^2$, whose vertices are the $m$th-roots of the unity $\{z \in \mathbb{C} \mid z^m = 1\}$. The corresponding simplicial complex is homeomorphic to the circle $S^1$ and is depicted in Fig. II.5.

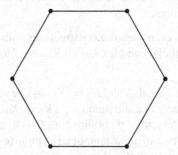

**Fig. II.5**

5. The Platonic solids can be subdivided by triangles; they give rise to simplicial complexes of $\mathbb{R}^3$. An example is given by the icosahedron of Fig. II.6.

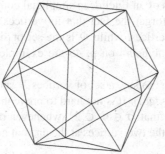

**Fig. II.6**

## II.2 Abstract Simplicial Complexes

In this section, we shall give the definition of the category **Csim** of simplicial complexes and simplicial maps; furthermore, we shall define two important functors with domain **Csim**, namely, the geometric realization functor and the homology functor.

An *(abstract) simplicial complex* is a pair $K = (X, \Phi)$ given by a *finite* set $X$ and a set of nonempty subsets of $X$ such that:

**K1** $(\forall x \in X)$, $\{x\} \in \Phi$,
**K2** $(\forall \sigma \in \Phi)(\forall \sigma' \subset \sigma, \sigma' \neq \emptyset)$, $\sigma' \in \Phi$.

The elements of $X$ are the *vertices* of $K$. The elements of $\Phi$ are the *simplexes* of $K$. If $\sigma$ is a simplex of $K$, every non-empty $\sigma' \subset \sigma$ is a *face* of $\sigma$. According to condition K2, we can say that all faces of a simplex are simplexes. A simplex $\sigma$ with $n+1$ elements ($n \geq 0$) is an *n-simplex* (we also say that $\sigma$ is a simplex of *dimension n*); we adopt the notation $\dim \sigma = n$. It follows that the 0-simplexes are *vertices*. The dimension of $K$ is the maximal dimension of its simplexes; if the dimensions of all simplexes of $K$ have a maximum $n$, we say that $K$ has *dimension n* or that $K$ is *n-dimensional*.

**(II.2.1) Remark.** We explicitly observe that in this book all simplicial complexes have a finite number of vertices.

Before we present some examples and constructions with simplicial complexes, we give a definition: a simplicial complex $L = (Y, \Psi)$ is a *subcomplex* of $K = (X, \Phi)$ if $Y \subset X$ and $\Psi \subset \Phi$.

**(II.2.2) Remark.** Let $K_0 = (X_0, \Phi_0)$ and $K_1 = (X_1, \Phi_1)$ be subcomplexes of a simplicial complex $K$; we observe that the union $K_0 \cup K_1 = (X_0 \cup X_1, \Phi_0 \cup \Phi_1)$ and the intersection $K_0 \cap K_1 = (X_0 \cap X_1, \Phi_0 \cap \Phi_1)$ (with $X_0 \cap X_1 \neq \emptyset$) are subcomplexes of $K$. In particular, the union of two *disjoint* simplicial complexes $K_0$ and $K_1$ (that is to say, such that $X_0 \cap X_1 = \emptyset$) is a simplicial complex.

Let us now give some examples.

1. Let $X$ be a finite set and let $\wp(X) = 2^X$ be the set of all subsets of $X$; clearly, the pair $K = (X, \wp(X) \smallsetminus \emptyset)$ is a simplicial complex.
2. The set of all simplexes of an Euclidean simplicial complex is an abstract simplicial complex if we forget the fact that its vertices are points of $\mathbb{R}^n$. The set $X$ is the set of all vertices, while $\Phi$ is the set of simplexes. Thus, the examples of Euclidean polyhedra on p. 45 are examples of abstract simplicial complexes.
3. Let $\Gamma$ be a graph (that is to say, a set of vertices $X$ and a symmetric subset $\Phi$ of $X \times X$, called *set of edges*). It is not hard to prove that $(X, \Phi)$ is a simplicial complex if we assume that $\sigma \in \Phi \subset 2^X$ whenever $\sigma$ is a set with just one element or is the set of the two vertices at the ends of an edge.

**(II.2.3) Definition** (generated complex). Let $K = (X, \Phi)$ be a simplicial complex; for every simplex $\sigma \in \Phi$, the pair

$$\overline{\sigma} = (\sigma, \wp(\sigma) \smallsetminus \emptyset)$$

is a simplicial complex; $\overline{\sigma}$ is the simplicial complex *generated* by $\sigma$ (sometimes also called *closure* of $\sigma$). More generally, let $B$ be a set of simplexes of $K$, that is to say, $B \subset \Phi$; then

$$\overline{B} = \bigcup_{\sigma \in B} \overline{\sigma}$$

is the simplicial complex *generated* by all the simplexes of the set $B$. Observe that $\overline{B}$ is a subcomplex of $K$.

**(II.2.4) Definition** (boundary of a simplex). For every simplex $\sigma$ of a simplicial complex,

$$\dot{\sigma} = (\sigma, \wp(\sigma) \smallsetminus \{\emptyset, \sigma\})$$

is a simplicial complex, called *boundary* of $\sigma$. By an abuse of notation, we write

$$\overline{\sigma} = \dot{\sigma} \cup \sigma.$$

**(II.2.5) Definition** (join and suspension). Given two simplicial complexes $K = (X, \Phi)$ and $L = (Y, \Psi)$, the *join* of $K$ and $L$ is the simplicial complex $K * L$ whose vertices are all the elements of the set $X \cup Y$, and whose simplexes are the elements of the sets $\Phi$, $\Psi$ and of the set

$$\Phi * \Psi = \{\{x_0, \ldots, x_n, y_0, \ldots, y_m\} \mid \{x_0, \ldots, x_n\} \in \Phi, \{y_0, \ldots, y_m\} \in \Psi\}.$$

In other words, a nonempty subset $\{x_0, \ldots, x_n, y_0, \ldots, y_m\}$ of $X \cup Y$ is a simplex of $K * L$ if and only if $\{x_0, \ldots, x_n\} \in \Phi \cup \{\emptyset\}$ and $\{y_0, \ldots, y_m\} \in \Psi \cup \{\emptyset\}$. In particular, if $L = (Y, \Psi)$ is the simplicial complex defined by a unique point $y$, $K * y = Ky$ is the *cone* (sometimes called *abstract cone*) of $K$ with vertex $y$ (of course, we can also define the cone $yK$). An $n$-simplex with $n \geq 1$ can be interpreted as the cone of any of its faces (of dimension $n - 1$).

If $L$ is the simplicial complex determined by exactly two points $x$ and $y$, that is to say,

$$L = (Y, \Psi) \text{ with } Y = \{x, y\}, \Psi = \{\{x\}, \{y\}\},$$

the join $K * L = \Sigma K$ is called *suspension* of $K$. Observe that $\Sigma K$ can be viewed as the union of the cones $K * x$ and $K * y$.

The category **Csim** of *simplicial complexes* is the category whose objects are all simplicial complexes, and whose morphisms $f : K = (X, \Phi) \to L = (Y, \Psi)$ are the functions (between sets) $f : X \to Y$ such that

$$(\forall \sigma = \{x_0, x_1, \ldots, x_n\} \in \Phi), f(\sigma) = \{f(x_0), f(x_1), \ldots, f(x_n)\} \in \Psi.$$

A morphism $f \in \mathbf{Csim}(K, L)$ is a *simplicial function* from $K$ to $L$.

## II.2.1 The Geometric Realization Functor

For a given simplicial complex $K = (X, \Phi)$, let $V(K)$ be the set of all functions $p: X \to \mathbb{R}_{\geq 0}$ (nonnegative real numbers); we define the *support* of an arbitrary $p \in V(K)$ to be the finite set

$$s(p) = \{x \in X \mid p(x) > 0\}.$$

Let $|K|$ be the set defined as follows:

$$|K| = \{p \in V(K) \mid s(p) \in \Phi \text{ and } \sum_{x \in s(p)} p(x) = 1\}.$$

We now define the function

$$d \colon |K| \times |K| \longrightarrow \mathbb{R}_{\geq 0}$$

which takes any pair $(p, q) \in |K| \times |K|$ into the real number

$$d(p, q) = \sqrt{\sum_{x \in X} (p(x) - q(x))^2}.$$

This function is a metric on $K$ (verify the conditions defining a metric given in Sect. I.1.5); hence, it defines a (metric) topology on $|K|$. The metric space $|K|$ is the *geometric realization* of $K$. Observe that $|K|$ is a *bounded* space, in the sense that $(\forall p, q \in |K|)$, $d(p, q) \leq \sqrt{2}$. Moreover, $|K|$ is a Hausdorff space.

We can write the elements of $K$ as finite linear combinations. In fact, for each vertex $x$ of $K$, with a slight abuse of language, let us denote with $x$ the function of $V(K)$, with value 1 at the vertex $x$ and 0 at any other vertex; in a more formal fashion,

$$(\forall y \in X) \, x(y) = \begin{cases} 0 \text{ if } y \neq x \\ 1 \text{ if } y = x \end{cases}$$

(in other words, we identify the vertex $x$ with the corresponding real function of $V(K)$, whose support coincides with the set $\{x\}$). Hence if $s(p) = \{x_0, x_1, \ldots, x_n\}$ is the support of $p \in |K|$, and assuming that $p(x_i) = \alpha_i$, $i = 0, 1, \ldots, n$, we can write $p$ as

$$p = \sum_{i=0}^{n} \alpha_i x_i.$$

The real numbers $\alpha_i$, $i = 0, \ldots, n$, are the *barycentric coordinates* of $p$ (in agreement with the barycentric coordinates defined by $n+1$ independent points of an Euclidean space).

**(II.2.6) Remark.** Because $K$ has a finite number of vertices, say $n$, we can embed the set of vertices $X$ in the Euclidean space $\mathbb{R}^n$, so that the images of the elements of $X$ coincide with the vectors of the standard basis. Then we can take the convex hulls in $\mathbb{R}^n$ of the vectors corresponding to the simplexes of $K$, to obtain an Euclidean simplicial complex $K' \subset \mathbb{R}^n$ associated with $K$. We shall see in a short while that $K'$ is isomorphic to the geometric realization $|K|$ (actually, there exists an isometry between these two metric spaces; furthermore, the set of all functions $X \to \mathbb{R}_{\geq 0}$ coincides with the positive quadrant of $\mathbb{R}^n$).

The following statement holds true: two points $p, q \in |K|$ coincide if and only if they have the same barycentric coordinates.

The *geometric realization functor*

$$| \ |: \mathbf{Csim} \longrightarrow \mathbf{Top}$$

is defined over an object $K \in \mathbf{Csim}$ as the geometric realization $|K|$, and over a morphism $f \in \mathbf{Csim}(K, L)$ as

$$|f|: \ |K| \to |L|, \ |f|(\textstyle\sum \alpha_i x_i) = \sum \alpha_i f(x_i).$$

To prove that $| \ |$ is indeed a functor, we need the following result.

**(II.2.7) Theorem.** *The function $|f|$ induced from a simplicial function $f: K \to L$ is continuous.*

*Proof.* It is enough to prove that, for every $p \in |K|$, there exists a constant $c(p) > 0$ which depends on $p$ and such that, for every $q \in |K|, d(|f|(p), |f|(q)) \leq c(p)d(p,q)$.

Assume that

$$s(p) = \{x_0, \ldots, x_n\} \ \text{ and } \ s(q) = \{y_0, \ldots, y_m\}$$

and also that $p(x_i) = \alpha_i$ for $i = 0, \ldots, n$, and $q(y_j) = \beta_j$ for $j = 0, \ldots, m$. We consider three cases.

*Case 1:* $s(p) \cap s(q) = \emptyset$ - In this situation

$$d(p,q) = \sqrt{\sum_{i=0}^{n} \alpha_i^2 + \sum_{j=0}^{m} \beta_j^2} \geq \sqrt{\sum_{i=0}^{n} \alpha_i^2} \ ;$$

because $\sum_{i=0}^{n} \alpha_i = 1$, $\sum_{i=0}^{n} \alpha_i^2$ has its minimum value only when $\alpha_i = 1/(n+1)$, for every $i = 0, \ldots, n$. It follows that $d(p,q) \geq 1/\sqrt{n+1}$ and

$$\frac{d(|f|(p), |f|(q))}{d(p,q)} \leq \frac{\sqrt{2}}{1/\sqrt{n+1}}$$

(recall that $d(|f|(p), |f|(q)) \leq \sqrt{2}$); so,

$$d(|f|(p), |f|(q)) \leq \sqrt{2(n+1)}d(p,q);$$

thus, we define $c(p) = \sqrt{2(n+1)}$.

*Case 2:* $s(p) \cap s(q) \neq \emptyset$, but $s(p) \not\subset s(q)$ and $s(q) \not\subset s(p)$ –

Let us rewrite the indices of the elements of $s(p)$ and $s(q)$ to have the following common elements:

$$x_r = y_0, x_{r+1} = y_1, \ldots, x_n = y_{n-r}.$$

Notice that the set $s(p) \cup s(q)$ has exactly $m+r+1$ common elements. Now consider the elements

$$z_i = \begin{cases} x_i, & 0 \le i \le r-1 \\ x_i = y_{i-r}, & r \le i \le n \\ y_{i-r}, & n+1 \le i \le m+r \end{cases}$$

together with the real numbers

$$\gamma_i = \begin{cases} -\alpha_i, & 0 \le i \le r-1 \\ -\alpha_i + \beta_{i-r}, & r \le i \le n \\ \beta_{i-r}, & n+1 \le i \le m+r. \end{cases}$$

Notice that $\gamma_i < 0$ for $i = 0, \ldots, r-1$ and $\gamma_i > 0$ for $i = n+1, \ldots, m+r$, because $\alpha_i > 0$ for every $i = 0, 1, \ldots, n$ and $\beta_i > 0$ for $i = 1, \ldots, m$. Let us order the numbers $\gamma_i$ in such a way that $\gamma_0 \le \gamma_1 \le \ldots \le \gamma_{m+r}$ (if necessary, we make a permutation of the indices). Let $l$ be the largest index for which $\gamma_l < 0$ (because of the assumptions we made, such a set of indices cannot be empty - thus $l$ exists - nor can it be the set of all indices - thus $r \le l \le n$); moreover, the vertices (viewed as functions) $z_0, z_1, \ldots, z_l$ are summands of $p$ (the numbers $\gamma_i$ are negative), while the vertices $z_{l+1}, \ldots, z_{m+r}$ are part of $q$ (the corresponding numbers $\gamma_i$ are non-negative). At this point, take $\lambda = \sum_{i=0}^{l} \gamma_i < 0$ and the two finite successions of real positive numbers

$$\{\frac{\gamma_0}{\lambda}, \ldots, \frac{\gamma_l}{\lambda}\} \text{ and } \{\frac{\gamma_{l+1}}{-\lambda}, \ldots, \frac{\gamma_{m+r}}{-\lambda}\}.$$

The elements

$$p' = \sum_{i=0}^{l} \frac{\gamma_i}{\lambda} z_i \text{ and } q' = \sum_{i=l+1}^{m+r} \frac{\gamma_i}{-\lambda} z_i$$

are in $|K|$ because

$$\sum_{i=0}^{l} \frac{\gamma_i}{\lambda} = \sum_{i=l+1}^{m+r} \frac{\gamma_i}{-\lambda} = 1;$$

from what we proved above, it follows that $s(p') \subset s(p)$ and $s(q') \subset s(q)$. But $s(p') \cap s(q') = \emptyset$ and so, by *Case 1*,

$$d(|f|(p'), |f|(q')) \le \sqrt{2(l+1)} d(p', q').$$

The equalities

$$d(p', q') = \frac{1}{-\lambda} d(p, q),$$

$$d(|f|(p'), |f|(q')) = \frac{1}{-\lambda} d(|f|(p), |f|(q)),$$

and the fact that $\sqrt{2(l+1)} \le \sqrt{2(n+1)}$ allow us to conclude that

$$d(|f|(p),|f|(q)) \leq \sqrt{2(n+1)}d(p,q).$$

*Case 3*: Let us assume that $s(p) \subset s(q)$. Rewrite the indices of the elements of $s(p)$ and $s(q)$ in such a way that, $x_i = y_i$ for every $i = 0, \ldots, n$. Similar to the previous case, we consider the elements

$$z_i = \begin{cases} x_i = y_i, & 0 \leq i \leq n \\ y_j, & n+1 \leq j \leq m \end{cases}$$

and the real numbers

$$\gamma_i = \begin{cases} -\alpha_i + \beta_i, & 0 \leq i \leq n \\ \beta_j, & n+1 \leq j \leq m. \end{cases}$$

If $-\alpha_i + \beta_i \geq 0$ for every $i = 0, \ldots, n$, then $s(p) = s(q)$ and $p = q$, because $\sum_{i=0}^{n} \alpha_i = 1$. Hence, there exists a number $0 \leq i \leq n$ such that $-\alpha_i + \beta_i < 0$. At this point, we argue as in the previous case. If $s(q) \subset s(p)$, we use an analogous procedure. ∎

In particular, the following result holds true:

**(II.2.8) Theorem.** *Any piecewise linear function (the simplicial realization of a simplicial function)*

$$F: |K| \to |L|, \quad F\left(\sum_{i=0}^{n} \alpha_i x_i\right) = \sum_{i=0}^{n} \alpha_i F(x_i)$$

*is continuous.*

Hence $|\ |$ is a functor.

We now investigate some of the properties of the geometric realization of a simplicial complex. Recall that it is possible to characterize a convex set $X$ of an Euclidean space as follows: for every $p, q \in X$, the segment $[p, q]$, with end-points $p$ and $q$, is contained in $X$. As we are going to see in the next theorem, this convexity property is valid for the geometric realization of the complex $\overline{\sigma}$ (called *geometric simplex*), for every simplex $\sigma$ of a simplicial complex $K$.

**(II.2.9) Theorem.** *Let $K - (X, \Phi)$ be a simplicial complex. The following results hold true:*

*(i) The geometric realization $\overline{\sigma}$ of any simplex $\sigma \in \Phi$ is convex.*
*(ii) For every two simplexes $\sigma, \tau \in \Phi$ we have*

$$|\overline{\sigma}| \cap |\overline{\tau}| = |\overline{\sigma \cap \tau}|.$$

*(iii) For every $\sigma \in \Phi$, $\overline{\sigma}$ is compact.*

*Proof.* (i) Assume that $\sigma = \{x_0, \ldots, x_n\}$ and let $p, q$ be arbitrary points of $|\overline{\sigma}|$; suppose that $p = \sum_{i=0}^{n} \alpha_i x_i$ and $q = \sum_{i=0}^{n} \beta_i x_i$. The segment $[p, q]$ is the set of all points $r = tp + (1-t)q$, for every $t \in [0, 1]$. Then

$$r = tp + (1-t)q = \sum_{i=0}^{n}(t\alpha_i + (1-t)\beta_i)x_i$$

with $\sum_{i=0}^{n}(t\alpha_i + (1-t)\beta_i) = 1$ and so, $r \in |\overline{\sigma}|$.

(*ii*) Let us first observe that if $p \in |\overline{\sigma}|$, then $s(p) \subset \sigma$. Now if $p \in |\overline{\sigma}| \cap |\overline{\tau}|$, $s(p) \subset \sigma \cap \tau$, and thus $p \in |s(p)| \subset |\overline{\sigma \cap \tau}|$. Conversely, if $p \in |\overline{\sigma \cap \tau}|$ then $p \in |s(p)| \subset |\overline{\sigma}|$, $p \in |s(p)| \subset |\overline{\tau}|$, and therefore $p \in |\overline{\sigma}| \cap |\overline{\tau}|$.

(*iii*) Take the standard $n$-simplex

$$\Delta^n = \{(z_0, \ldots, z_n) \in \mathbb{R}^{n+1} \mid 0 \le z_i \le 1, \sum_i z_i = 1\}$$

endowed with a system of barycentric coordinates with respect to the vertices

$$e_0 = (1,0,\ldots,0), \ldots e_n = (0,0,\ldots,1);$$

we can write the elements of $\Delta^n$ as linear combinations with nonnegative real coefficients $\sum_{i=0}^{n}\alpha_i e_i$ where $\sum_{i=0}^{n}\alpha_i = 1$. Furthermore, we observe that $\Delta^n$ is compact as a bounded and closed subset of $\mathbb{R}^{n+1}$ (see Theorem (I.1.36)). Let $f: \Delta^n \to |\overline{\sigma}|$ be the function taking any $p = \sum_{i=0}^{n}\alpha_i e_i \in \Delta^n$ to the point $f(p) = \sum_{i=0}^{n}\alpha_i x_i$. This function is bijective, continuous, and takes a compact space to a Hausdorff space; hence, $f$ is a homeomorphism (see Theorem (I.1.27)). It follows that $|\overline{\sigma}|$ is compact.  ■

As we have observed before, the geometric realization $|K|$ can be viewed as a subspace of $\mathbb{R}^n$, where $n$ is the number of vertices of $K$. Thus, it is possible to consider an affine structure on the ambient space $\mathbb{R}^n$, and again analyze the convexity of the various parts of $K$ and the linear combinations of elements with barycentric coordinates. For every $p \in |K|$, let $B(p)$ be the set of all $\sigma \in \Phi$ such that $p \in |\overline{\sigma}|$; now take the space

$$D(p) = \bigcup_{\sigma \in B(p)} |\overline{\sigma}|.$$

The *boundary* $S(p)$ of $D(p)$ is the union of the geometric realizations of the complexes generated by the faces $\tau \subset \sigma$, with $\sigma \in B(p)$ and $p \notin |\overline{\tau}|$. Intuitively, $D(p)$ is the "disk" defined by all geometric simplices, which contain $p$ and $S(p)$ is its bounding "sphere". Observe that $D(p)$ and $S(p)$ are closed subsets of $|K|$; finally, leaving out the geometric realization, $D(p)$ and $S(p)$ are subcomplexes of $K$.

**(II.2.10) Theorem.** *Let $K$ be a simplicial complex; the following properties are valid.*

(*i*) *For every $p \in |K|$, $D(p)$ is compact.*

(*ii*) *For every $q \in D(p) \smallsetminus \{p\}$ and every $t \in I$, the point $r = (1-t)p + tq$ belongs to $D(p)$.*

(*iii*) *Every ray in $D(p)$ with origin $p$ intersects $S(p)$ at a unique point.*

*Proof.* (*i*): The compactness of $D(p)$ follows from Theorem (II.2.9), (*iii*).

(*ii*): Because $q \in D(p) \smallsetminus \{p\}$, there exists a simplex $\sigma \in B(p)$ such that $q \in |\overline{\sigma}|$, a convex space; it follows that the segment $[p,q]$ is entirely contained in $|\overline{\sigma}| \subset D(p)$.

(*iii*): Let $\ell$ be a ray with origin $p$, and let $q$ be the point of $\ell$ determined by the condition

$$d(p,q) = \sup\{d(p,q') \mid q' \in \ell \cap D(p)\}.$$

Then, $|s(q)| \in S(p)$; otherwise, we could extend $\ell$ in $D(p)$ beyond $q$ and thus we would have $q \in S(p)$. On the other hand, because $s(q)$ is a face of a simplex containing $p$, the vertices of $s(p)$ and $s(q)$ define a simplex of which $s(p)$ is a face (in fact, $s(p) \cup s(q)$ is a simplex of the simplicial complex $D(p)$). The points of $\ell$ beyond $q$ cannot be in $S(p)$ and the open segment $(p,q)$ is contained in $D(p) \smallsetminus S(p)$. Hence, $\ell$ intersects $S(p)$ in one point only. ∎

Notice that $D(p)$ is not necessarily convex; at any rate, as we have seen in part (*ii*) of the previous theorem, $D(p)$ is endowed with a certain kind of convexity in the sense that, for every $q \in D(p)$, the segment $[p,q]$ is entirely contained in $D(p)$. We say that $D(p)$ is *p-convex* (star convex). Theorem (II.2.10) allows us to define a map

$$\pi_p \colon D(p) \smallsetminus \{p\} \to S(p) \, , \ q \mapsto \ell_{p,q} \cap S(p)$$

where $\ell_{p,q}$ is the ray with origin $p$ and containing $q$; the function $\pi_p$ is the *radial projection with center p* from $D(p)$ onto $S(p)$. Let $i \colon S(p) \to D(p) \smallsetminus \{p\}$ be the inclusion map; then $\pi_p i = 1_{S(p)}$, and $i\pi_p$ is homotopic to the identity map of $D(p) \smallsetminus \{p\}$ onto itself with homotopy given by the map

$$H \colon (D(p) \smallsetminus \{p\}) \times I \to D(p) \smallsetminus \{p\} \, , \ (q,t) \mapsto (1-t)q + t\pi_p(q).$$

Hence, $S(p)$ is a deformation retract of $D(p) \smallsetminus \{p\}$ (see Exercise 2, Sect. I.2). This shows another similarity between the spaces $D(p)$, $S(p)$ and, respectively, the $n$-dimensional Euclidean disk and its boundary.

The next result (cf. [24]) will be used only when studying triangulable manifolds (Sect. V.1); the reader could thus leave it for later on.

**(II.2.11) Theorem.** *Let $f \colon |K| \to |L|$ be a homeomorphism. Then, for every $p \in |K|$, $S(p)$ and $S(f(p))$ are of the same homotopy type.*

*Proof.* Assume that $s(f(p)) = \{y_0, \ldots, y_n\}$ and let $U = |\overline{s(f(p))}| \smallsetminus |s(f(p))|$ be the interior of $|s(f(p))|$, that is to say, the set of all $q \in |L|$ such that $q(y_i) > 0$, for every $i = 0, \ldots, n$. Notice that $U$ is an open set of $D(f(p))$; moreover, $f^{-1}(U)$ is an open set of $|K|$ containing $p$. The bounded, compact set $D(p)$ can be shrunk at will: in fact, for any real number $0 < \lambda \leq 1$ we define the *compression* $\lambda D(p)$ as the set of all points $r = (1-\lambda)p + \lambda q$, for every $q \in D(p)$; observe that $\lambda D(p)$ is a closed subset of $|K|$, and is homeomorphic to $D(p)$. Let $\lambda \in (0,1]$ be such that

$$p \in \lambda D(p) \subset f^{-1}(U);$$

then

$$f(p) \in f(\lambda D(p)) \subset U \subset D(f(p)).$$

In a similar fashion, we can find two other real numbers $\mu, \nu \in (0, 1]$ such that

$$f(p) \in f(\nu D(p)) \subset \mu D(f(p)) \subset f(\lambda D(p)) \subset D(f(p)).$$

Because $f(\nu D(p)) \subset \mu D(f(p))$, we can define the radial projection with center $f(p)$

$$\psi \colon f(\nu S(p)) \to \mu S(f(p)) , \ f(q) \mapsto \pi_{f(p)}(f(q))$$

for every $q \in \nu S(p)$. We also define the map

$$\phi \colon \mu S(f(p)) \to f(\nu S(p)) , \ q \mapsto f(\nu(\pi_p(f^{-1}(q))))$$

where $\pi_p$ is the radial function with center $p$ in $\lambda D(p)$ (notice that $f^{-1}(q) \neq p$, for every $q \in \mu S(f(p))$ and moreover, $f^{-1}(\mu D(f(p)) \subset \lambda D(p))$.

Since the spaces $f(\nu S(p))$ and $\mu S(f(p))$ are contained in $D(f(p))$ and this last space is $f(p)$-convex, we can define the homotopy

$$H_1 \colon \mu S(f(p)) \times I \longrightarrow D(f(p))$$

$$H_1(q,t) = \begin{cases} (1-2t)\psi\phi(q) + 2tf(p) , & 0 \leq t \leq \frac{1}{2} \\ (2-2t)f(p) + (2t-1)\phi(q) , & \frac{1}{2} \leq t \leq 1 \end{cases}$$

for every $q \in \mu S(f(p))$. Strictly speaking, $H_1$ is a homotopy between $\phi$ composed with the inclusion map $f(\nu S(p) \subset D(f(p))$ and $\psi\phi$ composed with $\mu S(f(p)) \subset D(f(p))$. We now take the maps

$$f^{-1}\phi \colon \mu S(f(p)) \to \nu D(p) \text{ and } f^{-1} \colon \mu S(f(p)) \to \lambda D(p).$$

Because $D(p)$ is $p$-convex, we can construct the homotopy

$$H_2(q,t) = \begin{cases} (1-2t)f^{-1}\phi(q) + 2tp , & 0 \leq t \leq \frac{1}{2} \\ (2-2t)p + (2t-1)f^{-1}(q) , & \frac{1}{2} \leq t \leq 1. \end{cases}$$

which, when composed with the homeomorphism $f$, gives rise to a homotopy

$$fH_2 \colon \mu S(f(p)) \times I \to D(f(p));$$

finally, we consider the homotopy

$$F \colon \mu S(f(p) \times I \to D(f(p))$$

defined by the formula

$$F(q,t) = \begin{cases} H_1(q, 2t) , & 0 \leq t \leq \frac{1}{2} \\ fH_2(q, 2t-1) , & \frac{1}{2} \leq t \leq 1. \end{cases}$$

The map $F$ is a homotopy between $\psi\phi$ and the identity map of $\mu S(f(p))$. Similarly, we prove that $\phi\psi$ is homotopic to the corresponding identity map. Thus, $\mu S(f(p))$ and $f(vS(p))$ are of the same homotopy type. On the other hand, $\mu S(f(p))$ and $f(vS(p))$ are homeomorphic, respectively, to $S(f(p))$ and $S(p)$; hence, $S(f(p))$ and $S(p)$ are of the same homotopy type. ∎

The geometric realization $|K|$ of a simplicial complex $K$ is called *polyhedron*.[2] The next theorem gives a better understanding of the topology of $|K| = (X, \Phi)$.

**(II.2.12) Theorem.** *A set $F \subset |K|$ is closed in $|K|$ if and only if, for every $\sigma \in \Phi$, the subset $F \cap |\overline{\sigma}|$ is closed in $|\overline{\sigma}|$.*

*Proof.* Because $|\overline{\sigma}|$ is a compact subset of a Hausdorff space $|K|$, $|\overline{\sigma}|$ is closed in $|K|$ (see Theorem (I.1.25)); thus, if $F$ is closed in $|K|$, $F \cap |\overline{\sigma}|$ is closed in $|K|$ and therefore, $F$ is closed in $|\overline{\sigma}|$.

Conversely, if $F \cap |\overline{\sigma}|$ is closed in $|\overline{\sigma}|$ for every $|\overline{\sigma}| \subset |K|$, then $F = \bigcup_{|\overline{\sigma}|} (F \cap |\overline{\sigma}|)$ is closed in $|K|$ as a finite union of closed sets. ∎

A topological space $X$ is said to be *triangulable* if there exists a polyhedron $K$, which is homeomorphic to $X$; the simplicial complex $K$ is a *triangulation* of $X$. A triangulable space can have more than one triangulation. For example, it is easy to understand that $S^1$ has a triangulation given by a simplicial complex whose geometric realization is homeomorphic to the boundary of an equilateral triangle; but it can also be triangulated by a complex whose geometric realization is a regular polygon with vertices in $S^1$ (the homeomorphisms are given by a projection from the center of $S^1$). More generally, a disk $D^n$ and its boundary $S^{n-1}$ are examples of triangulable spaces; these spaces also have several possible triangulations. Next, we describe the *standard triangulation* of $S^n$.

Let $\Sigma^n$ be the set of all points $(x_1, x_2, \ldots, x_{n+1}) \in \mathbb{R}^{n+1}$ such that $\sum_i |x_i| = 1$. Let $X$ be the set of all vertices of $\Sigma_n$, that is to say, of the points $a_i = (0, \ldots, \overset{i}{1}, \ldots, 0)$ and $a_i' = (0, \ldots, \overset{i}{-1}, \ldots, 0)$ in $\mathbb{R}^{n+1}$, $i = 1, \ldots, n+1$. Now, let $\Phi$ be the set of all nonempty subsets of $X$ of the type $\{x_{i_0}, \ldots, x_{i_r}\}$ with $1 \le i_0 < i_1 < \ldots < i_r \le n+1$ and $x_{i_s}$ equal to either $a_{i_s}$ or $a_{i_s}'$. Since any set of vertices of this type is linearly independent, $K^n = (X, \Phi)$ is a simplicial complex. Its geometric realization is homeomorphic to $\Sigma^n$; on the other hand, $\Sigma^n$ and $S^n$ are homeomorphic by a radial projection from the center and therefore, $K^n$ is a triangulation of $S^n$. The simplicial complex $K^n$ is the so-called *standard triangulation* of $S^n$.

**(II.2.13) Remark.** As we have already notice, in this book we work exclusively with finite simplicial complexes. However, it is possible to give a more extended definition of simplicial complexes, which includes the infinite case. With this in mind, we define a simplicial complex as a pair $K = (X, \Phi)$ in which $X$ is a set

---

[2] In some textbooks, *polyhedra* are the geometric realizations of two-dimensional complexes; for the more general case, they use the word *polytopes*.

(not necessarily finite) and $\Phi$ is a set of nonempty, *finite* subsets of $X$ satisfying the following properties:

1. $(\forall x \in X)$, $\{x\} \in \Phi$,
2. $(\forall \sigma \in \Phi)(\forall \sigma' \subset \sigma$, $\sigma' \neq \emptyset)$, $\sigma' \in \Phi$.

The price we must pay is a strengthening of the topology of $K$. We keep the metric

$$d: |K| \times |K| \longrightarrow \mathbb{R}_{\geq 0}$$

$$(\forall p, q \in |K|)\, d(p,q) = \left\{ \sum_{x \in X} (p(x) - q(x))^2 \right\}^{\frac{1}{2}}$$

which defines a topology on $K$. While the necessary condition of Theorem (II.2.12) is still valid, the sufficient condition does not hold because, to prove it, we need the assumption that $X$ is finite. However, it is precisely the topology of Theorem (II.2.12) that we impose on $|K|$; in other words, we must exchange the metric topology of $K$ with a finer topology. We say that

$F \subset |K|$ is *closed* $\iff$ $(\forall \sigma \in \Phi)\, F \cap |\overline{\sigma}|$ is closed in $\sigma$.

This topology is normally called "weak topology"; this is somehow a strange name, considering the fact that the weak topology for $K$ is finer (that is to say, has more open sets) than the metric topology.

## II.2.2  Simplicial Complexes and Immersions

We have proved, aided by the geometric realization functor, that every abstract finite simplicial complex $K$ can be immersed in an Euclidean space and hence can be viewed as an Euclidean simplicial complex. The dimension of the Euclidean space in question is equal to the number of vertices, say $m$, of the complex. At this point, we ask ourselves whether it is possible to immerse $K$ in an Euclidean space of dimension lower than $m$. The next theorem answers that question. Before stating the theorem, we define Euclidean simplicial complexes in a different (but equivalent) fashion. Let $\mathfrak{K} \subset \mathbb{R}^n$ be a union of *finitely many* Euclidean simplexes of $\mathbb{R}^n$ such that

1. If $\sigma \subset \mathfrak{K}$, every face of $\sigma$ is in $\mathfrak{K}$.
2. The intersection of any two Euclidean simplexes of $\mathfrak{K}$ is a face of both.

It is not difficult to prove that a set of simplexes of $\mathbb{R}^n$ verifying the previous conditions is an Euclidean simplicial complex as defined in Sect. II.1. We also notice that if $F \subset \mathfrak{K}$ is closed, the intersection $F \cap \sigma$ is closed in $\sigma$ for every Euclidean simplex $\sigma$ of $\mathfrak{K}$; conversely, if $F$ is a subset of $\mathfrak{K}$ such that, for every Euclidean simplex $\sigma$ of $\mathfrak{K}$, $F \cap \sigma$ is closed in $\sigma$, then $F$ is closed in $\mathfrak{K}$ because $F$ is the finite union of the closed sets $F \cap \sigma$. Clearly, an Euclidean complex $\mathfrak{K}$ of $\mathbb{R}^n$ is compact and closed in $\mathbb{R}^n$. We now state the immersion theorem for simplicial complexes.

**(II.2.14) Theorem.** *Every $n$-dimensional polyhedron $|K|$ is homeomorphic to an Euclidean simplicial complex.*

*Proof.* Let $\mathfrak{N}$ be the set of all points $P^i = (i, i^2, \ldots, i^{2n+1}) \in \mathbb{R}^{2n+1}$, for every $i \geq 0$. We claim that the set $\mathfrak{N}$ has the following property: every $2n+2$ points $P^{i_0}, \ldots, P^{i_{2n+1}}$ are linearly independent. In fact, a linear combination

$$\sum_{j=1}^{2n+1} \alpha_j (P^{i_j} - P^{i_0}) = 0$$

gives rise to the equations

$$\sum_{j=1}^{2n+1} \alpha_j = 0,$$

$$\sum_{j=1}^{2n+1} \alpha_j i_j^1 = 0,$$

$$\ldots$$

$$\sum_{j=1}^{2n+1} \alpha_j i_j^{2n+1} = 0;$$

because the determinant of the system of linear homogeneous equations defined by the $2n+2$ equations written above is equal to $\prod_{k>j}(i_k - i_j) \neq 0$, the only solution for the system is the trivial one, $\alpha_1 = \alpha_2 = \ldots = \alpha_{2n+1} = 0$.

Assume that $K = (X, \Phi)$ with $X = \{a_0, \ldots, a_s\}$ and $s \leq n$. To each vertex $a_i$, we associate the point $P^i = (i^1, i^2, \ldots, i^{2n+1}) \in \mathbb{R}^{2n+1}$, and to each simplex $\{a_{j_0}, a_{j_1}, \ldots, a_{j_p}\} \in \Phi$ we associate the Euclidean simplex $\{P^{j_0}, P^{j_1}, \ldots, P^{j_p}\}$ (observe that the points $P^{j_i}$ with $j = 0, \ldots, p$ are linearly independent because $p \leq n < 2n+1$). Let $\mathfrak{K}$ be the set of vertices and Euclidean simplexes obtained in this way.

We begin by observing that $\mathfrak{K}$ clearly satisfies condition 1 of the definition of Euclidean simplicial complexes. Let us prove that condition 2 is also valid. Let $\sigma_p$ and $\sigma_q$ be two Euclidean simplexes of $\mathfrak{K}$ with $r$ common vertices; altogether $\sigma_p$ and $\sigma_q$ have $p + q - r + 2$ vertices. Because $p + q - r + 2 \leq 2n + 2$, these vertices form an Euclidean simplex of $\mathbb{R}^{2n+1}$ having $\sigma_p$ and $\sigma_q$ as faces; hence, $\sigma_p \cap \sigma_q$ is either empty (if $r = 0$) or a common face of $\sigma_p$ and $\sigma_q$.

Therefore, $\mathfrak{K}$ is an Euclidean simplicial complex homeomorphic to $|K|$. ∎

The reader could ask whether Theorem (II.2.14) is the best possible result or else, whether it is possible to realize all $n$-dimensional simplicial complexes in Euclidean spaces of dimension less than $2n + 1$. Clearly, a complex of dimension $n$ must be immersed in a space of dimension at least $n$. We shall now give two examples of one-dimensional simplicial complexes (that is to say, graphs) that cannot be immersed in $\mathbb{R}^2$.

Let $K_{3,3}$ be the *complete bipartite graph* over two sets of 3 vertices, also known as *utility graph*: $K_{3,3} = (X, \Phi)$ with $X = \{1,2,3,a,b,c\}$ and

$$\Phi = \{\{1\},\{2\},\{3\},\{a\},\{b\},\{c\},\{1,a\},\{1,b\},\{1,c\},\{2,a\},\{2,b\},$$
$$\{2,c\},\{3,a\},\{3,b\},\{3,c\}\}.$$

Figure II.7 shows its graphic representation (which however is *not* a geometric real-

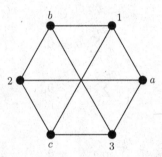

**Fig. II.7**

ization of $K_{3,3}$ because its distinct 1-simplexes have empty intersections). Another way to represent $K(3,3)$ is given in Fig. II.8. To ask whether or not $K(3,3)$ can be

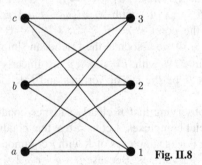

**Fig. II.8**

represented as a *planar* graph is a classical query; the answer would be affirmative if one could determine an immersion of $K(3,3)$ in $\mathbb{R}^2$ (but planarity is a weaker property: It is enough to show that $|K_{3,3}|$ is homeomorphic to a subspace of $\mathbb{R}^2$). Another example of a simplicial complex with an analogous property is given by the *complete graph* over 5 vertices $K_5$: $X = \{1,2,3,4,5\}$ and $\Phi$ is the set of all nonempty subsets of $X$ with at most 2 elements. Figure II.9 is a standard graphic representation of this graph. It is not difficult to prove that both $K_{3,3}$ and $K_5$ cannot be immersed in $\mathbb{R}^2$.

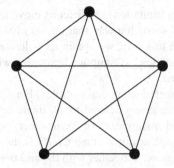

**Fig. II.9**

## II.2.3 The Homology Functor

We now define the *homology functor*

$$H_*(-;\mathbb{Z})\colon \textbf{Csim} \to \textbf{Ab}^{\mathbb{Z}},$$

another important functor with domain **Csim**.

Let $K = (X, \Phi)$ be an arbitrary simplicial complex. We begin our work by giving an *orientation* to the simplexes of $K$. Let $\sigma_n = \{x_0, x_1, \ldots, x_n\}$ be an $n$-simplex of $K$; the elements of $\sigma_n$ can be ordered in $(n+1)!$ different ways. We say that two orderings of the elements of $\sigma_n$ are *equivalent* whenever they differ by an even permutation; an *orientation* of $\sigma_n$ is an equivalence class of orderings of the vertices of $\sigma_n$, provided that $n > 0$. An $n$-simplex $\sigma_n = \{x_0, x_1, \ldots, x_n\}$ has two orientations. A 0-simplex has clearly only one ordering; its orientation is given by $\pm 1$.

If $\sigma_n = \{x_0, x_1, \ldots, x_n\}$ is oriented, the simplex $\{x_1, x_0, \ldots, x_n\}$ for example, is denoted with $-\sigma$. If $n \geq 1$, a given orientation of $\sigma_n = \{x_0, x_1, \ldots, x_n\}$ automatically defines an orientation in all of its $(n-1)$-faces: For example, if $\sigma_2 = \{x_0, x_1, x_2\}$ is oriented by the ordering $x_0 < x_1 < x_2$, its oriented 1-faces are

$$\{x_1, x_2\}\,, \ \{x_2, x_0\} = -\{x_0, x_2\} \text{ and } \{x_0, x_1\}.$$

More generally, if $\sigma_n = \{x_0, x_1, \ldots, x_n\}$ is oriented by the natural ordering of the indices of its vertices, its $(n-1)$-face

$$\sigma_{n-1,i} = \{x_0, x_1, \ldots, \widehat{x_i}, \ldots, x_n\} = \{x_0, x_1, \ldots, x_{i-1}, x_{i+1}, \ldots, x_n\}$$

(opposite to the vertex $x_i$ with $i = 0, \ldots, n$) has an orientation given by $(-1)^i \sigma_{n-1,i}$; we say that $\sigma_{n-1,i}$ is *oriented coherently* to $\sigma_n$ if $i$ is even, and is oriented coherently to $-\sigma_n$ if $i$ is odd. We observe explicitly that the symbol $\widehat{\phantom{x}}$ over the vertex $x_i$ means that such vertex has been eliminated.

We are now ready to order a simplicial complex $K = (X, \Phi)$. We recall that the technique used to give an orientation to a simplex was first to order its vertices in all possible ways, and then choose an ordering class (there are two possible classes: the class in which the orderings differ by an even permutation, and that in which the orderings differ by an odd permutation). Now let us move to $K$. Begin by taking a partial ordering of the set $X$ in such a way that the set of vertices of each simplex $\sigma \in \Phi$ is totally ordered; in this way, we obtain an ordering class – that is to say, an orientation – for each simplex. A simplicial complex whose simplexes are all oriented is said to be *oriented*.

Let $K = (X, \Phi)$ be an oriented simplicial complex. For every $n \in \mathbb{Z}$, with $n \geq 0$, let $C_n(K)$ be the free Abelian group defined by all linear combinations with coefficients in $\mathbb{Z}$ of the oriented $n$-simplexes of $K$; in other words, if $\{\sigma_n^i\}$ is the finite set of all oriented $n$-simplexes of $K$, then $C_n(K)$ is the set of all formal sums $\sum_i m_i \sigma_n^i$, $m_i \in \mathbb{Z}$ (called $n$-*chains*), together with the addition law

$$\sum_i p_i \sigma_n^i + \sum_i q_i \sigma_n^i := \sum_i (p_i + q_i) \sigma_n^i.$$

If $n < 0$, we set $C_n(K) = 0$. Now, for every $n \in \mathbb{Z}$, we define a homomorphism $\partial_n = \partial_n^K : C_n(K) \longrightarrow C_{n-1}(K)$ as follows: if $n \leq 0$, $\partial_n$ is the constant homomorphism 0; if $n \geq 1$, we first define $\partial_n$ over an oriented $n$-simplex $\{x_0, x_1, \ldots, x_n\}$ (viewed as an $n$-chain) as

$$\partial_n(\{x_0, x_1, \ldots, x_n\}) = \sum_{i=0}^{n} (-1)^i \{x_0, \ldots, \widehat{x_i}, \ldots, x_n\};$$

finally, we extend this definition by linearity over an arbitrary $n$-chain of oriented $n$-simplexes. The homomorphisms of degree $-1$, that we have just defined, are called *boundary homomorphisms*.

**(II.2.15) Lemma.** *For every $n \in \mathbb{Z}$, the composition $\partial_{n-1}\partial_n = 0$.*

*Proof.* The result is obvious if $n = 1$. Let $\{x_0, x_1, \ldots, x_n\}$ be an arbitrary oriented $n$-simplex with $n \geq 2$. Then

$$\partial_{n-1}\partial_n(\{x_0, x_1, \ldots, x_n\}) = \partial_{n-1} \sum_{i=0}^{n} (-1)^i \{x_0, \ldots, \widehat{x_i}, \ldots, x_n\}$$

$$= \sum_{j<i} (-1)^i (-1)^j \{x_0, \ldots, \widehat{x_j}, \ldots, \widehat{x_i}, \ldots, x_n\}$$

$$+ \sum_{j>i} (-1)^i (-1)^{j-1} \{x_0, \ldots, \widehat{x_i}, \ldots, \widehat{x_j}, \ldots, x_n\}.$$

This summation is 0 because its addendum $\{x_0, \ldots, \widehat{x_j}, \ldots, \widehat{x_i}, \ldots, x_n\}$ appears twice, once with the sign $(-1)^i(-1)^j$ and once with the sign $(-1)^i(-1)^{j-1}$. ∎

This important property of the boundary homomorphisms implies that, for every $n \in \mathbb{Z}$, the image of $\partial_{n+1}$ is contained in the kernel of $\partial_n$; using the notation $Z_n(K) = \ker \partial_n$ and $B_n(K) = \operatorname{im} \partial_{n+1}$, we conclude that $B_n(K) \subset Z_n(K)$ for every $n \in \mathbb{Z}$. Thus, to each integer $n \geq 0$, we can associate the quotient group

$$H_n(K; \mathbb{Z}) = Z_n(K)/B_n(K);$$

to each $n < 0$, we associate $H_n(K; \mathbb{Z}) = 0$.

**(II.2.16) Definition.** An $n$-chain $c_n \in C_n(K)$ is an *n-cycle* (or simply cycle) if $\partial_n(c_n) = 0$; thus $Z_n(K)$ is the set of all $n$-cycles. An $n$-chain $c_n$, for which we can find an $(n+1)$-chain $c_{n+1}$ such that $c_n = \partial_{n+1}(c_{n+1})$, is an *n-boundary*; thus, $B_n(K)$ is the set of all $n$-boundaries. Two $n$-chains $c_n$ and $c'_n$ are said to be *homologous* if $c_n - c'_n \in B_n(K)$.

What we have just described is a method to associate a graded Abelian group

$$H_*(K;\mathbb{Z}) = \{H_n(K;\mathbb{Z}) \mid n \in \mathbb{Z}\}$$

to any oriented simplicial complex $K \in \mathbf{Csim}$. To define a functor on $\mathbf{Csim}$ we must see what happens to the morphisms; we proceed as follows. Let $f\colon K = (X,\Phi) \to L = (Y,\Psi)$ be a simplicial function ($K$ and $L$ have a fixed orientation). We first define

$$C_n(f)\colon C_n(K) \longrightarrow C_n(L)$$

on the simplexes by

$$C_n(f)(\{x_0,\ldots,x_n\}) = \begin{cases} \{f(x_0),\ldots,f(x_n)\}, & (\forall i \neq j)\, f(x_i) \neq f(x_j) \\ 0, & \text{otherwise} \end{cases}$$

and then extend $C_n(f)$ linearly over the whole Abelian group $C_n(K)$. It is easy to prove that $\partial_n^L C_n(f) = C_{n-1}(f)\partial_n^K$, for every $n \in \mathbb{Z}$ (one can verify this on a single $n$-simplex). We now define

$$H_n(f)\colon H_n(K;\mathbb{Z}) \longrightarrow H_n(L;\mathbb{Z})$$
$$z + B_n(K) \mapsto C_n(f)(z) + B_n(L)$$

for every $n \geq 0$. We begin by observing that $C_n(f)(z)$ is a cycle in $C_n(L)$: in fact, since $z$ is a cycle,

$$\partial_n^L C_n(f)(z) = C_{n-1}(f)\partial_n^K(z) = 0 .$$

On the other hand, we note that $H_n(f)$ is well defined: let us assume that $z - z' = \partial_{n+1}^K(w)$; then

$$C_n(f)(z-z') = C_n(f)\partial_{n+1}^K(w) = \partial_{n+1}^L C_{n+1}(f)(w)$$

and thus $C_n(f)(z - z') \in B_n(L)$; from this, we conclude that $H_n(f)((z - z') + B_n(K)) = 0$.

If $n < 0$, we set $H_n(f) = 0$; in this way, we obtain a homomorphism $H_n(f)$ between Abelian groups, for every $n \in \mathbb{Z}$. The reader is invited to prove that

$$H_n(1_K) = 1_{H_n(K)} \text{ e } H_n(gf) = H_n(g)H_n(f)$$

for every $n \in \mathbb{Z}$.

**(II.2.17) Remark.** The construction of the *homology groups* $H_n(K;\mathbb{Z})$ is independent from the orientation of $K$, up to isomorphism. In fact, suppose that $O$ and $O'$

are two distinct orientations of $K$ and denote by $K^O$ and $K^{O'}$ the complex $K$ together with the orientations $O$ and $O'$, respectively.

The simplexes of $K^O$ are denoted by $\sigma$, and those of $K^{O'}$, by $\sigma'$. Now define $\phi_n : C_n(K^O) \to C_n(K^{O'})$ as the function taking a simplex $\sigma_n$ into the simplex $\sigma'_n$ if $O$ and $O'$ give the same orientation to $\sigma_n$, and taking $\sigma_n$ into $-\sigma'_n$ if $O$ and $O'$ give opposite orientations to $\sigma_n$; next, extend $\phi_n$ by linearity over the whole group $C_n(K^O)$. It is easy to prove that $\phi_n$ is a group isomorphism. Moreover, for every $n \in \mathbb{Z}$, $\partial_n \phi_n = \phi_{n-1} \partial_n$. For a given $n$-simplex $\sigma_n$ of $K^O$, we have two cases to consider:

Case 1: $O$ and $O'$ give the same orientation to $\sigma_n$; then

$$\partial_n \phi_n(\sigma_n) = \partial_n(\sigma'_n) = \sum_{i=0}^{n}(-1)^i \sigma'_{n-1,i}$$

$$\phi_{n-1}\partial_n(\sigma_n) = \phi_{n-1}(\sum_{i=0}^{n}(-1)^i \sigma_{n-1,i}) = \sum_{i=0}^{n}(-1)^i \sigma'_{n-1,i};$$

Case 2: $O$ and $O'$ give different orientations to $\sigma_n$; then

$$\partial_n \phi_n(\sigma_n) = \partial_n(-\sigma'_n) = \sum_{i=0}^{n}(-1)^{i+1} \sigma'_{n-1,i}$$

$$\phi_{n-1}\partial_n(\sigma_n) = \phi_{n-1}(\sum_{i=0}^{n}(-1)^{i+1} \sigma_{n-1,i}) = \sum_{i=0}^{n}(-1)^{i+1} \sigma'_{n-1,i}.$$

Similar to what we did to define the homomorphism $H_n(f)$, we can prove that $\phi_n$ induces a homomorphism

$$H_n(\phi_n) \colon H_n(K^O;\mathbb{Z}) \longrightarrow H_n(K^{O'};\mathbb{Z})$$

which is actually an isomorphism.

Therefore, up to isomorphism, the orientation given to a simplicial complex has no influence on the definition of the group $H_n(K;\mathbb{Z})$; thus, we forget the orientation (however, we note that in certain questions it cannot be ignored). With this, we define the covariant functor

$$H_*(-;\mathbb{Z}) \colon \mathbf{Csim} \longrightarrow \mathbf{Ab}^{\mathbb{Z}}$$

by setting

$$H_*(K;\mathbb{Z}) = \{H_n(K;\mathbb{Z}) \mid n \in \mathbb{Z}\} \text{ and } H_*(f) = \{H_n(f) \mid n \in \mathbb{Z}\}$$

on objects and morphisms, respectively. The graduate Abelian group $H_*(K;\mathbb{Z})$ is the *(simplicial) homology* of $K$ with coefficients in $\mathbb{Z}$.

We are going to compute the homology groups of the simplicial complex $T^2$ depicted in Fig. II.10 and whose geometric realization is the two-dimensional torus. We begin by orienting $T^2$ so that we go clockwise around the boundary of each 2-simplex. To simplify the notation, let us write $C_i(T^2)$ as $C_i$ (the same for the

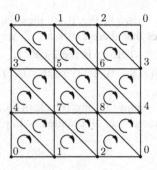

**Fig. II.10** A triangulation of the torus with oriented simplexes

groups of boundaries and cycles). We notice that $C_2 \cong \mathbb{Z}^{18}$, $C_1 \cong \mathbb{Z}^{27}$, $C_0 \cong \mathbb{Z}^9$. We represent the boundary homomorphisms in the next diagram

$$0 \longrightarrow C_2 \xrightarrow{\;\partial_2\;} C_1 \xrightarrow{\;\partial_1\;} C_0 \longrightarrow 0 .$$

Clearly, each vertex (and hence, each 0-chain) is a cycle; hence, $Z_0 = C_0$. The elements $\{0\}$, $\{1\} - \{0\}$, $\ldots \{8\} - \{0\}$ form a basis of $Z_0$. Any two vertices can be connected by a sequence of 1-simplexes and so the 0-cycles $\{1\} - \{0\}$, $\ldots$, $\{8\} - \{0\}$ are 0-boundaries. Since the boundary of a generic 1-simplex $\{i,j\}$ can be written as

$$\partial_1(\{i,j\}) = \{j\} - \{i\} = \{j\} - \{0\} - (\{i\} - \{0\}),$$

we have that $B_0 \subset Z_0$ is generated by $\{1\} - \{0\}, \ldots, \{8\} - \{0\}$ and thus,

$$H_0(T^2; \mathbb{Z}) \cong \mathbb{Z} .$$

The homology class of any vertex is a generator of this group.

Next, we compute $H_1(T^2; \mathbb{Z})$. The two 1-chains

$$z_1^1 = \{0,3\} + \{3,4\} + \{4,0\} \text{ and } z_1^2 = \{0,1\} + \{1,2\} + \{2,0\}$$

are cycles and generate (in $Z_1$) a free Abelian group of rank 2 which we denote by $S \cong \mathbb{Z} \oplus \mathbb{Z}$. Let $z \subset Z_1$ be a 1-cycle $z - \sum_i k_i \sigma_1^i$, in which $\sigma_1^i$ are the 1-simplexes and $k_i \in \mathbb{Z}$. By adding suitable multiples of 2-simplexes, it is possible to find a 1-boundary $b$ such that the 1-cycle $z - b$ does not contain the terms, which correspond to the diagonal 1-simplexes $\{0,5\}$, $\{1,6\}$, $\ldots$, $\{7,2\}$, $\{8,0\}$. Similarly, adding suitable pairs of adjacent 2-simplexes (those forming squares with a common diagonal) it is possible to find a 1-boundary $b'$ such that the cycle $z - b - b'$ contains only the terms corresponding to the 1-simplexes $\{0,3\}$, $\{3,4\}$, $\{4,0\}, \{0,1\}, \{1,2\}$, and $\{2,0\}$ (we leave the details to the reader, as an exercise). Because $z - b - b'$ is a 1-cycle, it follows that $z - b - b' \in S$. This argument shows that $B_1 + S = Z_1$. Let us now suppose that $B_1 \cap S \neq 0$; then there exists a linear combination of 2-simplexes $\sum_j h_j \sigma_2^j$ such that $\sum_j h_j \partial \sigma_2^j \in S$. If two 2-simplexes $\sigma_2^i$ and $\sigma_2^j$ have a common

1-simplex "internal" to the square of Fig. II.10, then they must have equal coefficients $h_i = h_j$. This implies that there exists $h \in \mathbb{Z}$ such that $h_j = h$ for each $j$, that is to say, $\sum_j h_j \sigma_2^j = h z_2$ where $z_2$ is the 2-chain $\sum_j \sigma_2^j$. It is easy to see that $\partial z_2 = \sum_j \partial \sigma_2^j = 0$ and so $B_1 \cap S = 0$, implying that $H_1(T^2; \mathbb{Z}) \cong S \cong \mathbb{Z}^2$, with free generators $z_1^1$ and $z_1^2$.

Finally, similar arguments show that any 2-cycle of $C_2$ is a multiple of the 2-chain $z_2$ defined above (given by the sum of all oriented 2-simplexes of $T^2$) and therefore, $H_2(T^2; \mathbb{Z}) \cong \mathbb{Z}$.

## Exercises

**1.** Let $\mathscr{U} = \{U_x | x \in X\}$ be a *finite* open covering of a topological space $B$, and take the set

$$\Phi = \{\sigma \subset X \,|\, \bigcap_{x \in \sigma} U_x \neq \emptyset\}.$$

Prove that $N(\mathscr{U}) = (X, \Phi)$ is a simplicial complex. This is the so-called *nerve* of $\mathscr{U}$.

**2.** Let $K = (X, \Phi)$ be a simplicial complex. For a given $x \in X$, let $St(x)$ be the complement in $|K|$ of the union of all $|\overline{\sigma}|$ such that $x \notin \sigma$, $\sigma \in \Phi$. $St(x)$ is called *star* of $x$ in $|K|$. Prove that $\mathscr{S} = \{St(x) \,|\, x \in X\}$ is an open covering of $|K|$, and $N(\mathscr{S}) = K$.

**3.** Let $X$ be a compact metric space and let $\varepsilon$ be a positive real number. Take the set $\Phi$ of all *finite* subsets of $X$ with diameter less than $\varepsilon$. Prove that $K = (X, \Phi)$ is a simplicial complex (infinite).

**4.** Exhibit a triangulation of the following spaces:

a) *Cylinder $C$* – recall that the cylinder $C$ is obtained from a rectangle by identification of two opposite sides;

b) *Möbius band $M$* obtained from a rectangle by identification of the "inverse" points of two opposite sides; more precisely, let $S$ be the rectangle with vertices $(0,0)$, $(0,1)$, $(2,0)$, and $(2,1)$ of $\mathbb{R}^2$; then

$$M = S/\{(0,t) \equiv (2, 1-t)\}, \; 0 \leq t \leq 1;$$

c) *Klein bottle $K$* obtained by identifying the "inverse" points of the boundary of the cylinder $C$;

d) *real projective plane $\mathbb{R}P^2$* obtained by the identification of the antipodal points of the boundary $\partial D^2 \cong S^1$ of the unit disk $D^2 \subset \mathbb{R}^2$;

e) $G_2$ obtained by attaching two handles to the sphere $S^2$; prove that $G_2$ is homeomorphic to the space obtained from an octagon with the suitable identifications of the edges of its border $a_1 b_1 a_1^{-1} b_1^{-1} a_2 b_2 a_2^{-1} b_2^{-1}$.

## II.3 Introduction to Homological Algebra

In the previous section, we have seen that we can associate a graded Abelian group $C(K) = \{C_n(K)\}$ with any simplicial complex $K$ and a homomorphism $\partial_n \colon C_n(K) \to C_{n-1}(K)$, such that $\partial_{n-1}\partial_n = 0$, to each integer $n$; these homomorphisms define a graded Abelian group $H_*(K;\mathbb{Z}) = \{H_n(K;\mathbb{Z})\}$. All this can be viewed in the framework of a more general and more useful context.

A *chain complex* $(C,\partial)$ is a graded Abelian group $C = \{C_n\}$ together with an endomorphism $\partial = \{\partial_n\}$ of degree $-1$, called *boundary homomorphism*[3] $\partial = \{\partial_n \colon C_n \to C_{n-1}\}$, such that $\partial^2 = 0$; this means that, for every $n \in \mathbb{Z}$, $\partial_n\partial_{n+1} = 0$. Hence

$$B_n = \operatorname{im} \partial_{n+1} \subset Z_n = \ker \partial_n$$

and so we can define the graded Abelian group

$$H_*(C) = \{H_n(C) = Z_n/B_n \mid n \in \mathbb{Z}\};$$

this is the *homology* of $C$.

A *chain homomorphism* between two chain complexes $(C,\partial)$ and $(C',\partial')$ is a graded group homomorphism $f = \{f_n \colon C_n \to C'_n\}$ of degree 0 commuting with the boundary homomorphism, that is to say, for every $n \in \mathbb{Z}$, $f_{n-1}\partial_n = \partial'_n f_n$.

Chain complexes and chain homomorphisms form a category $\mathfrak{C}$, the *category of chain complexes*.

It is costumary to visualize chain complexes as diagrams

$$\cdots \longrightarrow C_{n+1} \xrightarrow{\ \partial_{n+1}\ } C_n \xrightarrow{\ \partial_n\ } C_{n-1} \longrightarrow \cdots$$

and their morphisms as commutative diagrams

$$
\begin{array}{ccccccc}
\cdots \longrightarrow & C_{n+1} & \xrightarrow{\partial_{n+1}} & C_n & \xrightarrow{\partial_n} & C_{n-1} & \longrightarrow \cdots \\
& \downarrow{\scriptstyle f_{n+1}} & & \downarrow{\scriptstyle f_n} & & \downarrow{\scriptstyle f_{n-1}} & \\
\cdots \longrightarrow & C'_{n+1} & \xrightarrow{\partial'_{n+1}} & C'_n & \xrightarrow{\partial'_n} & C'_{n-1} & \longrightarrow \cdots
\end{array}
$$

The previous definitions are clearly inspired by what we did to define the homology groups of a simplicial complex; indeed, we emphasize the fact that, for every simplicial complex $X$, the graded Abelian group $\{C_n(K) \mid n \in \mathbb{Z}\}$ together with its boundary homomorphism $\partial^K = \{\partial_n^K \mid n \in \mathbb{Z}\}$ is a chain complex $(C(K),\partial^K)$. The chain complex $C(K)$ is said to be *positive* because its terms of negative index are 0. In particular, for every simplicial function $f \colon K \to M$, the homomorphism

$$C(f) \colon C(K) \to C(M)$$

is a chain homomorphism.

---

[3] In some textbooks, it is called *differential operator*.

An infinite sequence of Abelian groups

$$\cdots \longrightarrow G_{n+1} \xrightarrow{f_{n+1}} G_n \xrightarrow{f_n} G_{n-1} \longrightarrow \cdots$$

is said to be *exact* if and only if, for every $n \in \mathbb{Z}$, $\operatorname{im} f_{n+1} = \ker f_n$.

The exact sequences with only three consecutive nontrivial groups

$$\cdots \longrightarrow 0 \longrightarrow G_{n+1} \xrightarrow{f_{n+1}} G_n \xrightarrow{f_n} G_{n-1} \longrightarrow 0 \longrightarrow \cdots$$

are particularly important; in that case, $f_{n+1}$ is injective and $f_n$ is surjective. These sequences are called *short exact sequences*. The previous short exact sequence is also written up in the form

$$G_{n+1} \xrightarrow{f_{n+1}} G_n \xrightarrow{f_n} G_{n-1}.$$

The concept of short exact sequence of groups can be easily exported to the category $\mathfrak{C}$ of chain complexes: a sequence of chain complexes

$$(C,\partial) \xrightarrow{f} (C',\partial') \xrightarrow{g} (C'',\partial'')$$

is *exact* if every horizontal line of its representative diagram

$$
\begin{array}{ccccc}
\vdots & & \vdots & & \vdots \\
\downarrow{\scriptstyle\partial_{n+2}} & & \downarrow{\scriptstyle\partial'_{n+2}} & & \downarrow{\scriptstyle\partial''_{n+2}} \\
C_{n+1} & \xrightarrow{f_{n+1}} & C'_{n+1} & \xrightarrow{g_{n+1}} & C''_{n+1} \\
\downarrow{\scriptstyle\partial_{n+1}} & & \downarrow{\scriptstyle\partial'_{n+1}} & & \downarrow{\scriptstyle\partial''_{n+1}} \\
C_n & \xrightarrow{f_n} & C'_n & \xrightarrow{g_n} & C''_n \\
\downarrow{\scriptstyle\partial_n} & & \downarrow{\scriptstyle\partial'_n} & & \downarrow{\scriptstyle\partial''_n} \\
C_{n-1} & \xrightarrow{f_{n-1}} & C'_{n-1} & \xrightarrow{g_{n-1}} & C''_{n-1} \\
\downarrow{\scriptstyle\partial_{n-1}} & & \downarrow{\scriptstyle\partial'_{n-1}} & & \downarrow{\scriptstyle\partial''_{n-1}} \\
\vdots & & \vdots & & \vdots
\end{array}
$$

is exact and each square is commutative.

The next result is very important; it is the so-called **Long Exact Sequence Theorem**.

**(II.3.1) Theorem.** *Let*

$$(C, \partial) \rightarrowtail^{f} (C', \partial') \xrightarrow{g} \twoheadrightarrow (C'', \partial'')$$

*be a short exact sequence of chain complexes. For every $n \in \mathbb{Z}$, there exists a homomorphism*

$$\lambda_n \colon H_n(C'') \to H_{n-1}(C)$$

*(called connecting homomorphism) making exact the following sequence of homology groups*

$$\cdots \longrightarrow H_n(C) \xrightarrow{H_n(f)} H_n(C') \xrightarrow{H_n(g)} H_n(C'') \xrightarrow{\lambda_n} H_{n-1}(C) \longrightarrow \cdots$$

*Proof.* The proof of this theorem is not difficult. However, it is very long; we shall divide it into several steps, leaving some of the proofs to the reader, as exercises.

1. *Definition of $\lambda_n$.* Take the following portion of the short exact sequence of chain complexes:

$$
\begin{array}{ccccc}
C_n & \xrightarrow{f_n} & C'_n & \xrightarrow{g_n} & C''_n \\
\downarrow{\partial_n} & & \downarrow{\partial'_n} & & \downarrow{\partial''_n} \\
C_{n-1} & \xrightarrow[f_{n-1}]{} & C'_{n-1} & \xrightarrow[g_{n-1}]{} & C''_{n-1}
\end{array}
$$

Let $z$ be a cycle of $C''_n$; since $g_n$ is surjective, there exists a chain $\bar{z} \in C'_n$ such that $g_n(\bar{z}) = z$. Because the diagram is commutative,

$$g_{n-1} \partial'_n(\bar{z}) = \partial''_n g_n(\bar{z}) = \partial''_n(z) = 0$$

and thus, $\partial'_n(\bar{z}) \in \ker g_{n-1} = \operatorname{im} f_{n-1}$, hence, there exists a unique chain $c \in C_{n-1}$ such that

$$f_{n-1}(c) = \partial'_n(\bar{z}).$$

Actually, $c$ is a cycle because

$$f_{n-2}\partial_{n-1}(c) = \partial'_{n-1} f_{n-1}(c) = \partial'_{n-1} \partial'_n(\bar{z}) = 0$$

and $f_{n-2}$ is a monomorphism. It follows that we can define

$$\lambda_n \colon H_n(C'') \to H_{n-1}(C)$$

by setting $\lambda_n[z] := [c]$.

2. $\lambda_n$ *is well defined.* We must verify that $\lambda_n$ is independent from both the choice of the cycle $z$ representing the homology class and the chain $\tilde{z}$ mapped into $z$. Let $z' \in C''$ be a cycle such that $[z] = [z']$, and let $\tilde{z}' \in C_n'$ be such that $g_n(\tilde{z}') = z'$; moreover, take a cycle $c'$ in $C_{n-1}'$ satisfying the property $f_{n-1}(c') = \partial_n'(\tilde{z}')$. The definition of homology classes implies that there exists a chain $b \in C_{n+1}''$ such that $\partial_{n+1}''(b) = z - z'$. Since $g_{n+1}$ is an epimorphism, we can find a $\tilde{b} \in C_{n+1}'$ such that $g_{n+1}(\tilde{b}) = b$. Hence,

$$g_n(\tilde{z} - \tilde{z}' - \partial_{n+1}'(\tilde{b})) = z - z' - \partial_{n+1}''(b) = 0$$

and thus, there exists $a \in C_n$ such that

$$f_n(a) = \tilde{z} - \tilde{z}' - \partial_{n+1}'(\tilde{b}).$$

At this point, we have that

$$f_{n-1}(\partial_n(a)) = \partial_n'(\tilde{z} - \tilde{z}' - \partial_{n+1}'(\tilde{b})) = f_{n-1}(c - c')$$

and because $f_{n-1}$ is injective, we conclude that $c - c' = \partial_n(a)$. Therefore, $c$ and $c'$ represent the same homology class in $H_n(C)$.

**(II.3.2) Remark.** The previous items 1. and 2. are typical examples of the so-called *"diagram chasing"* technique. We suggest the reader to draw the diagrams indicating the maps without their indices which, although necessary for precision, are sometimes difficult to read; all this will help in following up the arguments.

3. *The sequence is exact.* To prove the exactness of the sequence of homology groups, we must show the following:

   a.  $\operatorname{im} H_n(f) = \ker H_n(g)$;
   b.  $\operatorname{im} H_n(g) = \ker \lambda_n$;
   c.  $\operatorname{im} \lambda_n = \ker H_{n-1}(f)$.

We shall only prove (b), leaving the proof of the other cases to the reader. We pick a class $[z] \in H_n(C'')$ and compute $\lambda_n H_n(g)([z]) = \lambda_n[g_n(z)]$. Since we can take any (!) element of $C_n'$ which is projected onto $g_n(z)$, we choose $z$ itself; given that $\partial_n''(z) = 0$, we conclude that $\lambda_n[g_n(z)] = 0$, that is to say, $\operatorname{im} H_n(g) \subseteq \ker \lambda_n$. Conversely, let $[z]$ be a homology class of $H_n(C'')$ such that $\lambda_n[z] = 0$; the definition of $\lambda_n$ implies that there exist $\tilde{z} \in C_n'$ and a cycle $c \in C_{n-1}$ such that

$$g_n(\tilde{z}) = z \text{ and } C_{n-1}(f)(c) = \partial_n'(\tilde{z}).$$

Because $\lambda_n[z] = 0$, there exists $\tilde{c} \in C_n$ such that $c = \partial_n(\tilde{c})$. Notice that

$$\partial_n'(C_n(f)(\tilde{c}) - \tilde{z}) = 0$$

and moreover, $H_n(g)(C_n(g)(\tilde{c}) - \tilde{z}) = z$; hence, $\ker \lambda_n \subseteq \operatorname{im} H_n(g)$.

■

The connecting homomorphisms $\lambda_n$ are *natural* in the following sense:

**(II.3.3) Theorem.** *Let*

$$
\begin{array}{ccccc}
(C,\partial) & \xrightarrow{\;f\;} & (C',\partial') & \xrightarrow{\;g\;} & (C'',\partial'') \\
\downarrow{\scriptstyle h} & & \downarrow{\scriptstyle k} & & \downarrow{\scriptstyle \ell} \\
(\bar{C},\bar{d}) & \xrightarrow[\;\bar{f}\;]{} & (\bar{C}',\bar{d}') & \xrightarrow[\;\bar{g}\;]{} & (\bar{C}'',\bar{d}'')
\end{array}
$$

*be a commutative diagram of chain complexes in which the horizontal lines are short exact sequences. Then, for every $n \in \mathbb{Z}$, the next diagram commutes.*

$$
\begin{array}{ccc}
H_n(C'') & \xrightarrow{\;\lambda_n\;} & H_{n-1}(C) \\
\downarrow{\scriptstyle H_n(\ell)} & & \downarrow{\scriptstyle H_{n-1}(h)} \\
H_n(\bar{C}'') & \xrightarrow[\;\bar{\lambda}_n\;]{} & H_{n-1}(\bar{C})
\end{array}
$$

The proof of this theorem is easy and is left to the reader.

Let $f,g\colon (C,\partial) \to (C',\partial')$ be chain complex morphisms. We say that $f$ and $g$ are *chain homotopic* if there is a graded group morphism of degree $+1$, $s\colon C \to C'$ such that $f - g = d's + sd$; more precisely

$$(\forall n \in \mathbb{Z})\; f_n - g_n = \partial'_{n+1}s_n + s_{n-1}\partial_n \;.$$

The morphism $s\colon C \to C'$ is a *chain homotopy* between $f$ and $g$ (or from $f$ to $g$). Notice that the chain homotopy relation just defined is an equivalence relation in the set

$$\mathfrak{C}((C,\partial),(C',\partial')) \;.$$

In particular, a morphism $f \in \mathfrak{C}((C,\partial),(C',\partial'))$ is *chain null-homotopic* if there exists a chain homotopy $s$ such that $f = d's + sd$ (it follows that $f$ and $g$ are chain homotopic if and only if $f - g$ is chain null-homotopic).

**(II.3.4) Proposition.** *If $f,g \in \mathfrak{C}((C,\partial),(C',\partial'))$ are chain homotopic, then*

$$(\forall n)\; H_n(f) = H_n(g)\colon H_nC \to H_nC'.$$

*Proof.* For any cycle $z \in Z_nC$, we have that

$$H_nf[z] = [f_n(z)] = [g_n(z)] + [\partial'_{n+1}s_n(z)] + [s_{n-1}\partial_n(z)] = H_ng[z];$$

we now notice that $\partial_n z = 0$ and that $\partial'_{n+1}s_n(z)$ is a boundary and thus, homologous to zero. ∎

Please notice that if $f$ is chain null-homotopic, then $H_n(f) = 0$ for every $n$.

A chain complex $(C, \partial)$ is *free* if all of its groups are free Abelian; it is *positive* if $C_n = 0$ for every $n < 0$. A positive chain complex $(C, \partial)$ is *augmented* (to $\mathbb{Z}$) if there exists an epimorphism

$$\varepsilon \colon C_0 \to \mathbb{Z}$$

such that $\varepsilon \partial_1 = 0$. The homomorphism $\varepsilon$ is the *augmentation (homomorphism)*.

**(II.3.5) Remark.** The chain complex $C(K)$ associated with a simplicial complex $K$ is free and positive. Moreover, the function

$$\varepsilon \colon C_0(K) \to \mathbb{Z}, \ \Sigma_{i=1}^n a_i\{x_i\} \mapsto \Sigma_{i=1}^n a_i$$

is an augmentation.

A chain complex $(C, \partial)$ is *acyclic* if, for every $n \in \mathbb{Z}$, $\ker \partial_n = \operatorname{im} \partial_{n+1}$, that is to say, if the sequence

$$\cdots \longrightarrow C_{n+1} \xrightarrow{\partial_{n+1}} C_n \xrightarrow{\partial_n} C_{n-1} \longrightarrow \cdots$$

is exact. A positive chain complex $(C, \partial)$ with augmentation is *acyclic* if the sequence

$$\cdots \longrightarrow C_n \xrightarrow{\partial_n} \cdots \xrightarrow{\partial_1} C_0 \xrightarrow{\varepsilon} \mathbb{Z}$$

is exact.

Let $(C, \partial)$ and $(C', \partial')$ be two positive augmented chain complexes. A morphism $f \in \mathfrak{C}((C, \partial), (C', \partial'))$ is an *extension* of a homomorphism $\bar{f} \colon \mathbb{Z} \to \mathbb{Z}$ if the next diagram commutes.

$$
\begin{array}{ccccccc}
\cdots \longrightarrow & C_1 & \xrightarrow{\partial_1} & C_0 & \xrightarrow{\varepsilon} & \mathbb{Z} \\
& \downarrow{\scriptstyle f_1} & & \downarrow{\scriptstyle f_0} & & \downarrow{\scriptstyle \bar{f}} \\
\cdots \longrightarrow & C_1' & \xrightarrow{\partial_1'} & C_0' & \xrightarrow{\varepsilon'} & \mathbb{Z}
\end{array}
$$

**(II.3.6) Theorem.** *Let $(C, \partial)$ and $(C', \partial')$ be positive augmented chain complexes; assume that $(C, \partial)$ is free and $(C', \partial')$ is acyclic. Then any homomorphism $\bar{f} \colon \mathbb{Z} \to \mathbb{Z}$ admits an extension $f \colon (C, \partial) \to (C', \partial')$, unique up to chain homotopy.*

*Proof.* Since the augmentation $\varepsilon' \colon C_0' \to \mathbb{Z}$ is surjective, for every basis element $x_0$ of $C_0$, we choose an element of $C_0'$ which is taken onto $\bar{f}\varepsilon(x_0)$ by $\varepsilon'$; in this way, we obtain a homomorphism $f_0 \colon C_0 \to C_0'$ such that $\bar{f}\varepsilon = \varepsilon' f_0$. We now take an arbitrary basis element $x_1$ of $C_1$; because

$$\varepsilon' f_0 \partial_1(x_1) = f\varepsilon \partial_1(x_1) = 0$$

and $\operatorname{im} \partial_1' = \ker \varepsilon'$, there exists $y_1' \in C_1'$ such that $\partial_1'(y_1') = f_0 \partial_1(x_1)$. This defines a homomorphism $f_1: C_1 \to C_1'$ such that $f_0 \partial_1 = \partial_1' f_1$.

Assume that we have inductively constructed the homomorphisms $f_i: C_i \to C_i'$ commuting with the boundary homomorphisms for $i \le n$; now, take the commutative diagram

For every basis element $x_{n+1}$ of $C_{n+1}$

$$\partial_n' f_n \partial_{n+1}(x_{n+1}) = f_{n-1} \partial_n \partial_{n+1}(x_{n+1}) = 0$$

that is to say, $f_n \partial_{n+1}(x_{n+1}) \in \ker \partial_n'$. It follows that $f_n \partial_{n+1}(x_{n+1})$ is an $n$-cycle of $(C', \partial')$; but this chain complex is acyclic and so there exists $y_{n+1} \in C_{n+1}'$ such that $\partial'(y_{n+1}) = f_n \partial(x_{n+1})$. By extending linearly $x_{n+1} \mapsto y_{n+1}$, we obtain a homomorphism

$$f_{n+1}: C_{n+1} \to C_{n+1}', \quad d' f_{n+1} = f_n d.$$

This concludes the inductive construction.

Suppose that $g: (C, \partial) \to (C', \partial')$ is another extension of $\bar{f}$. Then, for any arbitrary generator $x_0$ of $C_0$,

$$\varepsilon'(f_0 - g_0)(x_0) = 0;$$

since $\ker \varepsilon' = \operatorname{im} \partial_1'$, there exists an element $y_1 \in C_1'$ such that $\partial_1'(y) = (f_0 - g_0)(x_0)$. We define $s_0: C_0 \to C_1'$ by $s_0(x_0) = y_1$ on the generators and extend this function linearly over the entire group $C_0$; in this way, we obtain a homomorphism $s_0: C_0 \to C_1'$ such that $\partial_1' s_1 = f_0 - g_0$. Let us assume that, for every $i = 1, \cdots, n$, we have defined the homomorphisms $s_i: C_i \to C_{i+1}'$ satisfying the condition

$$\partial_{i+1}' s_i + s_{i-1} \partial_i = f_i - g_i.$$

For any generator $x_{n+1}$ of $C_{n+1}$

$$\partial_{n+1}'(f_{n+1} - g_{n+1} - s_n \partial_{n+1})(x_{n+1}) = 0$$

(because $\partial_{n+1}' s_n + s_{n-1} \partial_n = f_n - g_n$); thus, there exists $y_{n+2} \in C_{n+2}'$ such that

$$(f_{n+1} - g_{n+1} - s_n \partial_{n+1})(x_{n+1}) = \partial_{n+2}'(y_{n+2}).$$

In this fashion, we construct a homomorphism $s_{n+1}: C_{n+1} \to C_{n+2}'$ and, in the end, we obtain a chain homotopy from $f$ to $g$. ∎

**(II.3.7) Corollary.** *Let $(C, \partial)$ and $(C', \partial')$ be two positive augmented chain complexes; assume $(C, \partial)$ to be free and $(C', \partial')$ to be acyclic. If $f: (C, \partial) \to (C', \partial')$ is an extension of the trivial homomorphism $0: \mathbb{Z} \to \mathbb{Z}$, then $f$ is chain null-homotopic.*

The next result gives a good criterion to check if a positive, free, augmented chain complex is acyclic.

**(II.3.8) Lemma.** *Let $(C, \partial)$ be a positive, free, augmented chain complex with augmentation homomorphism*

$$\varepsilon: C_0 \longrightarrow \mathbb{Z}.$$

*Then, $(C, \partial)$ is acyclic if and only if the following conditions hold true:*

    **I.** *There exists a function $\eta: \mathbb{Z} \to C_0$ such that $\varepsilon\eta = 1$.*
    **II.** *There exists a chain homotopy $s: C \to C$ such that*

      *1. $\partial_1 s_0 = 1 - \eta\varepsilon$,*
      *2. $(\forall n \geq 1)\ \partial_{n+1} s_n + s_{n-1} \partial_n = 1$.*

*Proof.* Suppose that $C$ is acyclic. Since $\varepsilon: C_0 \to \mathbb{Z}$ is a surjection, there exists $x \in C_0$ such that $\varepsilon(x) = 1$. Define

$$\eta: \mathbb{Z} \to C_0,\ n \mapsto nx.$$

Clearly $\varepsilon\eta = 1$.

The homomorphisms $1: \mathbb{Z} \to \mathbb{Z}$ and $0: \mathbb{Z} \to \mathbb{Z}$ can be extended trivially to the chain homomorphisms $1, 0: (C, \partial) \to (C, \partial)$; because of Theorem (II.3.6), there exists a chain homotopy $s: C \to C$ satisfying conditions 1 and 2.

Conversely, assume that there exists a chain homotopy $s: C \to C'$, and a homomrrophism $\eta$ with properties 1. and 2.; because of Noether's Homomorphism Theorem,

$$C_0/\ker\varepsilon \cong \mathbb{Z};$$

since $\varepsilon\partial_1 = 0$, we have that $\operatorname{im}\partial_1 \subset \ker\varepsilon$ and, from $\partial_1 s_0 = 1 - \eta\varepsilon$, we conclude that every $x \in C_0$ can be written as

$$x = \partial_1 s_0(x) + \eta\varepsilon(x).$$

Hence for every $x \in \ker\varepsilon$, we have that $x = \partial_1 s_0(x)$, that is to say, $\ker\varepsilon \subset \operatorname{im}\partial_1$, and therefore

$$H_0(C) \cong \mathbb{Z}.$$

Finally, because

$$\partial_{n+1} s_n + s_{n-1} \partial_n = 1$$

in all positive dimensions, it follows that $H_n(C) = 0$ for every $n > 0$. Thus, $(C, \partial)$ is acyclic. ∎

Theorem (II.3.6) and Corollary (II.3.7) require $(C', \partial')$ to be acyclic; this requirement can be replaced by a more interesting condition within the framework of the so-called *acyclic carriers*. The following definition is needed: given $(C, \partial), (C', \partial') \in \mathfrak{C}$, we say that $(C', \partial')$ is a *(chain) subcomplex* of $(C, \partial)$ if, for every $n \in \mathbb{Z}$, $C'_n$ is a subgroup of $C_n$ and $\partial'_n = \partial_n | C'_n$; we use the notation $(C', \partial') \leq (C, \partial)$ to indicate that $(C', \partial')$ is a subcomplex of $(C, \partial)$. Let $(C, \partial)$ be a free chain complex; for each $n \in \mathbb{Z}$, let $\{x_\lambda^{(n)} \mid \lambda \in \Lambda_n\}$ be a basis of $C_n$. Now, let $(C', \partial')$ be an arbitrary chain complex. A *chain carrier* from $(C, \partial)$ to $(C', \partial')$ (relative to the choice of basis) is a function $S$, which associates with each basis element $x_\lambda^{(n)}$ a subcomplex $(S(x_\lambda^{(n)}), \partial_S) \leq (C', \partial')$ satisfying the following properties:

1. $(S(x_\lambda^{(n)}), \partial_S)$ is an acyclic chain complex.
2. If $x$ is a basis element of $C_n$ such that $\partial x = \sum a_\lambda x_\lambda^{(n-1)}$ and $a_\lambda \neq 0$, then

$$(S(x_\lambda^{(n-1)}), \partial_S) \leq (S(x), \partial_S).$$

We say that a morphism $f \in \mathfrak{C}((C, \partial), (C', \partial'))$ *has an acyclic carrier* $S$ if $f(x_\lambda^{(n)}) \in S(x_\lambda^{(n)})$ for every index $\lambda$ and every $n \in \mathbb{Z}$. In this case, if $x$ is a basis element, then $f(\partial(x)) \in S(x)$. The next result is the **Acyclic Carrier Theorem**.

**(II.3.9) Theorem.** *Let $(C, \partial), (C', \partial') \in \mathfrak{C}$ be positive augmented chain complexes; suppose that $(C, \partial)$ is free and let $S$ be an acyclic carrier from $(C, \partial)$ to $(C', \partial')$. Then, any homomorphism $\bar{f} \colon \mathbb{Z} \to \mathbb{Z}$ has an extension $f \colon (C, \partial) \to (C', \partial')$ with chain carrier $S$. The chain homomorphism $f$ is uniquely defined, up to chain homotopy.*

*Proof.* Take any generator $x_0$ of $C_0$; let $S(x_0) \leq (C', \partial')$ be the acyclic subcomplex defined by $S$. Notice that the restriction of $\varepsilon'$ to $S(x_0)_0$ is an augmentation homomorphism for $S(x_0)$. Since such a restriction is a surjection, there exists $y_0 \in S(x_0)_0$ such that $\varepsilon'(y_0) = \bar{f}\varepsilon(x_0)$; the usual argument determines $f_0$ with carrier $S$.

We continue the proof using an induction procedure as in Theorem (II.3.6). Assume that, for every $i \leq n$, we have constructed the homomorphisms $f_i : C_i \to C'_i$ that commute with the boundary homomorphisms. Notice that if $x_{n+1}$ is an arbitrary generator of $C_{n+1}$

$$\partial'_n f_n \partial_{n+1}(x_{n+1}) = f_{n-1} \partial_n \partial_{n+1}(x_{n+1}) = 0;$$

on the other hand, $f_n \partial_{n+1}(x_{n+1})$ belongs to the acyclic subcomplex

$$S(x_{n+1}) \leq (C', \partial')$$

and thus there exists $y_{n+1} \in C'_{n+1} \cap S(x_{n+1})$ such that $d'(y_{n+1}) = f_n d(x_{n+1})$. The function $x_{n+1} \mapsto y_{n+1}$ can be linearly extended to the homomorphism

$$f_{n+1} \colon C_{n+1} \to C'_{n+1}$$

such that $d' f_{n+1} = f_n d$.

Also the proof of the second part follows the steps of the proof given in Theorem (II.3.6). ∎

**(II.3.10) Corollary.** *Let $(C, \partial)$ and $(C', \partial')$ be given positive augmented chain complexes and let $(C, \partial)$ be free. Then any chain homomorphism $f: (C, \partial) \to (C', \partial')$, extending the trivial homomorphism $0: \mathbb{Z} \to \mathbb{Z}$ and having an acyclic carrier $S$, is null-homotopic.*

We prove now an important result known as **Five Lemma**.

**(II.3.11) Lemma.** *Let the diagram of Abelian groups and homomorphisms*

$$
\begin{array}{ccccccccc}
A & \xrightarrow{\;f\;} & B & \xrightarrow{\;g\;} & C & \xrightarrow{\;h\;} & D & \xrightarrow{\;k\;} & E \\
\downarrow{\alpha} & & \downarrow{\beta} & & \downarrow{\gamma} & & \downarrow{\delta} & & \downarrow{\varepsilon} \\
A' & \xrightarrow{\;f'\;} & B' & \xrightarrow{\;g'\;} & C' & \xrightarrow{\;h'\;} & D' & \xrightarrow{\;k'\;} & E'
\end{array}
$$

*be commutative and with exact lines. If the homomorphisms $\alpha$, $\beta$, $\delta$, and $\varepsilon$ are isomorphisms, so is $\gamma$.*

*Proof.* Let $c \in C$ be such that $\gamma(c) = 0$; then $\delta h(c) = h'\gamma(c) = 0$ and because $\delta$ is an isomorphism, $h(c) = 0$. In view of the exactness condition, there is a $b \in B$ with $g(b) = c$ and $g'\beta(b) = \gamma g(b) = 0$; thus, there exists $a' \in A'$ such that $f'(a') = \beta(b)$. But

$$
c = g(b) = g\beta^{-1}f'(a') = gf\alpha^{-1}(a') = 0
$$

and so $\gamma$ is injective. For an arbitrary $c' \in C'$,

$$
k\delta^{-1}h'(c') = \varepsilon^{-1}k'h'(c') = 0
$$

and, hence, there exists $c \in C$ such that $h(c) = \delta^{-1}h'(c')$. Moreover,

$$
h'(c' - \gamma(c)) = h'(c') - \delta\delta^{-1}h'(c') = 0
$$

and hence, there exists $b' \in B'$ such that $g'(b') = c' - \gamma(c)$. It follows that

$$
\gamma(c + g\beta^{-1}(b')) = \gamma(c) + g'\beta\beta^{-1}(b') = c'
$$

and so $\gamma$ is also surjective. ∎

## Exercises

**1.** A short exact sequence of Abelian groups

$$
G_{n+1} \xrightarrow{\;f_{n+1}\;} G_n \xrightarrow{\;f_n\;} G_{n-1}
$$

*splits* (or *is split*) if there exists a homomorphism $h_{n-1} : G_{n-1} \to G_n$ such that $f_n h_{n-1} = 1_{G_{n-1}}$ (or if there exists a homomorphism $k_n : G_n \to G_{n+1}$ such that $k_n f_{n+1} = 1_{G_{n+1}}$). Prove that if the short exact sequence

$$G_{n+1} \overset{f_{n+1}}{\rightarrowtail} G_n \overset{f_n}{\twoheadrightarrow} G_{n-1}$$

splits, then

$$G_n \cong G_{n+1} \oplus G_{n-1}.$$

**2.** Prove Theorem (II.3.3).


## II.4 Simplicial Homology

In this section, we give some results which allow us to study more in depth the homology of a simplicial complex. We begin with some important remarks on the homology of a simplicial complex $K$. The groups $C_n(K)$ of the $n$-chains are free, with rank equal to the (finite) number of $n$-simplexes of $K$; hence, also the subgroups $Z_n(K)$ and $B_n(K)$ of $C_n(K)$ are free, with a finite number of generators. Finally, the homology groups $H_n(K)$ are Abelian and finitely generated; therefore, by the decomposition theorem for finitely generated Abelian groups, they are isomorphic to direct sums

$$\mathbb{Z}^{\beta(n)} \oplus \mathbb{Z}_{n(1)} \oplus \ldots \oplus \mathbb{Z}_{n(k)}$$

where $\mathbb{Z}_{n(i)}$ is cyclic of order $n(i)$. The number $\beta(n)$ – equal to the *rank* of the Abelian group $H_n(K)$ – is the *nth-Betti number* of the complex $K$.

Let $p$ be the dimension of the simplicial complex $K$; for each $0 \le n \le p$, let $s(n)$ be the number of $n$-simplexes of $K$ (remember that $K$ is finite). Hence, the rank of the free Abelian group $C_n(K)$ is $s(n)$. We indicate with $z(n)$ and $b(n)$ the ranks of the groups $Z_n(K)$ and $B_n(K)$, respectively, where $n = 0, \ldots, p$. Since the boundary homomorphism $\partial_n : C_n(K) \to C_{n-1}(K)$ is a surjection on $B_{n-1}(K)$, by Nöther's Homomorphism Theorem, for each $n \ge 1$

$$(1) \qquad s(n) - z(n) = b(n-1);$$

if $n = 0$, we have $s(0) = z(0)$ because $C_0(K) = Z_0(K)$ and $B_{-1}(K) = 0$; on the other hand, $H_n(K, \mathbb{Z}) = Z_n(K)/B_n(K)$ and so

$$(2) \qquad \beta(n) = z(n) - b(n)$$

for $n \ge 0$. Subtracting (2) from (1) (when $n \ge 1$) it follows that

$$(3) \qquad s(n) - \beta(n) = b(n) - b(n-1).$$

If we do the alternate sum of the equalities (3) together with $s(0) - \beta(0) = b(0)$, we obtain

$$\sum_{n=0}^{p} (-1)^n (s(n) - \beta(n)) = \pm \beta(p);$$

since $C_{p+1}(K) = 0$, we have that $\beta(p) = 0$ and so the equality

$$\sum_{n=0}^{p} (-1)^n s(n) = \sum_{n=0}^{p} (-1)^n \beta(n)$$

holds true. The number

$$\chi(K) = \sum_{n=0}^{p} (-1)^n \beta(n)$$

is the *Euler-Poincaré characteristic* of $K$; this may be useful in determining the homology of some finite simplicial complexes.

Let $L$ be a simplicial subcomplex of a simplicial complex $K$; we now ask whether it is possible to compare the homology of a subcomplex $L \subset K$ with the homology of $K$. The (positive) answer lies with the *exact homology sequence* of the pair $(K, L)$. Let us see how we may find this exact sequence. For every $n \geq 0$, consider the quotient of the chain groups $C_n(K)/C_n(L)$ and define

$$\partial_n^{K,L} \colon C_n(K)/C_n(L) \to C_{n-1}(K)/C_{n-1}(L)$$

by

$$\partial_n^{K,L}(c + C_n(L)) = (\partial_n^K(c)) + C_{n-1}(L).$$

This is a well-defined formula because, if $c'$ is another representative of $c + C_n(L)$, then, $c - c' \in C_n(L)$ and

$$\partial_n^K(c - c') = \partial_n^L(c - c') \in C_{n-1}(L) ;$$

hence, $\partial_n^{K,L}(c + C_n(L)) = \partial_n^{K,L}(c' + C_n(L))$. The reader can easily verify that the homomorphisms $\partial_n^{K,L}$ are boundary homomorphisms and so that

$$C(K,L) = \{C_n(K)/C_n(L), \partial_n^{K,L}\}$$

is a chain complex whose homology groups $H_n(K,L;\mathbb{Z})$ are the so-called *relative homology groups* of the pair $(K, L)$. We point out that

$$H_n(K,L;\mathbb{Z}) = Z_n(K,L)/B_n(K,L)$$

where

$$Z_n(K,L) = \ker \partial_n^{K,L} \text{ and } B_n(K,L) = \operatorname{im} \partial_{n+1}^{K,L}.$$

Let **CCsim** be the category whose objects are pairs $(K, L)$, where $K$ is a simplicial complex, $L$ is one of its subcomplexes, and whose morphisms are pairs of simplicial functions

$$(k, \ell) \colon (K, L) \longrightarrow (K', L')$$

such that $k\colon K \to K'$ and $\ell\colon L \to L'$ is the restriction of $k$ to $L$. The reader can easily verify that the relative homology determines a covariant functor

$$H(-,-;\mathbb{Z})\colon \mathbf{CCsim} \longrightarrow \mathbf{Ab}^{\mathbb{Z}}.$$

The next result, which is an immediate application of the Long Exact Sequence Theorem (II.3.1), is called **Long Exact Homology Sequence Theorem**; it relates the homology groups of $L$, $K$, and $(K,L)$ to each other.

**(II.4.1) Theorem.** *Let $(K,L)$ be a pair of simplicial complexes. For every $n > 0$, there is a homomorphism*

$$\lambda_n\colon H_n(K,L;\mathbb{Z}) \to H_{n-1}(L;\mathbb{Z})$$

*(connecting homomorphism) that causes the following sequence of homology groups*

$$\ldots \to H_n(L;\mathbb{Z}) \xrightarrow{H_n(i)} H_n(K;\mathbb{Z}) \xrightarrow{q_*(n)} H_n(K,L;\mathbb{Z}) \xrightarrow{\lambda_n} H_{n-1}(L;\mathbb{Z}) \to \ldots,$$

*to be exact; here, $H_n(i)$ is the homomorphism induced by the inclusion $i\colon L \to K$ and $q_*(n)$ is the homomorphism induced by the quotient homomorphism $q_n\colon C_n(K) \to C_n(K)/C_n(L)$.*

*Proof.* For every $n > 0$, let

$$q_n\colon C_n(K) \to C_n(K)/C_n(L)$$

be the quotient homomorphism. With the given definitions, it is easily proved that

$$(\forall n \geq 0)\ \partial_n^{K,L} q_n = q_{n-1}\partial_n^K$$

and therefore,

$$q = \{q_n\}\colon C(K) \to C(K,L)$$

is a homomorphism of chain complexes. We note furthermore that for each $n \geq 0$, the sequence of Abelian groups

$$C_n(L) \xrightarrow{\ C_n(i)\ } C_n(K) \xrightarrow{\ q_n\ } C_n(K)/C_n(L)$$

is a short exact sequence and therefore, we have a short exact sequence of chain complexes

$$C(L) \xrightarrow{\ C(i)\ } C(K) \xrightarrow{\ q\ } C(K,L);$$

the result follows from Theorem (II.3.1).  ∎

The exact sequence of homology groups described in the statement of Theorem (II.4.1) is the *exact homology sequence of the pair $(K,L)$.*

In the context of the categories **Csim** and **CCsim**, the naturality of the connecting homomorphism

$$\lambda_n \colon H_n(K,L;\mathbb{Z}) \to H_{n-1}(L;\mathbb{Z})$$

can be explained as follows. We start with a result whose proof is easily obtained from the given definitions and is left to the reader.

**(II.4.2) Theorem.** *Let* $(k,\ell) \colon (K,L) \to (K',L')$ *be a given simplicial function. Then, for every* $n \geq 1$, *the following diagram commutes.*

$$
\begin{array}{ccc}
H_n(K,L;\mathbb{Z}) & \xrightarrow{\ \lambda_n\ } & H_{n-1}(L;\mathbb{Z}) \\
\Big\downarrow{\scriptstyle H_n(k,\ell)} & {\scriptstyle H_{n-1}(\ell)} & \Big\downarrow \\
H_n(K',L';\mathbb{Z}) & \xrightarrow[\ \lambda_n\ ]{} & H_{n-1}(L';\mathbb{Z})
\end{array}
$$

Let

$$pr_2 \colon \textbf{CCsim} \to \textbf{Csim}$$

be the functor defined by

$$(\forall (K,L) \in \textbf{CCsim})\ pr_2(K,L) = L$$

and

$$(\forall (k,\ell) \in \textbf{CCsim}((K,L),(K',L')))\ pr_2(k,\ell) = \ell .$$

For each $n \geq 0$, take the covariant functors

$$H_n(-,-) \colon \textbf{CCsim} \to Gr$$

and

$$H_{n-1}(-) \circ pr_2 \colon \textbf{CCsim} \to Gr .$$

Theorem (II.3.3) states that

$$\lambda_n \colon H_n(-,-;\mathbb{Z}) \to H_{n-1}(-;\mathbb{Z}) \circ pr_2$$

is a *natural transformation* (see the definition of natural transformation of functors in Sect. I.2).

Computing the homology of a complex $K$ can be made easier by the exact homology sequence, provided that we can compute the homology of $L$ and the relative homology of $(K,L)$. Another very useful technique for computing the homology of a simplicial complex is using the *Mayer–Vietoris sequence*. Consider two simplicial complexes $K_1 = (X_1, \Phi_1)$ and $K_2 = (X_2, \Phi_2)$ such that $K_1 \cap K_2$ and $K_1 \cup K_2$ are simplicial complexes; in addition, $K_1 \cap K_2$ must be a subcomplex of both $K_1$ and $K_2$. The inclusions

$$\Phi_1 \cap \Phi_2 \hookrightarrow \Phi_\alpha \ , \ \Phi_\alpha \hookrightarrow \Phi_1 \cup \Phi_2 \ , \ \alpha = 1,2$$

define simplicial functions

$$i_\alpha \colon K_1 \cap K_2 \longrightarrow K_\alpha \,, \ j_\alpha \colon K_\alpha \longrightarrow K_1 \cup K_2 \,, \ \alpha = 1,2$$

which, in turn, define the homomorphisms

$$\bar{\imath}(n) \colon C_n(K_1 \cap K_2) \to C_n(K_1) \oplus C_n(K_2)$$
$$c \mapsto (C_n(i_1)(c), C_n(i_2)(c)) \,,$$

$$\bar{\jmath}(n) \colon C_n(K_1) \oplus C_n(K_2) \to C_n(K_1 \cup K_2)$$
$$(c, c') \mapsto C_n(j_1)(c) - C_n(j_2)(c').$$

These homomorphisms have the following properties:

1. $\bar{\imath}(n)$ is injective;
2. $\bar{\jmath}(n)$ is surjective;
3. $\operatorname{im} \bar{\imath}(n) = \ker \bar{\jmath}(n)$;
4. $(\partial_n^{K_1} \oplus \partial_n^{K_2}) \bar{\imath}(n) = \bar{\imath}(n-1) \partial_n^{K_1 \cap K_2}$;
5. $\bar{\jmath}(n-1)(\partial_n^{K_1} \oplus \partial_n^{K_2}) = \partial_n^{K_1 \cup K_2} \bar{\jmath}(n)$.

In this way, the chain complex sequence

$$0 \to C(K_1 \cap K_2) \xrightarrow{\ \bar{\imath}\ } C(K_1') \oplus C(K_2) \xrightarrow{\ \bar{\jmath}\ } C(K_1 \cup K_2) \to 0$$

is short exact.

Theorem (II.3.1) enables us to state the next theorem, known as **Mayer–Vietoris Theorem**:

**(II.4.3) Theorem.** *For every $n \in \mathbb{Z}$, there is a homomorphism*

$$\lambda_n \colon H_n(K_1 \cup K_2; \mathbb{Z}) \longrightarrow H_{n-1}(K_1 \cap K_2; \mathbb{Z})$$

*such that the infinite sequence of homology groups*

$$\ldots \to H_n(K_1 \cap K_2; \mathbb{Z}) \xrightarrow{H_n(\bar{\imath})} H_n(K_1; \mathbb{Z}) \oplus H_n(K_2; \mathbb{Z})$$
$$\xrightarrow{H_n(\bar{\jmath})} H_n(K_1 \cup K_2; \mathbb{Z}) \xrightarrow{\lambda_n} H_{n-1}(K_1 \cap K_2; \mathbb{Z}) \to \ldots$$

*is exact.*

We now give some results on the homology of certain simplicial complexes. Given a simplicial complex $K = (X, \Phi)$, we say that two vertices $x, y \in X$ are *connected* if there is a sequence of 1-simplexes

$$\{\{x_0^i, x_1^i\} \in \Phi, i = 0, \ldots, n\}$$

where $x_0^0 = x, x_1^n = y$, and $x_1^i = x_0^{i+1}$; we then have an equivalence relation on the set $X$, braking it down into a union of disjoint subsets $X = X_1 \sqcup X_2 \sqcup \ldots \sqcup X_k$. The sets

$$\Phi_i = \{\sigma \in \Phi | (\exists x \in X_i | x \in \sigma)\} \, , \, i = 1, \ldots, k$$

are disjoint; moreover, the pairs $K_i = (X_i, \Phi_i)$, $i = 1, \ldots, k$, called *connected components* of $K$, are simplicial subcomplexes of $K$. Hence, the relation of connectedness subdivides the complex $K$ into a union of disjoint simplicial subcomplexes of $K$. From this point of view, a complex $K$ is *connected* if and only if it has a unique connected component.

**(II.4.4) Lemma.** *A simplicial complex $K$ is connected if and only if $|K|$ is connected.*

*Proof.* Suppose $K$ to be connected and let $p$ and $q$ be any two points of $|K|$. Join $p$ to a vertex $x$ of its carrier $s(p)$ by means of the segment with end points $p$ and $x$; this segment is contained in $|s(p)|$ and is therefore a segment of $|K|$; similarly, join $q$ to a vertex $y$ of its carrier $s(q)$. However, the vertices $x$ and $y$ are also vertices of $K$ and since $K$ is connected, there is a path of 1-simplexes of $K$ which links $x$ to $y$. In this manner, we obtain a path of $|K|$ that links $p$ to $q$; hence, $|K|$ is path-connected and so, $|K|$ is connected (see Theorem (I.1.21)).

Conversely, suppose $|K|$ to be connected and let $K_i$ be a connected component of $K$; since $K_i$ and $K \smallsetminus K_i$ are subcomplexes of $K$, we have that $|K_i|$ is open and closed in $|K|$; since $|K|$ is connected, $|K_i| = |K|$, that is to say, $K_i = K$ and so, $K$ is connected. ∎

The reader is encouraged to review the results on connectedness and path-connectedness in Sect. I.1; note that these two concepts are equivalent for polyhedra.

**(II.4.5) Lemma.** *The following properties regarding a simplicial complex $K = (X, \Phi)$ are equivalent:*

1. *$K$ is connected;*
2. *$H_0(K; \mathbb{Z}) \simeq \mathbb{Z}$;*
3. *the kernel of the augmentation homomorphism*

$$\varepsilon \colon C_0(K) \to \mathbb{Z} \, , \, \sum_{i=1}^{n} g_i\{x_i\} \mapsto \sum_{i=1}^{n} g_i$$

*coincides with the group $B_0(K)$.*

*Proof.* $1 \Rightarrow 3$: We first notice that the inclusion

$$B_0(K) \subset \ker \varepsilon$$

is always true: indeed,

$$\varepsilon \left( \partial_1 \left( \sum_{i=0}^{k} g_i\{x_0^i, x_1^i\} \right) \right) = \sum_{i=0}^{k} g_i - \sum_{i=0}^{k} g_i = 0.$$

Let $x$ be a fixed vertex of $K$. The connectedness of $K$ means that, for every vertex $y$ of $K$, the 0-cycles $\{x\}$ and $\{y\}$ are homological and so, for every 0-chain $c_0 = \sum_{i=0}^{k} g_i\{x_i\}$, there exists a 1-chain $c_1$ such that

$$\sum_{i=0}^{k} g_i\{x_i\} - \left(\sum_{i=0}^{k} g_i\right)\{x\} = \partial_1(c_1).$$

Therefore, it is clear that $c_0 \in \ker \varepsilon$ implies $c_0 \in B_0(K)$.

$3 \Rightarrow 2$: Given two homological 0-cycles $z_0$ and $z'_0$, it follows from the property $z_0 - z'_0 \in B_0(K)$ that $\varepsilon(z_0) = \varepsilon(z'_0)$ and so we may define the homomorphism

$$\theta \colon H_0(K;\mathbb{Z}) \to \mathbb{Z}, \ z_0 + B_0(K) \mapsto \varepsilon(z_0)$$

which is easily seen (by hypothesis 3) to be injective. The surjectivity of $\theta$ follows immediately; in fact, for every $g \in \mathbb{Z}$, we have $\theta(g\{x\} + B_0(K)) = g$, where $x \in X$ is a fixed vertex.

$2 \Rightarrow 1$: Let $K = K_1 \sqcup K_2 \sqcup \ldots \sqcup K_k$ be the decomposition of $K$ into its connected components. We obtain

$$H_0(K;\mathbb{Z}) \simeq \sum_{i=1}^{k} H_0(K_i;\mathbb{Z}) \simeq \sum_{i=1}^{k} \mathbb{Z}$$

from the given definitions and from what we have proved so far; however, since $H_0(K;\mathbb{Z}) \simeq \mathbb{Z}$, we must have $k = 1$, which means that $K$ is connected. ∎

The next three examples are examples of abstract simplicial complexes called *acyclic* because they induce chain complexes which are acyclic (see Sect. II.3).

**Homology of $\overline{\sigma}$** – Let $\overline{\sigma}$ be the simplicial complex generated by a simplex $\sigma = \{x_0, x_1, \ldots, x_n\}$. Since $\overline{\sigma}$ is connected, Lemma (II.4.5) ensures that $H_0(\overline{\sigma},\mathbb{Z}) = \mathbb{Z}$. We wish to prove that $H_i(\overline{\sigma},\mathbb{Z}) = 0$ for every $i > 0$. With this in mind, we begin to order the set of vertices, assuming that $x_0$ is the first element. Then, for any integer $0 < j < n$ and any ordered simplex $\{x_{i_0}, \ldots, x_{i_j}\}$, we define

$$k_j(\{x_{i_0}, \ldots, x_{i_j}\}) = \begin{cases} \{x_0, x_{i_0}, \ldots, x_{i_j}\} & \text{for } i_0 > 0 \\ 0 & \text{for } i_0 = 0 \end{cases}$$

and linearly extend it to all $j$-chain of $\overline{\sigma}$ and therefore, to a homomorphism

$$k_j \colon C_j(\overline{\sigma}) \longrightarrow C_{j+1}(\overline{\sigma}).$$

A simple computation (on the simplexes of $\overline{\sigma}$) shows that for every chain $c \in C_j(\overline{\sigma})$

$$\partial_{j+1} k_j(c) + k_{j-1} \partial_j(c) = c$$

and so any $z_j \in Z_j(\overline{\sigma})$ is a boundary, that is to say, $H_j(\overline{\sigma},\mathbb{Z}) \cong 0$. Regarding $H_n(\overline{\sigma},\mathbb{Z})$, we note that $\sigma$, being the only $n$-simplex of $\overline{\sigma}$, cannot be a cycle; consequently, $Z_n(\overline{\sigma}) \cong 0$.

**Homology of a simplicial cone** – Since $\sigma = \{x_0, x_1, \ldots, x_{n+1}\}$, we call the simplicial complex

$$C(\sigma) = \dot{\sigma} \setminus \{x_1, \ldots, x_{n+1}\},$$

obtained by removing the $n$-face opposite to the vertex $x_0$ from the simplicial complex $\overset{\bullet}{\sigma}$, *n-simplicial cone with vertex* $\{x_0\}$ . Clearly

$$H_0(C(\sigma), \mathbb{Z}) \cong \mathbb{Z}$$

because $C(\sigma)$ is connected. A similar proof to the one used for $\overline{\sigma}$ shows that $H_j(C(\sigma), \mathbb{Z}) \cong 0$ for every $0 < j < n$. We note that, when $j = n$, the vertex $x_0$ belongs to every $n$-simplex of $C(\sigma)$ and so

$$(\forall c \in C_n(C(\sigma)))c = k_{n-1}\partial_n(c),$$

allowing us to conclude that the trivial cycle 0 is the only $n$-cycle of $C(\sigma)$; in other words, $H_n(C(\sigma); \mathbb{Z}) \cong 0$.

In the next example we refer to the construction of an acyclic carrier.

**Homology of the (abstract) cone** – Let $vK = v * K$ be the join of a simplicial complex $K = (X, \Phi)$ and of a simplicial complex with a single vertex (and simplex) $v$.

**(II.4.6) Lemma.** *The cones $vK$ are acyclic simplicial complexes.*

*Proof.* Let $v$ be the simplicial complex with the single vertex $v$ and no other simplex; it is clear that $v$ (considered as a simplicial complex) is an acyclic simplicial complex. The chain complex $C(v)$ is a subcomplex of $C(vK)$; let $\iota\colon C(v) \to C(vK)$ be the inclusion. Consider the simplicial function

$$c\colon vK \to v \,,\, y \in v\Phi \mapsto \{v\} \,.$$

It is readily seen that the chain morphism

$$C(c)\iota\colon C(v) \longrightarrow C(v)$$

coincides with the identity homomorphism of $C(v)$; then, for every $n \in \mathbb{Z}$, the composite $H_n(c)H_n(\iota)$ equals the identity. Let us prove that $\iota C(c)$ and the identity homomorphism $1_{C(vK)}$ of $C(vK)$ are homotopic. We define

$$s_n\colon C_n(vK) \to C_{n+1}(vK)$$

on the oriented $n$-simplexes $\sigma \in v\Phi$ (understood as a chain) by the formula

$$s_n(\sigma) = \begin{cases} 0 & \text{if } v \in \sigma, \\ v\sigma & \text{if } v \notin \sigma; \end{cases}$$

$s_n$ may be linearly extended to a homomorphism of $C_n(vK)$. Let us take a look into the properties of these functions.

*Case 1: $n = 0$* – Let $x$ be any vertex of $vK$.

$$(1_{C_0(vK)} - \iota C_0(c))(x) = \begin{cases} x - v & \text{if } x \neq v, \\ 0 & \text{if } x = v. \end{cases}$$

$$\partial_1 s_0(x) = \begin{cases} x - v & \text{if } x \neq v, \\ 0 & \text{if } x = v. \end{cases}$$

*Case 2:* $n > 0$ – Let $\sigma$ be any oriented $n$-simplex of $vK$. We first observe that $\iota C_n(c)(\sigma) = 0$; moreover, if $v$ is not a vertex of $\sigma$, we have

$$\partial_{n+1}(v\sigma) = \sigma - v\partial_n(\sigma).$$

Consequently, $v \notin \sigma$ implies

$$s_{n-1}\partial_n(\sigma) + \partial_{n+1}s_n(\sigma) = v\partial_n(\sigma) + \partial_{n+1}(v\sigma) = \sigma.$$

We now suppose that $v \in \sigma$. Then,

$$s_{n-1}\partial_n(\sigma) + \partial_{n+1}s_n(\sigma) = s_{n-1}\partial_n(\sigma) = \sigma.$$

It follows from these remarks that $\iota C(c)$ and the identity homomorphism $1_{C(vK)}$ are homotopic and we conclude from Proposition (II.3.4) that, for every $n \in \mathbb{Z}$, $H_n(\iota)H_n(c)$ coincides with the identity homomorphism. ∎

We now seek a better understanding of the relative homology $H_*(K, L; \mathbb{Z})$ of a pair of simplicial complexes $(K, L)$. As usual, $K = (X, \Phi)$ and $L = (Y, \Psi)$ with $Y \subset X$ and $\Psi \subset \Phi$. Let $v$ be a point which is not in the set of vertices of either $K$ or $L$. Let $CL$ be the abstract cone $vL$. It follows from the definitions that $K \cap CL = L$.

**(II.4.7) Theorem.** *The homology groups $H_n(K, L; \mathbb{Z})$ and $H_n(K \cup CL; \mathbb{Z})$ are isomorphic for each $n \geq 1$.*

*Proof.* The central idea in this proof is to compare the exact homology sequence of the pair $(K, L)$ and the exact sequence of Mayer–Vietoris of $K$ and $CL$, before using the Five Lemma; the notation is the one already adopted for the Mayer–Vietoris Theorem.

Let us consider the simplicial function $f \colon K \to CL$ defined on the vertices by

$$f(x) = \begin{cases} x & \text{if } x \in Y \\ v & \text{if } x \in X \smallsetminus Y. \end{cases}$$

For each nonnegative integer $n$, we now define the homomorphisms

$$\tilde{k}_n \colon C_n(K) \to C_n(K) \oplus C_n(CL), \ c \mapsto (c, C_n(f)(c))$$

and

$$\tilde{h}_n \colon C_n(K)/C_n(L) \longrightarrow C_n(K \cup CL),$$
$$c + C_n(L) \mapsto C_n(j_1)(c) - C_n(j_2)C_n(f)(c)$$

(that is to say, $\tilde{h}_n(c + C_n(L)) = \tilde{\jmath}_n \tilde{k}_n(c)$). The function $\tilde{h}_n$ is well defined; in fact, had $c' \in C_n(K)$ been such that $c - c' \in C_n(L)$, we would have $c - c' \in C_n(K \cap CL)$ and so

$$\tilde{\jmath}_n \tilde{k}_n(c - c') = \tilde{\jmath}_n \tilde{\imath}_n(c - c') = 0.$$

The homomorphism sequences $\tilde{h} = \{\tilde{h}_n | n \geq 0\}$ and $\tilde{k} = \{\tilde{k}_n | n \geq 0\}$ are homomorphisms of chain complexes giving rise to a commutative diagram

$$
\begin{array}{ccccc}
C(L) & \xrightarrow{\ \ C(i)\ \ } & C(K) & \xrightarrow{\ \ \tilde{q}\ \ } & C(K,L) \\
\downarrow{\scriptstyle 1} & & \downarrow{\scriptstyle \tilde{k}} & & \downarrow{\scriptstyle \tilde{h}} \\
C(K \cap CL) & \xrightarrow{\ \ \tilde{\imath}\ \ } & C(K) \oplus C(CL) & \xrightarrow{\ \ \tilde{\jmath}\ \ } & C(K \cup CL)
\end{array}
$$

Since $CL$ is an acyclic simplicial complex, we obtain, for every $n \geq 2$, the commutative diagram of Abelian groups

$$
\begin{array}{ccccccccc}
H_n(L;\mathbb{Z}) & \rightarrow & H_n(K;\mathbb{Z}) & \rightarrow & H_n(K,L;\mathbb{Z}) & \rightarrow & H_{n-1}(L;\mathbb{Z}) & \rightarrow & H_{n-1}(K;\mathbb{Z}) \\
\downarrow{\scriptstyle 1} & & \downarrow{\scriptstyle \cong} & & \downarrow{\scriptstyle \gamma} & & \downarrow{\scriptstyle 1} & & \downarrow{\scriptstyle \cong} \\
H_n(L;\mathbb{Z}) & \rightarrow & H_n(K;\mathbb{Z}) & \rightarrow & H_n(K \cup CL;\mathbb{Z}) & \rightarrow & H_{n-1}(L;\mathbb{Z}) & \rightarrow & H_{n-1}(K;\mathbb{Z})
\end{array}
$$

and by the Five Lemma, we conclude that $\gamma$ is an isomorphism; when $n = 1$, the last vertical arrow is an injective homomorphism

$$H_0(K;\mathbb{Z}) \longrightarrow H_0(K;\mathbb{Z}) \oplus \mathbb{Z}$$

and again with an argument similar to the Five Lemma, we conclude that $\gamma$ is an isomorphism.  ∎

**(II.4.8) Remark.** We recall that we have defined the relative homology groups of a pair of simplicial complexes $(K,L)$ through the chain complex $C(K,L) = \{C_n(K)/C_n(L), \partial_n^{K,L}\}$; we now construct the relative groups $H_n(K,L;\mathbb{Z})$, $n \geq 0$, from a slightly different point of view which turns out to be very useful for computing homology groups.

For any $n \geq 0$, let $\overline{C}_n(K,L)$ be the Abelian group of formal linear combinations, with coefficients in $\mathbb{Z}$, of all $n$-simplexes of $K$ which are not in $L$; in other words, if $K = (X, \Phi)$, $L = (Y, \Psi)$ with $Y \subset X$ and $\Psi \subset \Phi$,

$$\overline{C}_n(K,L;\mathbb{Z}) = \{\sum_i m_i \sigma_n^i | \sigma_n^i \in \Phi \smallsetminus \Psi\}.$$

The inclusion $i \colon L \rightarrow K$ induces an injective homomorphism $C_n(i) \colon C_n(L) \rightarrow C_n(K)$ for each $n \geq 0$; we now take, for every $n \geq 0$, the following linear homomorphisms:

$$\beta_n \colon C_n(K) \rightarrow C_n(L)$$

defined on the $n$-simplexes of $K$ by the conditions

$$\beta_n(\sigma_n) = \begin{cases} 0 & \text{if } \sigma_n \in \Phi \smallsetminus \Psi \\ \sigma_n & \text{if } \sigma_n \in \Psi \end{cases}$$

$$\alpha_n \colon \overline{C}_n(K,L) \to C_n(K)\,, \; \sigma_n \in \Phi \smallsetminus \Psi \mapsto \sigma_n$$

$$\mu_n \colon C_n(K) \to \overline{C}_n(K,L)$$

such that

$$\mu_n(\sigma_n) = \begin{cases} \sigma_n & \text{if } \sigma_n \in \Phi \smallsetminus \Psi \\ 0 & \text{if } \sigma_n \in \Psi \,. \end{cases}$$

It is easy to check that $\beta_n C_n(i) = 1$, $\mu_n \alpha_n = 1$, $\mu_n C_n(i) = 0$, and $C_n(i)\beta_n + \alpha_n \mu_n = 1$ for each $n \geq 0$. Hence, for every $n \geq 0$, we have a short exact sequence

$$C_n(L) \xrightarrowtail{\;C_n(i)\;} C_n(K) \xrightarrow{\;\mu_n\;} \overline{C}_n(K,L).$$

We now consider the boundary homomorphism $\partial_n \colon C_n(K) \to C_{n-1}(K)$ and define

$$\overline{\partial}_n \colon \overline{C}_n(K,L) \to \overline{C}_{n-1}(K,L)$$

as the composite homomorphism $\overline{\partial}_n = \mu_{n-1}\partial_n \alpha_n$. We note that

$$\begin{aligned} \overline{\partial}_{n-1}\overline{\partial}_n &= (\mu_{n-2}\partial_{n-1}\alpha_{n-1})(\mu_{n-1}\partial_n \alpha_n) \\ &= (\mu_{n-2}\partial_{n-1})(1 - C_{n-1}(i)\beta_{n-1})\partial_n \alpha_n \\ &= \mu_{n-2}\partial_{n-1}\partial_n \alpha_n - \mu_{n-2}\partial_{n-1}C_{n-1}\beta_{n-1} = 0 \end{aligned}$$

since the factor $\partial_{n-1}\partial_n = 0$ appears in the first term and also because the second term is null on all $(n-1)$-simplex of $K$. The graded Abelian group $\{\overline{C}_n(K,L) \mid n \in \mathbb{Z}\}$, where $\overline{C}_n(K,L) = 0$ for every $n < 0$, has a boundary homomorphism $\{\overline{\partial}_n \mid n \in \mathbb{Z}\}$ with $\overline{\partial}_n = 0$ for $n \leq 0$; let

$$\overline{H}_*(K,L;\mathbb{Z}) = \{\overline{H}_n(K,L;\mathbb{Z})\}$$

be its homology. Let $\theta_n \colon \overline{C}_n(K,L) \to C_n(K)$ be the linear homomorphism defined on an $n$-simplex $\sigma_n \in \Phi \smallsetminus \Psi$ by $\theta_n(\sigma_n) = \sigma_n + C_n(L)$ (if $n < 0$, we define $\theta_n = 0$). We note that $\theta_n$ commutes with the boundary homomorphisms; it is sufficient to verify this statement for an $n$-simplex $\sigma_n \in \Phi \smallsetminus \Psi$:

$$\partial_n^{K,L}\theta_n(\sigma_n) = \partial_n(\sigma_n) + C_{n-1}(L) = \sum_{\sigma_{n-1,i} \in \Phi \smallsetminus \Phi} (-1)^i \sigma_{n-1,i} + C_n(L);$$

$$\theta_{n-1}\overline{\partial}_n(\sigma_n) = \theta_{n-1}(\mu_{n-1}\sum_i (-1)^i \sigma_{n-1,i}) = \sum_{\sigma_{i,n-1} \in \Phi \smallsetminus \Phi} (-1)^i \sigma_{n-1,i} + C_n(L).$$

Therefore, the set $\{\theta_n \mid n \in \mathbb{Z}\}$ induces a homomorphism

$$H_n(\theta_n) \colon \overline{H}_n(K,L;\mathbb{Z}) \to H_n(K,L;\mathbb{Z}) \ .$$

On the other hand, $\theta_n$ is an isomorphism for each $n \geq 0$ (it is injective by definition and surjective because $\theta_n \mu_n = q_n$). Therefore, the two types of homology groups are isomorphic.

Let

$$\{K_i = (X_i, \Phi_i) \mid i = 1, \ldots, p\}$$

be a finite set of simplicial complexes; we choose a base vertex $x_0^i \in X_i$ for each $K_i$ and construct the *wedge sum* of all $K_i$ as the simplicial complex

$$\vee_{i=1}^p K_i : \ = \bigcup_{i=1}^n (\{x_0^1\} \times \ldots \times K_i \times \ldots \times \{x_0^p\})$$

that is to say

$$\vee_{i=1}^p K_i = (\vee_{i=1}^p X_i, \vee_{i=1}^p \Phi_i) \ .$$

The next theorem shows that the homology of the wedge sum of simplicial complexes acts in a special way.

**(II.4.9) Theorem.** *For every $q \geq 1$,*

$$H_q(\vee_{i=1}^p K_i; \mathbb{Z}) \cong \oplus_{i=1}^p H_q(K_i; \mathbb{Z}).$$

*Proof.* It is enough to prove this result for $p = 2$. The short exact sequence of chain complexes

$$C(K_1) \ \overset{i}{\rightarrowtail} \ C(K_1 \vee K_2) \ \overset{k}{\twoheadrightarrow} \ \overline{C}(K_1 \vee K_2, K_1; \mathbb{Z})$$

induces a long exact sequence of homology groups

$$\ldots \to H_n(K_1; \mathbb{Z}) \overset{H_n(i)}{\longrightarrow} H_n(K_1 \vee K_2; \mathbb{Z}) \overset{q_*(n)}{\longrightarrow} H_n(K_2; \mathbb{Z}) \overset{\lambda_n}{\longrightarrow} H_{n-1}(K_1; \mathbb{Z}) \to \ldots$$

(see Remark (II.4.8)). Let us now examine how the homomorphisms of chain complexes

$$C(K_1) \overset{i}{\longrightarrow} C(K_1 \vee K_2)$$

$$C(K_1 \vee K_2) \overset{k}{\longrightarrow} \overline{C}(K_1 \vee K_2, K_1; \mathbb{Z}) \cong C(K_2)$$

are defined on simplexes (in other words, the generators of the free groups that concern us):

$$(\forall \sigma_n^1 \in \Phi_1) \ i_n(\sigma_n) = \sigma_n^1 \times \{x_0^2\}$$
$$(\forall \sigma_n^1 \times \{x_0^2\} \in \Phi_1 \times \{x_0^2\}) \ k_n(\sigma_n^1 \times \{x_0^2\} = 0$$
$$(\forall \{x_0^1\} \times \sigma_n^2 \in \{x_0^1\} \times \Phi_2) \ k_n(\{x_0^1\} \times \sigma_n^2) = \sigma_n^2.$$

We now define the homomorphisms

$$j\colon C(K_1 \vee K_2) \to C(K_1) \text{ and } h\colon \overline{C}(K_1 \vee K_2, K_1; \mathbb{Z}) \to C(K_1 \vee K_2)$$

as follows:

$$(\forall \sigma_n^1 \times \{x_0^2\} \in \Phi_1 \times \{x_0^2\}) \, j_n(\sigma_n^1 \times \{x_0^2\}) = \sigma_n^1$$
$$(\forall \{x_0^1\} \times \sigma_n^2 \in \{x_0^1\} \times \Phi_2) \, j_n(\{x_0^1\} \times \sigma_n^2) = 0$$
$$(\forall \sigma_n^2 \in \Phi_2) \, h_n(\sigma_n^2) = \{x_0^1\} \times \sigma_n^2.$$

Morphisms $i, k, j$, and $h$ are induced by simplicial functions and so they commute with boundary operators. Moreover, $ji = 1_{C(K_1)}$ and $kh = 1_{C(K_2)}$, a property that extends to the respective homomorphisms regarding homology groups. Hence, for each $q \geq 1$, we have a splitting short exact sequence of homology groups

$$H_q(K_1; \mathbb{Z}) \xrightarrow{\ H_q(i)\ } H_q(K_1 \vee K_2; \mathbb{Z}) \xrightarrow{\ H_q(k)\ } H_q(K_2; \mathbb{Z}).$$
∎

## II.4.1 Reduced Homology

It is sometimes an advantage to introduce a little change to the simplicial homology, named *reduced homology*; the only difference between the two homologies lies on the group $H_0(-; \mathbb{Z})$. To obtain the reduced homology $\tilde{H}_*(K; \mathbb{Z})$ of a simplicial complex $K$, we consider the chain complex

$$\tilde{C}(K, \mathbb{Z}) = \{\tilde{C}_n(K), \tilde{d}_n\}$$

where

$$\tilde{C}_n(K) = \begin{cases} C_n(K), & n \geq 0 \\ \mathbb{Z}, & n = -1 \\ 0, & n \leq -2 \end{cases}$$

and define the boundary homomorphism

$$\tilde{d}_n = \begin{cases} \partial_n, & n \geq 1 \\ \varepsilon\colon C_0(K) \to \mathbb{Z}, & n = 0 \\ 0, & n \leq -1 \end{cases}$$

where $\varepsilon$ is the augmentation homomorphism (see Lemma (II.4.5)). We only need to verify that $\tilde{d}_0 \tilde{d}_1 = 0$; but this follows directly from the definition of $\varepsilon$.

We leave to the reader, as an exercise, to prove that if $K$ is a connected simplicial complex, then

$$(\forall n \neq 0) \, \tilde{H}_n(K; \mathbb{Z}) \cong H_n(K; \mathbb{Z})$$

and

$$\tilde{H}_0(K; \mathbb{Z}) \cong 0.$$

## Exercises

**1.** The *simplicial n-sphere* is the simplicial complex

$$\dot{\sigma}_{n+1} = (\sigma_{n+1}, \Phi)$$

where $\sigma_{n+1} = \{x_0, x_1, \ldots, x_{n+1}\}$ and $\Phi = \wp(X) \setminus \{\emptyset, \sigma_{n+1}\}$. Prove that

$$H_p(\dot{\sigma}_{n+1}) = \begin{cases} \mathbb{Z}, & p = 0, n \\ 0, & p \neq 0, n. \end{cases}$$

**2.** Prove that a subgroup of a free Abelian group is free (if this proves to be very difficult, refer to [17], Theorem 5.3.1f).

**3.** Compute the homology groups of the triangulations associated with the following spaces (see Exercise 4 on p. 64).

a) cylinder $C = S^1 \times I$;
b) Möbius band $M$;
c) Klein bottle $K$;
d) real projective plane $\mathbb{R}P^2$;
e) $G_2$, obtained by adding two handles to the sphere $S^2$.

**4.** Compute the Betti numbers and the Euler–Poincaré characteristic for the surfaces of the previous exercise.

**5.** Let $K$ be a given connected simplicial complex and $\Sigma K = K * \{x, y\}$ be the suspension of $K$ (see examples of simplicial complexes given in Sect. II.2). Prove that

$$(\forall n \geq 0) \ \tilde{H}_n(\Sigma K) \cong \tilde{H}_{n-1}(K)$$

by means of the Mayer–Vietoris sequence.

**6.** Let $K$ be a one-dimensional connected simplicial complex (namely, a graph), and $C(K) = 1 - \chi(K)$ its *cyclomatic number* (also called the *circuit rank*). Prove that $C(K) \geq 0$ and that the equality holds if and only if $|K|$ is contractible (that is to say, $K$ is a *tree*).

## II.5 Homology with Coefficients

In Sect. II.4, we have studied the homology of oriented simplicial complexes $K$ determined by the chain complex

$$(C(K), \partial) = \{C_n(K), \partial_n^K | n \in \mathbb{Z}\},$$

$C_n(K)$ being the free Abelian groups of formal linear combinations with coefficients in $\mathbb{Z}$ of the $n$-simplexes of $K$. We now wish to generalize our homology with co-efficients in the Abelian group $\mathbb{Z}$ to a homology with coefficients drawn from any Abelian group $G$.

We begin by reviewing the construction of the *tensor product* of two Abelian groups $A$, $B$: by definition, $A \otimes B$ is the Abelian group generated by the set of elements

$$\{a \otimes b \mid a \in A,\ b \in B\}$$

where ($\forall a, a' \in A,\ b, b' \in B$)

1. $(a + a') \otimes b = a \otimes b + a' \otimes b$,
2. $a \otimes (b + b') = a \otimes b + a \otimes b'$.

We notice that the function

$$A \otimes \mathbb{Z} \to A,\ a \otimes n \mapsto na$$

is a group isomorphism, that is to say, $A \otimes \mathbb{Z} \cong A$ (similarly, $\mathbb{Z} \otimes A \cong A$). The reader may easily prove that

$$(A \oplus B) \otimes C \cong (A \otimes C) \oplus (B \otimes C)$$

for any three Abelian groups $A$, $B$, and $C$. Finally, given two group homomorphisms $\phi \colon A \to A'$ and $\psi \colon B \to B'$, the function $\phi \otimes \psi \colon A \otimes B \to A' \otimes B'$ defined by $\phi \otimes \psi(a \otimes b) = \phi(a) \otimes \psi(b)$ is a homomorphism of Abelian groups.

In this way, by fixing an Abelian group $G$ we are able to construct a covariant functor

$$- \otimes G \colon \mathbf{Ab} \to \mathbf{Ab}$$

that transforms a group $A$ into $A \otimes G$ and a morphism $\phi \colon A \to B$ into the morphism $\phi \otimes 1_G$.

We extend this functor to chain complexes. We transform a given chain complex $(C, \partial) \in \mathfrak{C}$ in $(C \otimes G, \partial \otimes 1_G)$, by setting

$$(C \otimes G)_n := C_n \otimes G$$

for every $n \in \mathbb{Z}$, and by defining the homomorphisms

$$(\partial \otimes 1_G)_n := \partial_n \otimes 1_G \colon C_n \otimes G \to C_{n-1} \otimes G.$$

Since

$$(\partial \otimes 1_G)_{n-1}(\partial \otimes 1_G)_n = (\partial_{n-1} \otimes 1_G)(\partial_n \otimes 1_G) = \partial_{n-1}\partial_n \otimes 1_G = 0,$$

we conclude that $(C \otimes G, \partial \otimes 1_G)$ is a chain complex whose homology groups are the *homology groups of $(C, \partial)$ with coefficients in $G$*. The $n$th-homology group of $(C, \partial)$ with coefficients in $G$ is defined by the quotient group

$$H_n(C;G) = \ker(\partial_n \otimes 1_G)/\operatorname{im}(\partial_{n+1} \otimes 1_G);$$

the graded Abelian group $H_*(C;G)$ is the graded *homology* group of $C$ with coefficients in $G$. In particular, if $(C,\partial) = (C(K),\partial)$, the chain complex of the oriented complex $K$, then $H_*(C(K);G)$ – simply denoted by $H_*(K;G)$ – is the homology of $K$ with coefficients in $G$.

We recall that the chain complex $(C(K),\partial)$ is positive, free, and has an augmentation homomorphism $\varepsilon\colon C_0(K) \to \mathbb{Z}$. To continue with our work, we only need one of these properties, namely, that $(C,\partial)$ be *free*.

For every free chain complex $(C,\partial)$ and for each $n \in \mathbb{Z}$, we have a short exact sequence of free Abelian groups

$$Z_n(C) \rightarrowtail C_n \xrightarrow{\ \partial_n\ } B_{n-1}(C) \ .$$

The main point is that, by taking the tensor product of each component of this exact sequence with $G$, we obtain again a short exact sequence.

**(II.5.1) Lemma.** *If*

$$A \overset{f}{\rightarrowtail} B \overset{g}{\twoheadrightarrow} C$$

*is a short exact sequence of free Abelian groups and $G$ is an Abelian group, then also the sequence*

$$A \otimes G \overset{f \otimes 1_G}{\rightarrowtail} B \otimes G \overset{g \otimes 1_G}{\twoheadrightarrow} C \otimes G$$

*is exact.*

*Proof.* We begin by noting that the group $C$ is free; therefore, we may define a map $s\colon C \to B$ simply by choosing for each element of a basis of $C$ an element of its anti-image under $g$, and by extending this operation linearly; through this procedure, we obtain a homomorphism of Abelian groups

$$s\colon C \longrightarrow B$$

such that $gs = 1_C$, the identity homomorphism of $C$ onto itself. It follows that $g(1_B - sg) = g - (gs)g = 0$, in other words, the image of $1_B - sg$ is contained in the $\ker g = \operatorname{im} f$; for this reason, we may define the map $r := f^{-1}(1_B - sg)\colon B \to A$ that also satisfies $rf = f^{-1}(1_B - sg)f = 1_A$. We thus obtain the relations

$$rf = 1_A , \quad gs = 1_C , \quad \text{and} \quad fr + sg = 1_B.$$

We know that the tensor product by $G$ is a functor and that it transforms sums of homomorphisms into sums of transformed homomorphisms; consequently, tensorization gives us the relations

$$(r \otimes 1_G)(f \otimes 1_G) = 1_{A \otimes G} \quad (g \otimes 1_G)(s \otimes 1_G) = 1_{C \otimes G}$$

and

$$(f \otimes 1_G)(r \otimes 1_G) + (s \otimes 1_G)(g \otimes 1_G) = 1_{B \otimes G}.$$

The first of these relations tells us that $(f \otimes 1_G)$ is injective; the second, that $(g \otimes 1_G)$ is surjective, and the third, that $\mathrm{im}(f \otimes 1_G) = \ker(g \otimes 1_G)$ because, if we take $x \in B \otimes G$ such that $(g \otimes 1_G)(x) = 0$, then

$$\begin{aligned}
x &= (f \otimes 1_G)(r \otimes 1_G)(x) + (s \otimes 1_G)(g \otimes 1_G)(x) \\
&= (f \otimes 1_G)((r \otimes 1_G)(x)) \in \mathrm{im}(f \otimes 1_G).
\end{aligned}$$

∎

Returning to our free chain complex $(C, \partial)$, we notice that for each integer $n$, the sequence

$$Z_n(C) \otimes G \rightarrowtail C_n \otimes G \xrightarrow{\partial_n \otimes 1_G} B_{n-1}(C) \otimes G$$

is short exact. In addition, we observe that the graded Abelian groups $Z(C) = \{Z_n(C) | n \in \mathbb{Z}\}$ and $B(C) = \{B_n(C) | n \in \mathbb{Z}\}$ may be viewed as chain complexes with trivial boundary operator 0; we then construct the chain complexes

1. $(Z(C) \otimes G, 0 \otimes 1_G)$;
2. $(C \otimes G, \partial \otimes 1_G)$;
3. $(\widetilde{B(C)} \otimes G, 0 \otimes 1_G)$, where $\widetilde{B(C)}_n = B_{n-1}(C)$

and observe that in view of the preceding short exact sequence of Abelian groups, we have a short exact sequence of chain complexes

$$(Z(C) \otimes G, 0 \otimes 1_G) \rightarrowtail (C \otimes G, \partial \otimes 1_G) \twoheadrightarrow (\widetilde{B(C)} \otimes G, 0 \otimes 1_G).$$

By the Long Exact Sequence Theorem (II.3.1), we obtain the long exact sequence of homology groups

$$\cdots \longrightarrow H_n(Z(C) \otimes G) \longrightarrow H_n(C \otimes G) \longrightarrow$$

$$H_n(\widetilde{B(C)} \otimes G) \longrightarrow H_{n-1}(Z(C) \otimes G) \longrightarrow \cdots$$

in other words, by considering the format of the boundary operators, we have the following exact sequence of Abelian groups:

$$\cdots \longrightarrow B_n(C) \otimes G \xrightarrow{i_n \otimes 1_G} Z_n(C) \otimes G \xrightarrow{j_n}$$

$$H_n(C; G) \xrightarrow{h_n} B_{n-1}(C) \otimes G \xrightarrow{i_{n-1} \otimes 1_G} Z_{n-1}(C) \otimes G \longrightarrow \cdots$$

Note that $i_n$ is the inclusion of $B_n(C)$ in $Z_n(C)$ and $j_n$ is the induced homomorphism by the inclusion of $Z_n(C)$ in $C_n$; the reader is also asked to notice that the connecting homomorphism $\lambda_{n+1}$ in Theorem (II.3.1) coincides with $i_n \otimes 1_G$.

Since $\operatorname{im} j_n = \ker h_n$, we conclude that, for every $n \geq 0$, the sequence

$$\operatorname{im} j_n \rightarrowtail H_n(C;G) \xrightarrow{\ h_n\ } \operatorname{im} h_n$$

is short exact.

**(II.5.2) Lemma.** *If the group $G$ is free, the short exact sequence*

$$\operatorname{im} j_n \rightarrowtail H_n(C;G) \xrightarrow{\ h_n\ } \operatorname{im} h_n$$

*splits,*[4] *and so*

$$H_n(C;G) \cong \operatorname{im} j_n \oplus \operatorname{im} h_n$$

*(however, one should note that this isomorphism is not canonic).*

*Proof.* Let us take the homomorphism of Abelian groups

$$h_n \colon H_n(C;G) \to B_{n-1}(C) \otimes G.$$

As a subgroup of the free Abelian group $C_{n-1}$, $B_{n-1}(C)$ is free and by hypothesis $G$ is also free; then $B_{n-1}(C) \otimes G$ is free and it follows that $\operatorname{im} h_n$ is free. We now choose, for every generator $x \in \operatorname{im} h_n$, an element $y \in H_n(C;G)$ such that $h_n(y) = x$; by linearity, we obtain a homomorphism

$$s \colon \operatorname{im} h_n \longrightarrow H_n(C;G)$$

such that $h_n s = 1_{\operatorname{im} h_n}$. Exercise 1 in Sect. II.3 completes the proof.

The homomorphism $s$ depends on the choice of the elements $y$ for the generators $x$; therefore, $s$ is not canonically determined. ∎

We now give another interpretation of the groups $\operatorname{im} j_n$ and $\operatorname{im} h_n$. Note that

$$\operatorname{im} j_n \cong Z_n(C) \otimes G / \ker j_n = Z_n(C) \otimes G / \operatorname{im}(i_n \otimes 1_G);$$

the quotient group

$$Z_n(C) \otimes G / \operatorname{im}(i_n \otimes 1_G) := \operatorname{coker}(i_n \otimes 1_G)$$

is called *cokernel* of $i_n \otimes 1_G$. Since $\operatorname{im} h_n = \ker(i_{n-1} \otimes 1_G)$, the exact sequence

$$\operatorname{im} j_n \rightarrowtail H_n(C;G) \xrightarrow{\ h_n\ } \operatorname{im} h_n$$

is written as

$$\operatorname{coker}(i_n \otimes 1_G) \rightarrowtail H_n(C;G) \twoheadrightarrow \ker(i_{n-1} \otimes 1_G)$$

---

[4] The definition of splitting short exact sequence can be found in Exercise 1, Sect. II.3.

where the first and the third terms may be viewed in another way; we begin with $\text{coker}(i_n \otimes 1_G)$.

**(II.5.3) Lemma.** $\text{coker}(i_n \otimes 1_G) \cong H_n(C) \otimes G$.

*Proof.* Since by definition $\text{coker}(i_n \otimes 1_G) = Z_n(C) \otimes G / \text{im}(i_n \otimes 1_G)$, there exists a homomorphism

$$\phi: \text{coker}(i_n \otimes 1_G) \to H_n(C) \otimes G$$

defined on the generators by $\phi[z \otimes g] := p(z) \otimes g$ (where $p: Z_n(K) \to H_n(C)$ is the natural projection). On the other hand for each $y \in H_n(C)$, we may choose an $x \in p^{-1}(y) \subseteq Z_n(C)$ and define

$$\psi: H_n(C) \otimes G \to \text{coker}(i_n \otimes 1_G)$$

on the generators, with $\psi(y \otimes g) := [x \otimes g]$. The homomorphism $\psi$ is well defined because, in view of the exactness of the long exact sequence, we have for each $x'$ such that $p(x') = y$

$$x \otimes g - x' \otimes g = (x - x') \otimes g \in \ker j_n \cong \text{im}(i_n \otimes 1_G),$$

that is to say, $[x \otimes g] = [x' \otimes g]$. The homomorphisms $\phi$ and $\psi$ are clearly the inverse of each other and so $\text{coker}(i_n \otimes 1_G) \cong H_n(C) \otimes G$. ∎

The preceding lemma shows that $\text{coker } i_n \otimes 1_G$ depends neither on $B_n(C) \otimes G$ nor on $Z_n(C) \otimes G$, but only on $H_n(C)$ (the cokernel of the monomorphism $i_n: B_n(C) \to Z_n(C)$) and on $G$. This fact suggests that the same may be true for $\ker(i_n \otimes 1_G)$ and indeed it is so.

**(II.5.4) Theorem.** *Let $H$ be the cokernel of the the monomorphism $i: B \to Z$ between free Abelian groups and let $G$ be any fixed Abelian group. Then, both the kernel and the cokernel of the homomorphism $i \otimes 1_G$ depend entirely on $H$ and $G$.*

*Moreover, $\text{coker}(i \otimes 1_G) \cong H \otimes G$, while $\ker(i \otimes 1_G)$ gives rise to a new covariant functor*

$$\text{Tor}(-, G): \mathbf{Ab} \longrightarrow \mathbf{Ab}$$

*called torsion product.*

*Proof.* Due to the fact that $H$ is the cokernel of the monomorphism $i: B \to Z$, the bases of $Z$ and $B$ represent $H$ with generators and relations; we then have a *free presentation* of $H$

$$B \overset{i}{\rightarrowtail} Z \overset{q}{\twoheadrightarrow} H.$$

Suppose that we had another free presentation of $H$

$$R \overset{j}{\rightarrowtail} F \overset{q'}{\twoheadrightarrow} H$$

and consider the following free chain complexes with augmentation to the Abelian group $H$ (viewed as a $\mathbb{Z}$-module):[5]

1. $(C, \partial)$, with $C_1 = B$, $C_0 = Z$, $\partial_1 = i$, $\varepsilon = q$, $C_i = 0$ for all $i \neq 0, 1$ and $\partial_i = 0$ for all $i \geq 2$;
2. $(C', \partial')$, with $C'_1 = R$, $C'_0 = F$, $\partial'_1 = j$, $\varepsilon' = q'$, $C'_i = 0$ for all $i \neq 0, 1$ and $\partial'_i = 0$ for all $i \geq 2$.

These chain complexes are free and acyclic; by Theorem (II.3.6), we obtain chain morphisms $f\colon C \to C'$ and $g\colon C' \to C$ whose composites $fg$ and $gf$ are chain homotopic to the respective identities. The tensor product with $G$ is a functor that preserves compositions of morphisms; therefore, their tensor products by $G$ produce the chain morphisms

$$f \otimes 1_G\colon C \otimes G \to C' \otimes G$$
$$g \otimes 1_G\colon C' \otimes G \to C \otimes G$$

and besides,

$$(f \otimes 1_G)(g \otimes 1_G) \text{ and } (g \otimes 1_G)(f \otimes 1_G)$$

are still chain homotopic to their respective identities. This means that the induced morphisms in homology

$$
\begin{array}{ccc}
 & H_1(f \otimes 1_G) & \\
\ker(i \otimes 1_G) & \rightleftarrows & \ker(j \otimes 1_G) \\
 & H_1(g \otimes 1_G) &
\end{array}
$$

are the inverse of each other and likewise for

$$
\begin{array}{ccc}
 & H_0(f \otimes 1_G) & \\
\operatorname{coker}(i \otimes 1_G) & \rightleftarrows & \operatorname{coker}(j \otimes 1_G) \\
 & H_0(g \otimes 1_G) &
\end{array}
$$

This implies that neither $\ker(i \otimes 1_G)$ nor $\operatorname{coker}(i \otimes 1_G)$ depends on the chosen presentation of $H$.

Hence, by following the argument in Lemma (II.5.3),

$$\operatorname{coker}(i \otimes 1_G) \cong H \otimes G$$

regardless of which free presentation of $H$ we take.

---

[5] Chain complexes can be constructed over $\Lambda$-modules, with $\Lambda$ a commutative ring with unit element.

We now focus our attention on functor $\text{Tor}(-, G)$. For any $H \in \mathbf{Ab}$, we define the group $\text{Tor}(H, G)$ as follows. Let $F(H)$ be the free group generated by all the elements of $H$; the function

$$q \colon F(H) \to H \,, \ h \mapsto h$$

is an epimorphism of $F(H)$ onto $H$. Let $i \colon \ker q \to F(H)$ be the inclusion homomorphism; we then have a representation of $H$ by free groups

$$\ker q \xrightarrow{\ \ i\ \ } F(H) \xrightarrow{\ \ q\ \ } H.$$

We define

$$\text{Tor}(H, G) := \text{coker}(i \otimes 1_G).$$

By the first part of the theorem, $\text{Tor}(H, G)$ does not depend on the presentation of $H$. As for the morphisms, for any $\bar{f} \in \mathbf{Ab}(H, H')$, we choose the presentations $R \rightarrowtail F \twoheadrightarrow H$ and $R' \rightarrowtail F' \twoheadrightarrow H'$; by Theorem (II.3.6), we obtain a chain morphism $f$ between the complexes $C$ and $C'$ (determined by $R \rightarrowtail F$ and $R' \rightarrowtail F'$, respectively) that extends $\bar{f}$ and is unique up to chain homotopy. By taking their tensor product by $G$ and computing the homology groups, we obtain

$$H_1(f \otimes 1_G) \colon \ \text{Tor}(H, G) \to \text{Tor}(H', G)$$

which is, by definition, the result of applying the torsion product on $\bar{f}$. ∎

When $(C, \partial)$ is the chain complex $(C(K), \partial)$ of an oriented simplicial complex $K$, the previous results prove the **Universal Coefficients Theorem in Homology**:

**(II.5.5) Theorem.** *The homology of a simplicial complex $K$ with coefficients in an Abelian group $G$ is determined by the following short exact sequences:*

$$H_n(K; \mathbb{Z}) \otimes G \rightarrowtail H_n(K; G) \twoheadrightarrow \text{Tor}(H_{n-1}(K; \mathbb{Z}), G).$$

*What is more, if $G$ is free,*

$$H_n(K; G) \cong H_n(K; \mathbb{Z}) \otimes G \oplus \text{Tor}(H_{n-1}(K; \mathbb{Z}), G).$$

Let us now see what happens when $G = \mathbb{Q}$, the additive group of rational numbers. This group is not free, but it is *locally free*: we say that an Abelian group $G$ is *locally free* if every finitely generated subgroup of $G$ is free; in particular, due to the Finitely Generated Abelian Groups Decomposition Theorem (see p. 75), a finitely generated Abelian group is locally free if and only if it is torsion free. We now state the following

**(II.5.6) Lemma.** *If $i \colon A \to A'$ is a monomorphism and $G$ is locally free, then*

$$i \otimes 1_G \colon A \otimes G \longrightarrow A' \otimes G$$

*is a monomorphism.*

In particular, the monomorphism

$$i_{n-1}\colon B_{n-1}(K) \longrightarrow Z_{n-1}(K)$$

determines the monomorphism

$$i_{n-1} \otimes 1_{\mathbb{Q}}\colon B_{n-1}(K) \otimes \mathbb{Q} \longrightarrow Z_{n-1}(K) \otimes \mathbb{Q}$$

and so

$$\mathrm{Tor}(H_{n-1}(K;\mathbb{Z}),\mathbb{Q}) = \ker(i_{n-1} \otimes 1_{\mathbb{Q}}) = 0 \,.$$

Theorem (II.5.5) allows us to afirm that

$$H_n(K;\mathbb{Q}) \cong H_n(K;\mathbb{Z}) \otimes \mathbb{Q}$$

and so, helped once more by the Finitely Generated Abelian Groups Decomposition Theorem, we say that $H_n(K;\mathbb{Q})$ is a rational vector space of dimension equal to the rank of $H_n(K;\mathbb{Z})$ (the $n$th- Betti number of $K$).

## Exercises

**1.** Prove that, if $A$ and $B$ are free Abelian groups, then, $A \otimes B$ is a free Abelian group.

**2.** Let $K$ be any simplicial complex. Prove that for every prime number $p$ the short exact sequence

$$0 \longrightarrow \mathbb{Z} \xrightarrow{\;p\cdot -\;} \mathbb{Z} \xrightarrow{\;\mathrm{mod}\,p\;} \mathbb{Z}_p \longrightarrow 0$$

creates an exact sequence of homology groups

$$\cdots \longrightarrow H_n(K;\mathbb{Z}) \xrightarrow{\;p\cdot -\;} H_n(K;\mathbb{Z}) \xrightarrow{\;\mathrm{mod}\,p\;}$$

$$\xrightarrow{\;\mathrm{mod}\,p\;} H_n(K;\mathbb{Z}_p) \xrightarrow{\;\beta_p\;} H_{n-1}(K;\mathbb{Z}) \longrightarrow \cdots$$

called *Bockstein long exact sequence*. The homomorphism of Abelian groups

$$H_n(K;\mathbb{Z}_p) \xrightarrow{\;\beta_p\;} H_{n-1}(K;\mathbb{Z})$$

is called *Bockstein operator*.

**3.** (*Snake Lemma*) Consider the following commutative diagram, whose rows are exact sequences of Abelian groups:

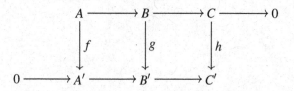

Prove that there exists a homomorphism $d$ that turns the sequence

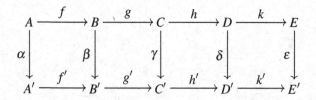

into an exact sequence; also, by using this Lemma, give an alternative proof to Theorem (II.3.1).

**4.** (*General form of the Five Lemma*) Let

$$
\begin{array}{ccccccccc}
A & \xrightarrow{f} & B & \xrightarrow{g} & C & \xrightarrow{h} & D & \xrightarrow{k} & E \\
\Big\downarrow{\alpha} & & \Big\downarrow{\beta} & & \Big\downarrow{\gamma} & & \Big\downarrow{\delta} & & \Big\downarrow{\varepsilon} \\
A' & \xrightarrow{f'} & B' & \xrightarrow{g'} & C' & \xrightarrow{h'} & D' & \xrightarrow{k'} & E'
\end{array}
$$

be a commutative diagram of Abelian groups with exact rows. Prove that:

- If $\alpha$ is surjective and $\beta$, $\delta$ are injective, then $\gamma$ is injective;
- If $\varepsilon$ is injective and $\beta$, $\delta$ are surjective, then $\gamma$ is surjective.

It follows directly from these results that, if $\alpha$, $\beta$, $\delta$, and $\varepsilon$ are isomorphisms, then also $\gamma$ is an isomorphism.

# Chapter III
# Homology of Polyhedra

## III.1 The Category of Polyhedra

In Sect. II.2, we have defined polyhedra as geometric realizations of (finite) simplicial complexes. We have also seen that polyhedra are compact topological spaces. We shall indicate with **P** the category of polyhedra and continuous functions between them; clearly, **P** is a subcategory of **Top**.

We start by proving that the category **P** is *closed under finite products*, that is to say, the Cartesian product of a finite number of polyhedra is (homeomorphic to) a polyhedron; it is enough to prove this result for two polyhedra.

**(III.1.1) Theorem.** *Let two simplicial complexes*

$$K = (X, \Phi) \ and \ L = (Y, \Psi),$$

*be given. There exists a simplicial complex $K \times L$ such that*

$$|K \times L| \cong |K| \times |L|.$$

*Proof.* The simplicial complex $K \times L$ is defined as follows:

1. The vertices of $K \times L$ are the elements of $X \times Y$.
2. To define the simplexes of $K \times L$, we first order the sets $X$ and $Y$; we then give $X \times Y$ the lexicographical order.
3. A $n$-simplex of $K \times L$ is given by a set $\{(x_{i_0}, y_{j_0}), \ldots, (x_{i_n}, y_{j_n})\}$ of elements of $X \times Y$ satisfying the following conditions:

$$(a) \ (x_{i_0}, y_{j_0}) < \ldots < (x_{i_n}, y_{j_n})$$

$$(b) \ \{x_{i_0}, \ldots, x_{i_n}\} \in \Phi$$

$$(c) \ \{y_{j_0}, \ldots, y_{j_n}\} \in \Psi$$

D.L. Ferrario and R.A. Piccinini, *Simplicial Structures in Topology*,
CMS Books in Mathematics, DOI 10.1007/978-1-4419-7236-1_III,
© Springer Science+Business Media, LLC 2011

(the vertices $x_{i_k}$ and $y_{j_k}$ are not necessarily different). As a consequence, the projections

$$pr_1 \colon X \times Y \to X , \; pr_2 \colon X \times Y \to Y$$

are morphisms between simplicial complexes determining therefore a map

$$\phi = |pr_1| \times |pr_2| \colon |K \times L| \to |K| \times |L| .$$

We wish to prove that $\phi$ is a homeomorphism.

Let $p \in |K|$ and $q \in |L|$ be given by

$$p = \sum_{i=0}^{m} \alpha_i x_i , \; \alpha_i > 0 , \; \sum_{i=0}^{m} \alpha_i = 1 , \; x_0 < \cdots < x_m$$

and

$$q = \sum_{j=0}^{n} \beta_j y_j , \; \beta_j > 0 , \; \sum_{j=0}^{n} \beta_j = 1 , \; y_0 < \cdots < y_n .$$

We define

$$a_s = \sum_{i=0}^{s} \alpha_i , \; s = 0, 1, \ldots, m$$

and

$$b_t = \sum_{j=0}^{t} \beta_j , \; t = 0, 1, \ldots, n ;$$

we then take the set $\{0, a_0, \ldots, a_{m-1}, b_0, \ldots, b_n = 1\}$, rename and reorder its $m + n + 2$ elements so to obtain the ordered set $\{c_{-1}, c_0, \ldots, c_{m+n}\}$ such that

$$0 = c_{-1} < c_0 < c_1 < \ldots < c_{m+n}, c_{m+n} = 1 .$$

For every $r = 0, 1, \ldots, m + n$, take $z_r = (x_i, y_j)$ where the indices $i$ (respectively, $j$) equal the number of real numbers $a_s$ (respectively, $b_t$) in the set $\{c_0, c_1, \ldots, c_{r-1}\}$. We note that

$$z_0 = (x_0, y_0) , \; z_{m+n} = (x_m, y_n) .$$

We also note that if $z_r = (x_i, y_j)$, the element $z_{r+1}$ is equal either to $(x_{i+1}, y_j)$ or to $(x_i, y_{j+1})$; in either case $z_r < z_{r+1}$ and so

$$z_0 = (x_0, y_0) < z_1 < \ldots < z_{m+n} = (x_m, y_n).$$

Since

$$\sum_{i=0}^{m+n} (c_i - c_{i-1}) = c_{m+n} - c_{-1} = 1 ,$$

we may conclude that

$$w = \sum_{i=0}^{m+n} (c_i - c_{i-1}) z_i \in |K \times L| .$$

This line of reasoning allows us to define a function

$$\psi: |K| \times |L| \longrightarrow |K \times L| , \ \psi(p,q) = \sum_{i=0}^{m+n} (c_i - c_{i-1})z_i$$

which is the inverse of $\phi$. In fact, if we take $p \in |K|$ and $q \in |L|$ as before,

$$|pr_1|\psi(p,q) = \sum_{i=0}^{m} \gamma_i x_i .$$

Let $z_r < z_{r+1} < \ldots < z_{r+t}$ be the elements $z_i$ of $\psi(p,q)$ with $x_s$ for first coordinate, as described by the following table:

| vertices | coefficients in $\psi(p,q)$ |
|---|---|
| $z_{r-1} = (x_{s-1}, y_\alpha)$ | $c_{r-1} - c_{r-2}$ |
| $z_r = (x_s, y_\alpha)$ | $c_r - c_{r-1}$ |
| $\ldots$ | $\ldots$ |
| $\ldots$ | $\ldots$ |
| $\ldots$ | $\ldots$ |
| $z_{r+t} = (x_s, y_{\alpha+t})$ | $c_{r+t} - c_{r+t-1}$ |
| $z_{r+t+1} = (x_{s+1}, y_{\alpha+t})$ | $c_{r+t+1} - c_{r+t} .$ |

We note that the coefficient $\gamma_s$ in $|pr_1|\psi(p,q)$ is equal to $c_{r+t} - c_{r-1}$; moreover, by the definitions of $z_{r-1}$ and $z_r$, we may conclude that $c_{r-1} = a_{s-1}$ (similarly, $c_{r+t} = a_s$). It follows that

$$\gamma_s = c_{r+t} - c_{r-1} = a_s - a_{s-1} = \alpha_s$$

and so, $|pr_1|\psi(p,q) = p$. The proof of

$$|pr_2|\psi(p,q) = q$$

is completely similar; these two results show that $\phi\psi = 1_{|K| \times |L|}$.

Let us now take an element

$$u = \sum_{r=0}^{s} \zeta_r z_r \in |K \times L|$$

where $\sum_{r=0}^{s} \zeta_r = 1$ and $z_0 < z_1 < \ldots < z_s$. Let $x_0, \ldots, x_m$ (respectively, $y_0, \ldots, y_n$) be the distinct vertices that appear as first (respectively, second) coordinates of $z_0, \ldots, z_s$; then

$$\{x_0, \ldots, x_m\} \in \Phi , \ \{y_0, \ldots, y_n\} \in \Psi$$

and besides, $z_0 = (x_0, y_0)$ and $z_s = (x_m, y_n)$. By the definitions given here, we are able to write the equality

$$|pr_1|(u) = \sum_{i=0}^{m} \alpha_i x_i ,$$

where $\alpha_i$ is the sum of the coefficients $\zeta_r$ of the elements $z_r$ whose first coordinate equals $x_i$; now, we can easily see that $\sum_{i=0}^m \alpha_i = 1$. After making similar remarks for $|pr_2|(u)$, we conclude that

$$\psi\phi = 1_{|K\times L|} .$$

Since $|K \times L|$ is compact, $\phi$ is a homeomorphism.                                   ∎

**(III.1.2) Example.** The diagram in Fig. III.1 illustrates the case in which $K$ is

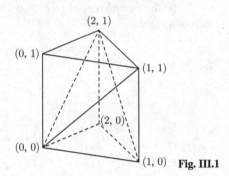

**Fig. III.1**

a 2-simplex with vertices $\{0,1,2\}$ and $L$ is the 1-simplex with vertices $\{0,1\}$. The resulting prism consists of the three 3-simplexes $\{(0,0),(1,0),(2,0),(2,1)\}$, $\{(0,0),(1,0),(1,1),(2,1)\}$ and $\{(0,0),(0,1),(1,1),(2,1)\}$, respectively.

Let $K = (X,\Phi)$ be a simplicial complex; we construct an infinite sequence of simplicial complexes $\{K^{(r)} | r \geq 0\}$ as follows:

1. $K^{(0)} = K$
2. $K^{(1)} = (X^{(1)}, \Phi^{(1)})$ is the simplicial complex defined by:
     (i) $X^{(1)} = \Phi$
     (ii) $\Phi^{(1)}$ is the set of all nonempty subsets of $\Phi$ such that

$$\{\sigma_{i_0},\ldots,\sigma_{i_n}\} \in \Phi^{(1)} \Leftrightarrow \sigma_{i_0} \subset \ldots \subset \sigma_{i_n};$$

3. $K^{(r)}$ is defined iteratively:

$$K^{(r)} = (K^{(r-1)})^{(1)} .$$

The simplicial complex $K^{(r)}$ is the *rth- barycentric subdivision* of $K$.

**(III.1.3) Remark.** The definition we gave of the first barycentric subdivision $K^{(1)}$ of a simplicial complex is, perhaps, a little convoluted, and so we give here another, which is equivalent but refers to the geometric realization $|K|$. For every $n$-simplex $\sigma = \{x_0,\ldots,x_n\}$ of $K$, we define the *barycenter* of $\sigma$ to be the point

$$b(\sigma) = \sum_{i=0}^n \frac{1}{n+1} x_i ;$$

then, the vertices of $K^{(1)}$ are represented abstractly by the elements of the set

$$X^{(1)} = \{b(\sigma) | \sigma \in \Phi\}$$

(in other words, $X^{(1)}$ contains all vertices of $K$ together with those vertices represented by barycenters of all simplexes of $K$ that are not vertices). From this point of view, a set $\{b(\sigma_{i_0}), \ldots, b(\sigma_{i_n})\}$ of $n+1$ vertices of $K^{(1)}$ is an $n$-simplex if and only if $\sigma_{i_0} \subset \sigma_{i_1} \subset \ldots \subset \sigma_{i_n}$.

From the simplicial complexes point of view, the elements of the sequence $\{K^{(r)} | r \geq 0\}$ are all different; but, from the geometric point of view, they are not distinct; actually, we have the following result:

**(III.1.4) Theorem.** *Let $K = (X, \Phi)$ be a simplicial complex and $r$ any positive integer. Then the polyhedra $|K|$ and $|K^{(r)}|$ are homeomorphic.*

*Proof.* It is sufficient to prove the result for $r = 1$. For instance, Fig. III.2 shows

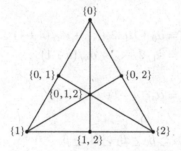

**Fig. III.2**

the barycentric subdivision of a 2-simplex. The function that associates with each $\sigma \in K^{(1)}$ the barycentre $b(\sigma)$ may be extended linearly to a continuous function

$$F: |K^{(1)}| \longrightarrow |K| , \ F\left(\sum_{i=0}^{n} \alpha_i \sigma^i\right) = \sum_{i=0}^{n} \alpha_i b(\sigma^i)$$

(all linear functions are continuous. cf. Theorem (II.2.8)). We wish to prove that the function $F$ is a bijection. Let $p = \sum_{i=0}^{n} \alpha_i \sigma^i$ be an arbitrary point of $|K^{(1)}|$; note that $\sigma^0, \ldots \sigma^n$ are simplexes of $K$ such that

$$\sigma^0 \subset \sigma^1 \subset \ldots \subset \sigma^n .$$

Let us suppose that dim $\sigma^0 = r$; we may assume that dim $\sigma^1 = r+1$ (otherwise, we could take intermediate simplexes

$$\sigma^0 = \tau^0 \subset \tau^1 \subset \ldots \subset \tau^k = \sigma^1$$

such that dim $\tau^{j+1} = $ dim $\tau^j + 1$ and to which we could assign coefficients equal to zero in the sum representing $p$). Consequently, we may assume that

$$\sigma^0 = \{x_0\},$$
$$\sigma^1 = \{x_0, x_1\},$$
$$\ldots$$
$$\sigma^n = \{x_0, x_1, \ldots, x_n\}$$

and so

$$F(p) = \alpha_0 x_0 + \alpha_1 \left( \frac{x_0 + x_1}{2} \right) + \ldots + \alpha_n \left( \frac{x_0 + x_1 + \ldots + x_n}{n+1} \right).$$

Therefore, if

$$q = \sum_{i=0}^{n} \beta_i x_i \in |K|$$

and $F(p)$ coincide, we have

$$\beta_0 = \alpha_0 + \alpha_1/2 + \ldots + \alpha_n/(n+1)$$
$$\beta_1 = \alpha_1/2 + \ldots + \alpha_n/(n+1)$$
$$\ldots$$
$$\beta_n = \alpha_n/(n+1)$$

and

$$1 \geq \beta_0 \geq \beta_1 \geq \ldots \geq \beta_n \geq 0.$$

On the other hand, given a sequence of real numbers

$$1 \geq \beta_0 \geq \beta_1 \geq \ldots \geq \beta_n \geq \beta_{n+1} = 0$$

the real numbers

$$\alpha_i = (i+1)(\beta_i - \beta_{i+1}), \ i = 0, 1, \ldots, n$$

satisfy the preceding equalities and so we see that the coefficients $\alpha_i$ and $\beta_i$ are determined by each other; this means that $F$ is a bijection. The proof is completed by recalling that $F$ is a continuous bijection from a compact space to a Hausdorff space. ∎

Although the spaces $|K^{(r)}|$ are homeomorphic, they may differ by the length of the geometric realizations of their 1-simplexes. Indeed, let us consider $|K^{(r)}|$ as a subspace of $|K|$ (the topology on $|K|$ being defined by the metric $d$) and define the *diameter* of $|K^{(r)}|$ to be the maximum length of the geometric realizations of all 1-simplexes of $K^{(r)}$; the notation diam $|K^{(r)}|$ indicates the diameter of $|K^{(r)}|$. We note that diam $|K| = \sqrt{2}$. The next result shows that the diameter decreases with the successive barycentric subdivisions.

**(III.1.5) Theorem.** *For every real number $\varepsilon > 0$, there is a positive integer $r$ such that* diam $|K^{(r)}| < \varepsilon$.

*Proof.* Suppose that $\dim K = n$ and take a 1-simplex $\{\sigma_0, \sigma_1\}$ of $K^{(1)}$ arbitrarily; we may assume that

$$\sigma_0 = \{x_{i_0}, \ldots, x_{i_q}\}$$

and

$$\sigma_1 = \{x_{i_0}, \ldots, x_{i_q}, x_{j_0}, \ldots, x_{j_p}\}$$

where $p + q + 1 \leq n$. Since $|K| \cong |K^{(1)}|$ (actually, it is useful to make the identification $|K| \equiv |K^{(1)}|$), the length of the 1-simplex of $|K^{(1)}|$ represented abstractly by $\{\sigma_0, \sigma_1\}$ is computed through the formula

$$d\left( \sum_{k=0}^{q} \frac{1}{q+1} x_{i_k}, \sum_{k=0}^{q} \frac{1}{p+q+2} x_{i_k} + \sum_{\ell=0}^{p} \frac{1}{p+q+2} x_{j_\ell} \right) =$$

$$= \left\{ \left( \frac{1}{q+1} - \frac{1}{p+q+2} \right)^2 (q+1) + \left( \frac{1}{p+q+2} \right)^2 (p+1) \right\}^{1/2} =$$

$$= \frac{1}{p+q+2} \sqrt{\frac{(p+1)(p+q+2)}{q+1}} =$$

$$= \frac{p+q+1}{p+q+2} \sqrt{\frac{(p+1)(p+q+2)}{(p+q+1)^2(q+1)}} < \frac{n}{n+1} \sqrt{2}$$

because

$$\frac{p+q+1}{p+q+2} < \frac{n}{n+1}, \quad \frac{(p+1)(p+q+2)}{(p+q+1)^2(q+1)} < 2.$$

Therefore,

$$\text{diam } |K^{(1)}| < \frac{n}{n+1} \sqrt{2}. \qquad \blacksquare$$

Note that the dimension of a simplicial complex is not affected by successive barycentric subdivisions. Besides, by definition, the *dimension* of a polyhedron $|K|$ is the dimension of the simplicial complex $K$, that is to say, $\dim |K| = \dim K$.

Let $L$ be a simplicial subcomplex of $K$; the pair of polyhedra $(|K|, |L|)$ has the **Homotopy Extension Property**: for every topological space $Z$ and every pair of maps

$$f\colon |K| \times \{0\} \longrightarrow Z$$
$$G\colon |L| \times I \longrightarrow Z$$

with the same restriction to $|L| \times \{0\}$, there is a map (not necessarily unique) $F\colon |K| \times I \to Z$ whose restrictions to $|K| \times \{0\}$ and $|L| \times I$ coincide with $f$ and $G$ (see the following commutative diagram).

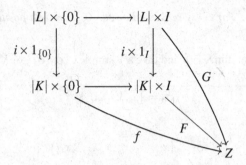

Before proving this property of pairs of polyhedra, we prove a lemma that characterizes the homotopy extension property.

**(III.1.6) Lemma.** *Let A be a closed subspace of X. Then, $(X,A)$ has the homotopy
extension property if and only if there exists a retraction*

$$r\colon\ X \times I \longrightarrow \widehat{X} = X \times \{0\} \cup A \times I$$

*(that is to say, a map r whose restriction to $\widehat{X}$ is the identity).*

*Proof.* We note that, since $A$ is closed in $X$, the following diagram is a pushout:

$$
\begin{array}{ccc}
A \times \{0\} & \xrightarrow{\;1_A \times i_0\;} & A \times I \\[2pt]
{\scriptstyle i \times 1_{\{0\}}}\big\downarrow & & \big\downarrow{\scriptstyle i \times 1_{\{0\}}} \\[2pt]
X \times \{0\} & \xrightarrow[\;1_A \times i_0\;]{} & \widehat{X}
\end{array}
$$

In the diagram, $i_0$ is the inclusion of $\{0\}$ in $I$ and $i$ is the inclusion of $A$ in $X$.

Let us suppose that the pair $(X,A)$ has the homotopy extension property. Then,
there is a map

$$r\colon\ X \times I \longrightarrow \widehat{X}$$

such that

$$r(i \times 1_I) = \overline{i \times 1_0} \text{ and } r(1_X \times i_0) = \overline{1_A \times i_0}\,.$$

Let $\iota$ be the inclusion of $\widehat{X}$ in $X \times I$. By the universal property of pushouts, $r\iota = 1_{\widehat{X}}$
and so $r$ is a retraction.

Conversely, let us suppose that $r\colon X \times I \to \widehat{X}$ is a retraction and let $f\colon X \times \{0\} \to$
$Z$ and $G\colon A \times I \to Z$ be maps such that $f(i \times i_0) = G(1_A \times i_0)$; by the definition
of pushout, there is a map $H\colon \widehat{X} \to Z$ that completes the following commutative
diagram:

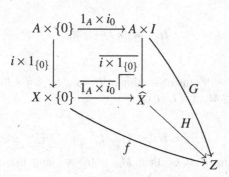

The function $F = Hr\colon X \times I \to Z$ is the extension of the homotopy that we were seeking.  ∎

**(III.1.7) Theorem.** *Let $L$ be a subcomplex of the simplicial complex $K$. Then, the pair of polyhedra $(|K|, |L|)$ has the homotopy extension property.*

*Proof.* We start the proof by assuming that $|K|$ and $|L|$ are the geometric realizations of the simplicial complexes $K = (X, \Phi)$ and $L = (Y, \Psi)$, respectively. By Lemma (III.1.6), we only need to prove that there exists a retraction of $|K| \times I$ onto $\widehat{|K|}$. Let $\sigma = \{x_0, \ldots, x_n\}$ be an $n$-simplex of $K$ and let $|\sigma|$ be the geometric realization of $(\sigma, \wp(\sigma) \smallsetminus \emptyset)$. Let $b(\sigma)$ be the barycenter of $|\sigma|$. For every $p \in |\sigma|$ and for every real number $t \in I$, let the point $tb(\sigma) + (1-t)p \in |\sigma|$ be denoted by $[p, t]$ (we recall that $|\sigma|$ is a convex space – see Theorem (II.2.9)). We now consider the function

$$H_\sigma \colon |\sigma| \times I \times I \longrightarrow |\sigma| \times I$$

defined by the equations:

$$H_\sigma([p,t], s, u) = \begin{cases} ([p, (1-u)t + \frac{u(2t-s)}{2-s}], (1-u)s), & s \leq 2t \\ ([p, (1-u)t], (1-u)s + \frac{u(s-2t)}{1-t}), & 2t \leq s. \end{cases}$$

To understand geometrically how we have come to this function, suppose that $|\sigma|$ is a 2-simplex that we have placed in the plane $(x, y)$ of $\mathbb{R}^3$ with its barycenter $b(\sigma)$ at the origin $(0, 0, 0)$; we then project $|\sigma| \times I$ on $|\sigma| \times \{0\} \cup |\dot\sigma| \times I$ from the point $(0, 0, 2)$ (here $|\dot\sigma|$ is the boundary of $|\sigma|$). Note that, in this way, we obtain a retraction of $|\sigma| \times I$ onto $|\sigma| \times \{0\} \cup |\dot\sigma| \times I$. In the general case, when $\sigma$ is an $n$-simplex, the function $H_\sigma(-, 1)$ is a retraction.[1]

For each integer $n \geq -1$, we define

$$M_n = |K| \times \{0\} \cup |K^n \cup L| \times I$$

where $K^n$ is the simplicial subcomplex determined by all simplexes of $K$ with dimension $\leq n$, and such that $K^{-1} = \emptyset$; then,

$$M_{-1} = \widehat{|K|} = |K| \times \{0\} \cup |L| \times I.$$

---

[1] $|\dot\sigma|$ is the geometric realization of the simplicial complex $\dot\sigma = \{\sigma' \subset \sigma \mid \dim \sigma' \leq n-1\}$.

We now define
$$H_n\colon M_n \times I \longrightarrow M_n$$

as follows:

1. For every $\sigma \in \Phi \smallsetminus \Psi$, the restriction of $H_n$ onto $|\sigma| \times I \times I$ coincides with $H_\sigma$
2. For every $(x,t) \in M_{n-1} \times I$, $H_n(x,t) = x$

The function
$$r_n = H_n(-,1)\colon M_n \longrightarrow M_{n-1}$$

is a retraction; if $\dim K = m$, then $M_m = |K| \times I$ and the composite function $r = r_0 \cdots r_m$ is a retraction of $|K| \times I$ onto $\widehat{|K|}$. ∎

## III.2  Homology of Polyhedra

We begin this section by recalling that the geometric realizations of a polyhedron and any one of its barycentric subdivisions are homeomorphic; one of our objectives is to prove that, given two simplicial complexes $K$ and $L$, and a map $f\colon |K| \to |L|$, there exists a simplicial function from a barycentric subdivision $K^{(r)}$ to $L$ whose geometric realization is homotopic to the composite of $f$ and the homeomorphism $F\colon |K^{(r)}| \to |K|$. Once this "simplicial approximation" of $f$ has been obtained, we may define a functor
$$H_*(-,\mathbb{Z})\colon \mathbf{P} \to \mathbf{Ab}^{\mathbb{Z}}$$

from the category of polyhedra and continuous functions to that of graded Abelian groups. Here is an overview of how this is done. We associate the graded Abelian group $H_*(K,\mathbb{Z})$ with a polyhedron $|K|$; the map $f$ induces a homomorphism, among the corresponding graded groups, which derives from the "approximation" of $f$.

With all this, it is not surprising that we may also give the concept of homotopy among chain morphisms, in parallel to the homotopy of maps among spaces. We shall see later some consequences of this important result.

Once again we remind the reader that an oriented simplicial complex $K$ defines a positive free augmented chain complex $(C(K), \partial^K)$ with augmentation homomorphism $\varepsilon\colon C_0(K) \to \mathbb{Z}$. In particular, if the simplicial complex $K$ is acyclic (for instance, $K$ is the simplicial complex $\overline{\sigma}$ generated by a simplex $\sigma$, a simplicial $n$-cone, or an abstract cone), then the chain complex $(C(K), \partial^K)$ is acyclic.

We define the functor
$$H_n(-,\mathbb{Z})\colon \mathbf{P} \to \mathbf{Ab}^{\mathbb{Z}}$$

on the objects $|K| \in \mathbf{P}$ simply as
$$H_n(|K|,\mathbb{Z}) = H_n(K,\mathbb{Z})\,.$$

Now the problem is to define $H_n(-,\mathbb{Z})$ on the morphisms of $\mathbf{P}$, in other words, on continuous functions between polyhedra. To solve this problem, we make here some initial remarks. Two simplicial functions $f,g\colon K=(X,\Phi)\to L=(Y,\Psi)$ are *contiguous* if for every $\sigma\in\Phi$ there exists $\tau\in\Psi$ such that $f(\sigma)\subset\tau$ and $g(\sigma)\subset\tau$; in symbols,

$$(\forall\sigma\in\Phi)(\exists\tau\in\Psi)\, f(\sigma)\subset\tau\, ,\, g(\sigma)\subset\tau\, .$$

Suppose that $f$ and $g$ are contiguous; for each $n$-simplex $\sigma\in\Phi$ (a generator of $C_n(K)$), let $\tau$ be the smallest simplex of $L$ which contains both simplexes $f(\sigma)$ and $g(\sigma)$. We define the chain complex $(S(\sigma),\partial^\sigma)$ as follows:

$$S(\sigma)_n=\begin{cases} C_n(\overline{\tau})\, , & n\geq 0\\ 0\, , & n<0\end{cases}$$

(recall that $\overline{\tau}$ is the simplicial complex $(\tau,\wp(\tau)\smallsetminus\emptyset)$). Since $\overline{\tau}$ is an acyclic simplicial complex, it follows that $S$ is an acyclic carrier between $C(K)$ and $C(L)$. Moreover, $S$ is an acyclic carrier of $C(f)-C(g)$: in fact, if $x$ is a vertex of $K$ such that $f(x)\neq g(x)$, the fact that $f$ and $g$ are contiguous ensures the existence of a simplex $\tau$ of $L$, which contains both vertices $f(x)$ and $g(x)$; therefore,

$$(C_0(f)-C_0(g))(x)\in C_0(\overline{\tau})=S(x)_0\, .$$

It is easy to prove that, for any generator $\sigma$ of $C(K)$,

$$(C(f)-C(g))(\sigma)\subset S(\sigma)$$

that is to say, $S$ is an acyclic carrier for the chain homomorphism $C(f)-C(g)$.

**(III.2.1) Theorem.** *If $f,g\colon K\to L$ are contiguous, $H_n(f,\mathbb{Z})=H_n(g,\mathbb{Z})$ for every $n\in\mathbb{Z}$.*

*Proof.* We first prove that the chain homomorphism

$$C(f)-C(g)\colon C(K)\to C(L)$$

extends the homomorphism $0\colon\mathbb{Z}\to\mathbb{Z}$. Indeed, let $x$ be a vertex of $K$; if $f(x)=g(x)$, then $(C_0(f)-C_0(g))(x)=0$; otherwise, if $f(x)\neq g(x)$,

$$\varepsilon(C_0(f)-C_0(g))(x)=\varepsilon(f(x)-g(x))=0$$

and so, $C(f)-C(g)$ extends $0$ (recall that $\varepsilon\colon C_0(K)\to\mathbb{Z}$ is the augmentation homomorphism). By Corollary (II.3.10), we conclude that $C(f)-C(g)$ is homologically null. We complete the proof with the help of Theorem (II.3.4). ∎

Theorem (III.1.4) shows that the geometric realizations of a simplicial complex and one of its barycentric subdivisions are homeomorphic; we now show that the homology groups of simplicial complexes remain unchanged (up to isomorphism) under barycentric subdivisions.

Let $K = (X, \Phi)$ be a simplicial complex. A *projection of $K^{(1)}$ on $K$* is a function

$$\pi \colon K^{(1)} \to K$$

that takes each vertex of $K^{(1)}$ (that is to say, a simplex of $K$) to one of its vertices. Any projection is a simplicial function; in fact, if

$$\{\sigma_{i_0}, \cdots, \sigma_{i_n}\} \in \Phi^{(1)} \text{ , with } \sigma_{i_0} \subset \cdots \subset \sigma_{i_n} \text{ ,}$$

$\pi(\{\sigma_{i_0}, \cdots, \sigma_{i_n}\}) \subset \sigma_{i_n}$ and, since the latter is a simplex of $K$, we conclude that $\pi(\{\sigma_{i_0}, \cdots, \sigma_{i_n}\}) \in \Phi$. From the homological point of view, the choice of vertex for each simplex is absolutely irrelevant because, if $\pi'$ were any other projection, we would have

$$\pi'(\{\sigma_{i_0}, \cdots, \sigma_{i_n}\}) \subset \sigma_{i_n} \supset \pi(\{\sigma_{i_0}, \cdots, \sigma_{i_n}\})$$

for every $\{\sigma_{i_0}, \cdots, \sigma_{i_n}\} \in \Phi^{(1)}$; therefore, the projections $\pi$ and $\pi'$ would be contiguous. It follows from these considerations that we may choose $\pi$ to be the function that associates to each simplex of $K$ its last vertex.

**(III.2.2) Theorem.** *Let* $\pi \colon K^{(1)} \to K$ *be a projection. Then, for every $n \in \mathbb{Z}$, $H_n(\pi, \mathbb{Z})$ is an isomorphism.*

*Proof.* The projection $\pi$ produces a chain complex homomorphism

$$C(\pi) \colon C(K^{(1)}) \longrightarrow C(K) \text{ ;}$$

we wish to find a chain complex homomorphism

$$\aleph \colon C(K) \longrightarrow C(K^{(1)})$$

such that $\aleph C(\pi)$ is chain homotopic to $1_{C(K^{(1)})}$ and $C(\pi)\aleph$ is chain homotopic to $1_{C(K)}$. If we reach this goal, from the homological point of view, the homomorphism

$$H_*(\aleph) \colon H_*(K, \mathbb{Z}) \longrightarrow H_*(K^{(1)}, \mathbb{Z})$$

induced by $\aleph$ is the inverse of $H_*(\pi, \mathbb{Z})$.

The morphism $\aleph$ does not come from a simplicial function and is defined by induction as follows. Since the vertices of $K$ are also vertices of $K^{(1)}$, we define $\aleph_0$ on the generators $\{x\}$ of $C_0(K)$ by $\aleph_0(\{x\}) = \{x\}$. Suppose that we have defined $\aleph_i$ for every $i = 1, \ldots, n - 1$ such that

$$\aleph_{i-1}\partial_i^K = \partial_i^{K^{(1)}} \aleph_i \text{ .}$$

We define $\aleph_n$ on the generators $\sigma_n$ of $C_n(K)$ (namely, the oriented $n$-simplexes of $K$) by the formula

$$\aleph_n(\sigma_n) := \aleph_{n-1}(\partial_n^K(\sigma_n)) * b(\sigma_n)$$

where $b(\sigma_n)$ is the barycenter of $\sigma_n$ (hence, a vertex of $K^{(1)}$) and

$$\aleph_{n-1}(\partial_n^K(\sigma_n)) * b(\sigma_n)$$

represents the $n$-chain obtained by taking the join of each component of the $(n-1)$-chain $\aleph_{n-1}(\partial_n^K(\sigma_n))$ and $b(\sigma_n)$ (abstract cone with vertex $b(\sigma_n)$). The only important fact to be proved here is that $\aleph_{n-1}\partial_n^K = \partial_n^{K^{(1)}}\aleph_n$. But this is immediate:

$$\partial_n^{K^{(1)}}(\aleph_n(\sigma_n)) = \aleph_{n-1}(\partial_n^K(\sigma_n)) - \partial_{n-1}^{K^{(1)}}(\aleph_{n-1}(\partial_n^K(\sigma_n))) * b(\sigma_n))$$
$$= \aleph_{n-1}(\partial_n^K(\sigma_n))$$

since $\partial_{n-1}^{K^{(1)}}\aleph_{n-1} = \aleph_{n-2}\partial_{n-1}^K$. We call $\aleph \colon C(K) \longrightarrow C(K^{(1)})$ *barycentric subdivision homomorphism.*

Arriving to the equality $C(\pi)\aleph = 1_{C(K)}$ is a straightforward procedure. We now prove that $1_{C(K^{(1)})} - \aleph C(\pi)$ is chain null-homotopic and so, by Proposition (II.3.4), we have $H_*(\aleph)H_*(\pi) = 1$. As a matter of simplicity, we shall indicate $1_{C(K^{(1)})}$ with 1, $K^{(1)}$ with $K'$, $\partial_n^{K^{(1)}}$ with $\partial_n'$, and all $n$-simplexes of $K'$ with the generic expression $\sigma_n'$.

Let $\varepsilon \colon C_0(K') \to \mathbb{Z}$ be the augmentation homomorphism of the chain complex $(C(K'), \partial')$; since

$$\varepsilon((1 - \aleph_0 C_0(\pi))(\sigma_n)) = \varepsilon(\sigma_n - \{x_n\}) = 0,$$

$1 - \aleph C(\pi))$ is a trivial extension of the homomorphism $\mathbb{Z}$ on itself.

On the other hand, to each generator

$$\sigma_n' = \{\sigma^0, \ldots, \sigma^n\}$$

of $C_n(K')$, we associate the chain complex $S(\sigma_n') \leq C(K')$ defined by the free groups

$$S(\sigma_n')_i = \begin{cases} C_i((\overline{\sigma^n})^{(1)}), & i \geq 0 \\ 0, & i < 0. \end{cases}$$

The simplicial complex $\overline{(\sigma^n)}^{(1)}$ comes from the first barycentric subdivision of $\sigma^n$ and is, therefore, an acyclic complex since it may be interpreted as the abstract cone with vertex at the barycenter of $\sigma^n$ relative to the boundary of $(\sigma^n)^{(1)}$ (see Sect. II.4); hence, the chain complex

$$S(\sigma_n') \leq C(K')$$

just defined is acyclic. We have thus obtained an acyclic carrier of $C(K')$ on itself. We maintain that $S$ is an acyclic carrier of $1 - \aleph C(\pi)$. Indeed, for any vertex $\sigma_n = \{x_0, \ldots, x_n\}$ of $K'$,

$$(1 - \aleph_0 C_0(\pi))(\sigma_n) = \sigma_n - \{x_n\} \in C_0(\sigma_n^{(1)}) = S(\sigma_n')_0 \,;$$

the given definitions ensure that, for every $n \geq 1$,

$$(1 - \aleph_n C_n(\pi))(\sigma'_n) \in S(\sigma'_n) .$$

We complete the proof by using Corollary (II.3.10).

Theorem (III.2.2) may be extended by iteration: Let $K^{(r)}$ be the $r$th-barycentric subdivision of $K$ and let

$$\pi^r \colon K^{(r)} \to K$$

be the composition of projections

$$K^{(r)} \xrightarrow{\ \pi\ } K^{(r-1)} \xrightarrow{\ \pi\ } \dots \xrightarrow{\ \pi\ } K .$$

Then for every $n \geq 0$,

$$H_n(\pi^r, \mathbb{Z}) \colon H_n(K^{(r)}, \mathbb{Z}) \to H_n(K, \mathbb{Z})$$

is an isomorphism whose inverse is induced by the (not simplicial) homomorphism

$$\aleph_n^r \colon C_n(K) \longrightarrow C_n(K^{(r)})$$

obtained from $\aleph_n$ by iteration.

We precede the important Simplicial Approximation Theorem by some remarks toward the characterization of the simplexes of a simplicial complex $K = (X, \Phi)$ by working with the topology on $|K|$. For each vertex $x$ of $K$, let $A(x)$ be the set of all points $p \in |K|$ such that $p(x) > 0$; in addition, we define the function

$$\delta_x \colon |K| \longrightarrow \mathbb{R} , \ \delta_x(p) = p(x) .$$

This function is continuous since, for every $q \in |K|$,

$$|\delta_x(p) - \delta_x(q)| < d(p,q) ;$$

since $A(x) = \delta_x^{-1}(0, \infty)$, we conclude that $A(x)$ is open in $|K|$. Note that in this way we obtain an open covering of $|K|$.

(III.2.3) Lemma.  *Given* $x_0, \dots, x_n \in X$ *arbitrarily,*

$$\sigma = \{x_0, \dots, x_n\} \in \Phi \iff \bigcap_{i=0}^{n} A(x_i) \neq \emptyset .$$

*Proof.* $\Rightarrow$: if $\sigma \in \Phi$, its barycenter

$$b(\sigma) = \sum_{i=0}^{n} \frac{1}{n+1} x_i \in \bigcap_{i=0}^{n} A(x_i) .$$

$\Leftarrow$: if $p \in \bigcap_{i=0}^{n} A(x_i)$, then $p(x_i) > 0$ for every $i = 0, \dots, n$; hence

$$\{x_0, \dots, x_n\} \subset s(p) \in \Phi .$$

A simplicial function $g\colon K^{(r)} \to L$ is called a *simplicial approximation* of a map $f\colon |K| \longrightarrow |L|$ if $|g|(p) \in |s(fF(p))|$ for every $p \in |K^{(r)}|$ (recall that $s(fF(p))$ is the carrier of the point $fF(p)$ and that $F\colon |K^{(r)}| \to |K|$ is the homeomorphism in Theorem (III.1.4)).

**(III.2.4) Theorem (Simplicial Approximation Theorem).** *Let $K = (X, \Phi)$ and $L = (Y, \Psi)$ be two simplicial complexes and let $f\colon |K| \to |L|$ be a map. Then, there are an integer $r \geq 0$ and a simplicial function $g\colon K^{(r)} \to L$, which is a simplicial approximation of $f$.*

*Proof.* Consider the finite open covering $\{A(y)|y \in Y\}$ of $|L|$ and let $\ell$ be the Lebesgue number (see Theorem (I.1.41)) of the finite open covering $\{f^{-1}(A(y)) \mid y \in Y\}$ of $|K|$. Let $r$ be a positive integer such that diam $|K^{(r)}| < \ell/2$ (see Theorem (III.1.5)). For every vertex $\sigma$ of $K^{(r)}$, there exists a vertex $y \in Y$ such that

$$A(\sigma) \subset f^{-1}(A(y)) .$$

We now define a function

$$g\colon K^{(r)} \longrightarrow L$$

by setting $g(\sigma) = y$. The function $g$ is simplicial: for each simplex $\{\sigma^0, \ldots, \sigma^n\}$ of $K^{(r)}$,

$$\bigcap_{i=0}^{n} A(\sigma^i) \neq \emptyset$$

(see Lemma (III.2.3)); moreover,

$$(\forall i = 0, \ldots, n)\, A(\sigma^i) \subset f^{-1}(A(g(\sigma^i)))$$

and so

$$\bigcap_{i=0}^{n} A(g(\sigma^i)) \neq \emptyset$$

follows from

$$\bigcap_{i=0}^{n} A(\sigma^i) \subset f^{-1}\left(\bigcap_{i=0}^{n} A(g(\sigma^i))\right);$$

again by Lemma (III.2.3) we conclude that

$$\{g(\sigma^0), \ldots, g(\sigma^n)\} \in \Psi .$$

Let us now prove that $g$ is a simplicial approximation of $f$: for any

$$p = \sum_{i=0}^{n} \alpha_i \sigma^i \in |K^{(r)}|$$

we have $p \in A(\sigma^i)$ for each $i = 0, \ldots, n$; then

$$f(p) \in f(A(\sigma^i)) \subset A(g(\sigma^i))$$

that is to say, $\{g(\sigma^0),\ldots,g(\sigma^n)\} \subset s(f(p))$. Since

$$|g|(p) = \sum_{i=0}^{n} \alpha_i g(\sigma^i) ,$$

we may conclude that $|g|(p) \in |\overline{s(fF(p))}|$. ∎

**(III.2.5) Theorem.** *The geometric realization of a simplicial approximation* $g\colon K^{(r)} \to L$ *of a map* $f\colon |K| \to |L|$ *is homotopic to* $fF$.

*Proof.* By definition, given any $p \in |K^{(r)}|$, the segment with ends $fF(p)$ and $|g|(p)$ is entirely contained in the convex space $|\overline{s(fF(p))}| \subset |L|$ (see Theorem (II.2.9)). The map

$$H\colon |K^{(r)}| \times I \to |L| ,\ H(p,t) = tfF(p) + (1-t)|g|(p)$$

is a homotopy between $fF$ and $|g|$. ∎

**(III.2.6) Theorem.** *Let* $g\colon K^{(r)} \to L$ *and* $g'\colon K^{(s)} \to L$ *be two simplicial approximations of a map* $f\colon |K| \to |L|$. *Suppose that* $s < r$ *and* $\pi^{(r,s)}\colon K^{(r)} \to K^{(s)}$ *is the composition of projections. Then the simplicial functions*

$$g\colon K^{(r)} \to L \text{ and } g'\pi^{(r,s)}\colon K^{(r)} \to L$$

*are contiguous.*

*Proof.* We have to prove that, for every simplex $\sigma \in \Phi^{(r)}$, there is a simplex $\tau \in \Psi$ such that $g(\sigma) \subset \tau$ and $g'\pi^{(r,s)}(\sigma) \subset \tau$.

For every $\sigma \in \Phi^{(r)}$, let us define $\tau = s(f(b(\sigma)))$ where $b(\sigma)$ is the barycenter of $\sigma$. Since $g$ is a simplicial approximation of $f$,

$$|g|(b(\sigma)) \in |\overline{s(f(b(\sigma)))}| ;$$

but $g(\sigma)$ is the smallest simplex of $L$ whose geometric realization contains the point $|g|(b(\sigma))$ and so, $g(\sigma) \subset \tau$.

On the other hand, let $P(\sigma)$ be the smallest simplex of $K^{(s)}$ such that $|\sigma| \subset |P(\sigma)|$. Clearly,

$$|\overline{\pi^{(r,s)}(\sigma)}| \subset |\overline{P(\sigma)}|$$

and

$$|\overline{g'(\pi^{(r,s)}(\sigma))}| \subset |\overline{g'(P(\sigma))}| .$$

From $b(\sigma) \in |\overline{\sigma}| \subset |\overline{P(\sigma)}|$, it follows that

$$|g'|(b(\sigma)) \in |\overline{g'(P(\sigma))}|$$

and because $|g'|(b(\sigma)) \in |\overline{s(f(b(\sigma)))}|$, it also follows that

$$|\overline{g'(P(\sigma))}| \subset |\overline{s(f(b(\sigma)))}|.$$

From this, we conclude that

$$|g'(\pi^{(r,s)}(\sigma))| \subset |s(f(b(\sigma)))|$$

and, consequently, $g'(\pi^{(r,s)}(\sigma)) \subset s(f(b(\sigma)))$. ∎

**(III.2.7) Corollary.** *A map* $f\colon |K| \to |L|$ *defines a unique homomorphism of graded groups*

$$H_*(f,\mathbb{Z})\colon H_*(|K|,\mathbb{Z}) \to H_*(|L|,\mathbb{Z}).$$

*Proof.* For any simplicial approximation $g\colon K^{(r)} \to L$ of $f$, let us set

$$H_*(f,\mathbb{Z}) = H_*(g,\mathbb{Z})(H_*(\pi^r))^{-1}$$

(recall that $(H_*(\pi^r))^{-1} = H_*(\aleph^r)$ ).

Suppose $g'\colon K^{(s)} \to L$ to be another simplicial approximation of $f$, with $s < r$. The contiguity of the simplicial functions $g$ and $g'\pi^{(r,s)}$ (which follows from the previous theorem) ensures that

$$H_*(g'\pi^{(r,s)}) = H_*(g)$$

(cf. Corollary (III.2.1)). But $\pi^r = \pi^s \pi^{(r,s)}$ and therefore

$$H_*(g',\mathbb{Z})(H_*(\pi^s,\mathbb{Z}))^{-1} = H_*(g,\mathbb{Z})(H_*(\pi^r,\mathbb{Z}))^{-1}.$$

∎

Corollary (III.2.7) completes the definition of functor $H_*(-,\mathbb{Z})$ on morphisms; we repeat it here: for every $n \geq 0$ and any map $f\colon |K| \to |L|$,

$$H_n(f,\mathbb{Z}) := H_n(g,\mathbb{Z})(H_n(\pi^r,\mathbb{Z}))^{-1}\colon H_n(|K|,\mathbb{Z}) \to H_n(|L|,\mathbb{Z})$$

where $g\colon K^{(r)} \to L$ is any simplicial approximation of $f$.

We now prove that the homomorphism $H_*(f,\mathbb{Z})$, defined by a map $f\colon |K| \to |L|$, is a *homotopy invariant*, meaning that the next result is true.

**(III.2.8) Theorem.** *Let* $f,g\colon |K| \to |L|$ *be homotopic maps. Then*

$$H_*(f,\mathbb{Z}) = H_*(g,\mathbb{Z}).$$

*Proof.* Let $H\colon |K| \times I \to |L|$ be the homotopy that links $f$ to $g$; suppose that $H\, i_0 = f$ and $H\, i_1 = g$, where $i_0$ and $i_1$ are the maps

$$i_\alpha\colon |K| \to |K| \times I,\ p \mapsto (p,\alpha),\ \alpha = 0,1.$$

We represent the intervall $I = [0,1]$ as the geometric realization of the complex

$$\mathscr{I} = (Y,\Upsilon),\ Y = \{0,1\},\ \Upsilon = \{\{0\},\{1\},\{0,1\}\}.$$

With the appropriate identification $|K \times \mathscr{I}| \equiv |K| \times I$ (cf. Theorem (III.1.1)), we may assume that $i_0$ and $i_1$ are the geometric realizations of the simplicial functions

$$\widetilde{i}_\alpha : K \to K \times \mathscr{I}$$

that take each 0-simplex $\{x\}$ of $K$ to the 0-simplex $\{(x,\alpha)\}$ of $K \times \mathscr{I}$. To verify this theorem, it is sufficient to prove that $\widetilde{i}_0$ and $\widetilde{i}_1$ are contiguous. We order the set $X = \{x_0, x_1 \ldots, x_r\}$ of all vertices of $K$ by setting $x_0 < x_1 < \ldots < x_r$. For each integer $n = 0, \ldots, r+1$, we define the simplicial functions

$$f_n : K \to K \times \mathscr{I}$$

$$f_n(\{x_i\}) = \begin{cases} \{(x_i,0)\} \, , \, i < n \, , \\ \{(x_i,1)\} \, , \, i \geq n \, . \end{cases}$$

These functions have the following properties:

1. $(\forall n = 0, \ldots, r-1)$ $f_n$ and $f_{n+1}$ are contiguous.
2. $f_{r+1} = \widetilde{i}_0$ and $f_0 = \widetilde{i}_1$.

The second property is a direct consequence of the definition of $f_n$; to prove the first property, we consider any

$$\sigma = \{x_{i_0}, \ldots, x_{i_k}\} \in \Phi$$

such that

$$x_{i_0} < x_{i_1} < \ldots < x_{n-1} < x_n < x_{n+1} < \ldots < x_{i_k} \, .$$

It follows from the definition of $f_n$ and $f_{n+1}$ that

$$f_n(\sigma) = \{(x_{i_0},0), \ldots, (x_{n-1},0), (x_n,1), \ldots, (x_{i_k},1)\}$$

and

$$f_{n+1}(\sigma) = \{(x_{i_0},0), \ldots, (x_n,0), (x_{n+1},1), \ldots, (x_{i_k},1)\};$$

notice that both sets $f_n(\sigma)$ and $f_{n+1}(\sigma)$ are subsets of

$$\{(x_{i_0},0), \ldots, (x_{n-1},0), (x_n,0), (x_n,1), \ldots, (x_{i_k},1)\} \, . \quad \blacksquare$$

**(III.2.9) Corollary.** *The homology groups of two polyhedra of the same homotopy type are isomorphic.*

*Proof.* Let $f \colon |K| \to |L|$ be a homotopy equivalence with homotopy inverse $f' \colon |L| \to |K|$. Then

$$H_*(f,\mathbb{Z})H_*(f',\mathbb{Z}) = 1 \text{ and } H_*(f',\mathbb{Z})H_*(f,\mathbb{Z}) = 1 \, . \quad \blacksquare$$

Corollary (III.2.7) and Theorem (III.2.8) allow us to define a functor

$$H_* \colon HP \longrightarrow \mathbf{Ab}^{\mathbb{Z}},$$

where $H\mathbf{P}$ is the category of polyhedra and of homotopy classes of maps among polyhedra.

**(III.2.10) Remark.** Corollary (III.2.9) tells us in particular that, for every polyhedron $|K|$ and whatever barycentric subdivision $|K^{(r)}|$, we have

$$H_*(|K|;\mathbb{Z}) \cong H_*(|K^{(r)}|;\mathbb{Z});$$

moreover, the homology $H_*(X;\mathbb{Z})$ of a compact and triangulable topological space is invariant under the chosen triangulation of $X$.

## Exercises

**1.** Prove that a projection $\pi\colon K^{(1)} \to K$ is a simplicial approximation of the identity map $1\colon |K| \to |K|$.

**2.** Prove (using the Simplicial Approximation Theorem) that the set $[|K|,|L|]$ of homotopy classes of maps from a polyhedron $|K|$ to a polyhedron $|L|$ is countable.

## III.3 Some Applications

In this section, we give some applications of simplicial homology to geometry. Our first important result is the Lefschetz Fixed Point Theorem. Before we state it, let us review some well-known results in linear algebra.

The *trace* of a rational square matrix $(a_{ij})$, $i = j = 1,\ldots,n$ is the rational number

$$\mathrm{Tr}\,((a_{ij})) = \sum_{i=1}^{n} a_{ii}\,.$$

We note that for two rational square matrices $A$ and $B$,

$$\mathrm{Tr}\,(AB) = \mathrm{Tr}\,(BA)\,.$$

Thus, we may define the trace of a linear transformation $\alpha\colon V \to V$ of an $n$-dimensional rational vector space. Indeed, if $A$ represents $\alpha$, then $B$ represents $\alpha$ if and only if there exists $C$ such that $B = CAC^{-1}$. We then define

$$\mathrm{Tr}\,(\alpha) = \mathrm{Tr}\,(B) = \mathrm{Tr}\,(A)\,.$$

**(III.3.1) Lemma.** *Let*

$$0 \to U \xrightarrow{\alpha} V \xrightarrow{\beta} W \to 0$$

*be an exact sequence of finite dimensional rational vector spaces and of linear transformations. Let $\gamma\colon V \to V$ be a linear transformation whose restriction to $U$ is a* *linear transformation* $\gamma_U : U \to U$. *Then* $\gamma$ *induces a unique linear transformation* $\gamma_W : W \to W$ *such that* $\gamma_W \beta = \beta \gamma$, *and therefore*

$$Tr(\gamma) = Tr(\gamma_U) + Tr(\gamma_W).$$

*Proof.* Let us consider a basis $\{\vec{u}_1, \ldots, \vec{u}_r\}$ of $U$ and the vectors $\vec{v}_i = \alpha(\vec{u}_i) \in V$, $i = 1, \ldots, r$. The latter are linearly independent in $V$ and may, therefore, be completed so that we have a basis $\{\vec{v}_1, \ldots, \vec{v}_n\}$ of $V$. Note that the vectors $\vec{w}_{r+1} = \beta(\vec{v}_{r+1}), \ldots, w_n = \beta(\vec{v}_n)$ constitute a basis for $W$. The matrix $(a_{ij})$, with $i, j = 1, \ldots, n$, obtained by the equalities

$$\gamma(\vec{v}_i) = \sum_{j=1}^{n} a_{ji} \vec{v}_j, \; i = 1, \ldots, n$$

represents $\gamma$. Transformations $\gamma_U$ and $\gamma_W$ are defined by the rules

$$\gamma_U(\vec{u}_i) = \sum_{j=1}^{r} a_{ji} \vec{u}_j, \text{ for every } i = 1, \ldots, r$$

and

$$\gamma_W(\vec{w}_i) = \sum_{j=r+1}^{n} a_{ji} \vec{w}_j, \text{ for every } j = r+1, \ldots, n. \qquad \blacksquare$$

We now concern ourselves with geometry. Our proofs will take place in the realm of rational homology, that is to say, homology with coefficients in $\mathbb{Q}$. We briefly recall that, for every polyhedron $|K|$, $H_n(|K|; \mathbb{Q})$ is a rational vector space of dimension $\beta(n)$. On the other hand, a map $f : |K| \to |K|$ produces a linear transformation

$$H_n(f, \mathbb{Q}) : H_n(|K|; \mathbb{Q}) \to H_n(|K|; \mathbb{Q})$$

for each integer $n \in \{0, \ldots, \dim K\}$. By definition, the **Lefschetz Number** of a map $f : |K| \to |K|$ is the rational number

$$\Lambda(f) = \sum_{n=0}^{\dim K} (-1)^n \mathrm{Tr}\, H_n(f, \mathbb{Q}).$$

We consider a simplicial approximation $g : K^{(r)} \to K$ to $f$ (see Theorem (III.2.4)) that produces a chain morphism

$$C(g) : C(K^{(r)}) \to C(K)$$

and therefore, a homomorphism

$$C_n(g) \otimes 1_{\mathbb{Q}} : C(K^{(r)}, \mathbb{Q}) \to C(K, \mathbb{Q});$$

we now consider the chain morphism (subdivision homomorphism)

$$\aleph : C(K) \to C(K^{(1)})$$

defined in Theorem (III.2.2); we remind the reader that the homomorphism

$$C(g)\aleph^r : C(K) \longrightarrow C(K)$$

is precisely the homomorphism that defines $H_n(f,\mathbb{Z}) = H_n(C(g)\aleph_n^r)$. Since for every $n = 0,\ldots,\dim K$,

$$(C_n(g) \otimes 1_{\mathbb{Q}})(\aleph_n^r \otimes 1_{\mathbb{Q}}) : C_n(K,\mathbb{Q}) \to C_n(K,\mathbb{Q})$$

is a linear transformation, we may define the number

$$\Lambda(C(g)\aleph) = \sum_{n=0}^{\dim K} (-1)^n \mathrm{Tr}\left[(C_n(g) \otimes 1_{\mathbb{Q}})(\aleph_n^r \otimes 1_{\mathbb{Q}})\right] ;$$

it is natural to wonder whether there is any relation between this number and the Lefschetz Number $\Lambda(f)$. The answer to this question is the next result, known as the **Hopf Trace Theorem**.

**(III.3.2) Lemma.** *For every map* $f : |K| \to |K|$, *we have*

$$\Lambda(C(g)\aleph) = \Lambda(f) .$$

*Proof.* For every $n \geq 0$, we consider two exact sequences

(1) $\quad 0 \to Z_n(K,\mathbb{Q}) \to C_n(K,\mathbb{Q}) \to B_{n-1}(K,\mathbb{Q}) \to 0$

(2) $\quad 0 \to B_n(K,\mathbb{Q}) \to Z_n(K,\mathbb{Q}) \to H_n(K,\mathbb{Q}) \to 0$

Then for any fixed $t \in \mathbb{Z}$, we define the numbers

$$c(t) = \sum t^n \mathrm{Tr}\left[(C_n(g) \otimes 1_{\mathbb{Q}})(\aleph_n^r \otimes 1_{\mathbb{Q}})\right],$$
$$z(t) = \sum t^n \mathrm{Tr}\left[(C_n(g) \otimes 1_{\mathbb{Q}})(\aleph_n^r \otimes 1_{\mathbb{Q}})\right]|Z_n(K,\mathbb{Q})),$$
$$b(t) = \sum t^n \mathrm{Tr}\left[(C_n(g) \otimes 1_{\mathbb{Q}})(\aleph_n^r \otimes 1_{\mathbb{Q}})\right]|B_n(K,\mathbb{Q})), \text{ and}$$
$$h(t) = \sum t^n \mathrm{Tr}\left(H_n(f,\mathbb{Q})\right).$$

By applying Lemma (III.3.1) to the exact sequences (1) and (2), we conclude that

$$c(t) = z(t) + tb(t)$$
$$z(t) = b(t) + h(t) ;$$

it follows from these two equalities that $c(t) - h(t) = (1+t)b(t)$; we arrive to the thesis by setting $t = -1$. ∎

We now turn to the **Lefschetz Fixed Point Theorem**.

**(III.3.3) Theorem.** *For every map* $f\colon |K| \to |K|$ *with no fixed points,* $\Lambda(f) = 0$.

*Proof.* Let $\Psi\colon |K| \to \mathbb{R}$ be a function defined by $\Psi(p) = d(p, f(p))$ for each $p \in |K|$. Since $f$ has no fixed points, $(\forall p \in |K|)\ \Psi(p) > 0$. On the other hand, since $|K|$ is compact, there exists an $\varepsilon > 0$ such that $\Psi(p) \geq \varepsilon$ for every $p \in |K|$. By Theorem (III.1.5), there is an integer $r$ such that the diameter of $|K^{(r)}|$ is less than $\frac{\varepsilon}{3}$. By Theorem (III.1.4), there exists a homeomorphism $F\colon |K^{(r)}| \to |K|$; we then define the map $f^{(r)} = F^{-1}fF\colon |K^{(r)}| \to |K^{(r)}|$. The next two assertions are now easily proved and we leave their proofs to the reader.

1. $\Lambda(f^{(r)}) = \Lambda(f)$.
2. $f$ has no fixed points if and only if $f^{(r)}$ has no fixed points.

So, we only need to prove that $\Lambda(f^{(r)}) = 0$.

Let the simplicial function $g\colon K^{(t)} \to K^{(r)}$, with $t \geq r$, be an approximation to $f^{(r)}$; then

$$(\forall p \in |K^{(t)}|)\ |g|(p) \in |s(f^{(r)}(p))|$$

and, since $s(f^{(r)}(p))$ is a simplex of $K^{(r)}$, we conclude that

$$(\forall p \in |K^{(t)}|)\ d(|g|(p), f^{(r)}(p)) < \frac{\varepsilon}{3}.$$

This last inequality, the metric triangular property, and condition

$$(\forall p \in |K^{(t)}|)\ \varepsilon < \Psi(p) = d(p, f^{(r)}(p))$$

enable us to conclude that

$$(\forall p \in |K^{(t)}|)\ d(p, |g|(p)) > 2\frac{\varepsilon}{3};$$

in other words, for every point $p \in |K^{(t)}|$, $|g|(p)$ and $p$ cannot coexist in the same simplex of $|K^{(r)}|$. Now, let $\sigma$ be any $n$-simplex of $K^{(r)}$ viewed as an element of the canonic basis of $C_n(K^{(r)}, \mathbb{Q})$, and suppose $\sigma$ to be one of the vectors in the composition of $C_n(g)\aleph_n^r(\sigma)$,[2] written as a linear combination of vectors of the canonic basis. However, $\aleph_n^r(\sigma)$ is a linear combination of $n$-simplexes of $K^{(t)}$, whose geometric realizations are contained in $|\sigma|$; since $\sigma$ is a component of $C_n(g)\aleph_n^r(\sigma)$, $C_n(g)$ must take to $\sigma$ one of the $n$-simplexes of $K^{(t)}$, among the components of the chain $\aleph_n^r(\sigma)$; on the other hand, since $C_n(\pi^{(t,r)})\aleph_n^r(\sigma) = \sigma$, there must be a point $p \in |K^{(t)}|$ such that $d(p, |g|(p)) < \frac{\varepsilon}{3}$, contradicting the preceeding results. Hence, $\mathrm{Tr}\,(C_n(g)\aleph_n^r) = 0$; the Hopf Trace Theorem allows us to conclude that $\Lambda(f^{(r)}) = 0$ and so, $\Lambda(f) = 0$.                                            ∎

The Lefschetz Fixed Point Theorem supplies a necessary but not sufficient condition for a map $f\colon |K| \to |K|$ to have no fixed points. Indeed,

$$f\colon S^1 \to S^1\ ,\ e^{2\pi i t} \mapsto e^{2\pi i(t+\frac{1}{12})}, t \in [0,1)$$

---

[2] We simplify the notation by writing $C_n(g)\aleph_n^r$ instead of $(C_n(g) \otimes 1_\mathbb{Q})(\aleph_n^r \otimes 1_\mathbb{Q})$.

has no fixed points and so $\Lambda(f) = 0$; but $f$ is homotopic to the identity $1_{S^1}$ through

$$H(e^{2\pi i t}, s) = e^{2\pi i(t + s\frac{1}{12})}$$

and therefore $\Lambda(1_{S^1}) = 0$ (because the Lefschetz Number is invariant up to homotopy); however, $1_{S^1}$ has only fixed points.

We now consider an important corollary to the Lefschetz Fixed Point Theorem, known as the **Brouwer Fixed Point Theorem**. A *self-map* of a space $X$ is a map of $X$ into itself.

**(III.3.4) Corollary.** *Every nonconstant self-map of a connected acyclic polyhedron has a fixed point.*

*Proof.* Let $|K|$ be such a polyhedron[3]; by hypothesis, the homology of $|K|$ is all trivial, except for $H_0(K; \mathbb{Q}) \cong \mathbb{Q}$. Let us suppose that $f$ is simplicially approximated by $g\colon K^{(r)} \to K$. Let $x_0$ be a vertex of $|K|$ and let us suppose that $\{x_0\} + B_0(K, \mathbb{Q})$ is a generator of $H_0(K, \mathbb{Q})$. On the other hand,

$$C_0(g)\aleph_0(\{x_0\}) = g(x_0)$$

and $g(x_0)$ is a vertex of $K$ because $g$ is simplicial. Since $|K|$ is connected, $g(x_0)$ is homological to $\{x_0\}$ and therefore $\Lambda(f) = 1$. Lefschetz theorem allows us to conclude that $f$ must have a fixed point. ∎

Before stating the next corollary, let us look into the definition of the *degree* of a self-map of a sphere. It is easy to see that the $n$-sphere $S^n$ ($n \geq 1$) is the geometric realization of a simplicial complex $\Sigma^n$. Let $f\colon S^n \to S^n$ be a given map and $g\colon (\Sigma^n)^{(r)} \to \Sigma^n$ be a simplicial approximation of $f$. $\Sigma^n$ has trivial homology in all dimensions except for $0$ and $n$, when it is isomorphic to the Abelian group $\mathbb{Z}$. We recall that $H_n((\Sigma^n)^{(r)}) \cong Z_n((\Sigma^n)^{(r)})$ has only two possible generators (differing by their orientation); let $z$ be one of them. Then, there exists an integer $d$ such that $H_n(g)(z) = d\,z$. This number, which is obviously independent from the homotopy class of the map $f$, the cycle $z$, and the simplicial approximation $g$ of $f$, is the *degree* of $f$ (notation: gr $(f)$).[4]

**(III.3.5) Lemma.** *For every map* $f\colon S^n \to S^n$,

$$\Lambda(f) = 1 + (-1)^n \, \text{gr}\,(f).$$

*Proof.* The Lefschetz number is defined for the homology with rational coefficients. Since $\Sigma^n$ is connected, $\text{Tr}\, H_0(f, \mathbb{Q}) = 1$; we only need to prove that $\text{Tr}\, H_n(f, \mathbb{Q}) = \text{gr}\,(f)$, to which end it is sufficient noting that

$$Z_n((\Sigma^n)^{(r)}, \mathbb{Q}) \cong Z_n((\Sigma^n)^{(r)}) \otimes \mathbb{Q}.$$ ∎

---

[3] See examples of acyclic complexes in Sect. II.4.

[4] Case $n = 1$ is particularly interesting: when we move along the cycle $z$, its image $g(z)$ wraps the corresponding generating cycle of $Z_1(S^1)$ $d$ times around $S^1$.

**(III.3.6) Corollary.** *If a map $f\colon S^2 \to S^2$ has no fixed point, there exists $p \in S^2$ such that $f(p) = -p$.*

*Proof.* Suppose that $f$ has no fixed point; by the preceeding lemma gr $(f) = -1$. In particular, the *antipodal* function

$$A\colon S^2 \to S^2 \, , \; p \mapsto -p$$

has degree $-1$. Since gr $(Af) =$ gr $(A)$gr $(f) = 1$, it follows that

$$\Lambda(Af) = 1 + (-1)^2 \text{gr} \, (Af) = 2$$

and so, there is $p \in S^2$ such that $Af(p) = p$; hence, $f(p) = -p$.  ■

**(III.3.7) Corollary.** *There is no map $f\colon S^{2n} \to S^{2n}$ with $n \geq 1$, such that the vectors $p$ and $f(p)$ are perpendicular for every $p \in S^{2n}$.*

*Proof.* Suppose there is such a function. Then

$$(\forall p \in S^{2n}) \; \|(1-t)f(p) + tp\| \neq 0$$

and we may therefore define a map

$$F\colon S^{2n} \times I \to S^{2n} \, , \; (p,t) \mapsto \frac{(1-t)f(p) + tp}{\|(1-t)f(p) + tp\|} \, .$$

This map is a homotopy between function $f$ and the identity function $1_{S^{2n}}$. It follows that

$$\Lambda(f) = \Lambda(1_{S^{2n}}) = 1 + (-1)^{2n} = 2$$

and so $(\exists p \in S^{2n}) \, f(p) = p$, against the hypothesis on $f$.  ■

**(III.3.8) Remark.** A *tangent vector field* on a sphere $S^n$ is a set of vectors of $\mathbb{R}^{n+1}$ tangent to $S^n$, one at each point $p \in S^n$, and such that the length and direction of the vector at $p$ vary continuously with $p$. Corollary (III.3.7) states that no sphere of even dimension has a nonvanishing tangent vector field. Odd-dimensional spheres have such fields; for instance,

$$S^{2n-1} \to S^{2n-1} \, , \; (x_1, \ldots, x_{2n}) \mapsto (-x_2, x_1, \ldots, -x_{2n}, x_{2n-1}) \, .$$

The important **Fundamental Theorem of Algebra** may be easily proved as a consequence of the Lefschetz Fixed Point Theorem:

**(III.3.9) Corollary.** *A polynomial*

$$f(z) = z^n + a_1 z^{n-1} + \ldots + a_{n-1} z + a_n \in \mathbb{C}[z]$$

*(with $n \geq 1$) has a complex root. Consequently, the equation $f(z) = 0$ has $n$ solutions, not necessarily distinct.*

*Proof.* We consider that polynomial as a map $P \colon \mathbb{R}^2 \to \mathbb{R}^2$ To prove that $P$ is surjective and that, as a consequence, there exists an $\alpha \in \mathbb{C}$ corresponding to $0 \in \mathbb{R}^2$ such that

$$\alpha^n + a_1 \alpha^{n-1} + \ldots + a_{n-1}\alpha + a_n = 0 \,.$$

We take the compactification $S^2$ of $\mathbb{R}^2$ obtained by adding a point $\infty$ to $\mathbb{R}^2$ and we extend $P$ to a map $P \colon S^2 \to S^2$ by setting $P(\infty) = \infty$. The map

$$H \colon S^2 \times I \to S^2 \,, \ (p,t) \mapsto z^n + (1-t)\left(a_1 z^{n-1} + \ldots + a_n\right) \,, \ \infty \mapsto \infty$$

is a homotopy between maps $P$ and

$$f \colon S^2 \to S^2 \,, \ z \mapsto z^n \,.$$

We wish to prove that $\mathrm{gr}\,(f) = n$. To this end, we construct two simplicial complexes $K$ and $L$ homeomorphic to $S^2$, and consider a simplicial approximation of $f$. Let us take the complex numbers

$$x_s = e^{\frac{2\pi i s}{n^2}} \,, \ s = 0, 1, \ldots, n^2 - 1$$

and construct the simplicial complex $K$, formed by
vertices:

$$\{0\}, \{x_0\}, \ldots, \{x_{n^2-1}\}, \{\infty\} \,;$$

1-simplexes:

$$\{0, x_0\}, \ldots, \{0, x_{n^2-1}\}$$
$$\{x_0, \infty\}, \ldots, \{x_{n^2-1}, \infty\}$$
$$\{x_0, x_1\}, \ldots, \{x_{n^2-1}, x_0\}$$

2-simplexes:

$$\{0, x_0, x_1\}, \ldots, \{0, x_1, x_2\}, \ldots, \{0, x_{n^2-1}, x_0\}$$
$$\{\infty, x_0, x_1\}, \ldots, \{\infty, x_1, x_2\}, \ldots, \{\infty, x_{n^2-1}, x_0\} \,,$$

whose geometric realization is homeomorphic to $S^2$.

We now take the complex numbers

$$y_t = e^{\frac{2\pi i t}{n}} \,, \ t = 0, 1, \ldots, n - 1$$

and consider the simplicial complex $L$, formed by
vertices:

$$\{0\}, \{y_0\}, \ldots, \{y_{n-1}\}, \{\infty\} \,;$$

1-simplexes:

$$\{0,y_0\},\ldots,\{0,y_{n-1}\}$$
$$\{y_0,\infty\},\ldots,\{y_{n-1},\infty\}$$
$$\{y_0,y_1\},\ldots,\{y_{n-1},y_0\}$$

2-simplexes:

$$\{0,y_0,y_1\},\ldots,\{0,y_1,y_2\},\ldots,\{0,y_{n-1},y_0\}$$
$$\{\infty,y_0,y_1\},\ldots,\{\infty,y_1,y_2\},\ldots,\{\infty,y_{n-1},y_0\}\,,$$

whose geometric realization is also homeomorphic to $S^2$. Note that

$$f(0) = 0\,,\ f(\infty) = \infty \text{ and } f(x_s) = y_s.$$

Let $z$ be a generator of $H_2(L;\mathbb{Z})$ (the sum of all oriented 2-simplexes of $L$) and $z'$ be the corresponding generator of $K$ (a subdivision of $L$); we can easily see that $H_2(f)(z') = nz$, which means that gr $(f) = n$.

Since the degree of a map is invariant up to homotopy, we conclude that gr $(P) = n$. Assuming that $P$ is not a surjection, there should exist a $p \in S^2$ such that $P(S^2) \subset S^2 \setminus p \approx \mathbb{R}^2$. We could then interpret $P$ as a map from $S^2$ to $\mathbb{R}^2$ and, therefore, homotopic to the constant map from $S^2$ to $0 \in \mathbb{R}^2$ (if $X$ is a topological space and $h\colon X \to \mathbb{R}^2$ is a continuous function, then $h$ is homotopic to the constant map to 0 through the homotopy $h_t(x) = tx$). But a constant map has degree 0, contradicting the fact that $P$ has degree $n > 0$.                                        ■

## Exercises

**1.** Let $A\colon S^1 \to S^1$ be the antipodal function. Prove that gr $A = 1$.

**2.** Prove that there is no retraction of the unit disk $D^n \subset \mathbb{R}^n$ onto its boundary $S^{n-1}$.

**3.** A polyhedron $|K|$ has the *fixed point property* if every map $f\colon |K| \to |K|$ has at least one fixed point. Prove that the fixed point property is invariant up to homeomorphism.

**4.** Let $|K|$ be a polyhedron with the fixed point property. Prove that if $|L|$ is a retraction of $|K|$ then also $|L|$ has the fixed point property.

**5.** Consider the space

$$X = \{(x_0,x_1,x_2) \in \mathbb{R}^3 \,|\, (\forall i = 0,1,2)x_i \geq 0\}$$

with the topology induced by $\mathbb{R}^3$ and let $f\colon X \to X$ be a given continuous function. Prove that it is possible to find a unit vector $\vec{v} \in X$ and a nonnegative real number $\lambda$ such that $f(\vec{v}) = \lambda\vec{v}$.

**6.** Let $M_{3\times 3}^+$ be the set of all real square matrices with no negative elements. Apply the preceeding exercise to prove that every matrix $A \in M_{3\times 3}^+$ has at least one nonnegative eigenvalue.

## III.4 Relative Homology for Polyhedra

In Sect. II.4, we have studied the relative homology of a pair of simplicial complexes $(K,L)$; specifically, we have proved that given two simplicial complexes $K = (X,\Phi)$ and $L = (Y,\Theta)$ with $Y \subset X$ and $\Theta \subset \Phi$, then

$$(\forall n \geq 1) \ H_n(K,L;\mathbb{Z}) \cong H_n(K \cup CL;\mathbb{Z})$$

where $CL$ is the abstract cone $vL$ Theorem (II.4.7). In Sect. III.2, we have studied the homology functor in the category of polyhedra, giving the definition of $H_*(|K|,\mathbb{Z}) := H(K,\mathbb{Z})$ for any polyhedron $|K|$ (the definition of functor $H_*(-,\mathbb{Z})$ on morphisms is more intricate and depends on the Simplicial Approximation Theorem). Nevertheless, we may say that

$$(\forall n \geq 1) \ H_n(|K|,|L|;\mathbb{Z}) \cong H_n(|K \cup CL|;\mathbb{Z})$$

for every pair of polyhedra $(|K|,|L|)$. In this section, we prove the following result:

**(III.4.1) Theorem.** *For every pair of polyhedra $(|K|,|L|)$ (with $L$ a subcomplex of $K$) and every integer $n \geq 1$,*

$$H_n(|K|,|L|;\mathbb{Z}) \cong H_n(|K|/|L|;\mathbb{Z}) \ .$$

This theorem is a direct consequence of a more general result in the category of topological spaces:

**(III.4.2) Theorem.** *Let $(X,A)$ be a pair of topological spaces with $A$ closed in $X$; suppose that $(X,A)$ has the Homotopy Extension Property. Let $CA$ be the cone $(A \times I)/(A \times \{0\})$. Then the adjunction space $X \cup CA$ and the quotient space $X/A$ are of the same homotopy type.*

*Proof.* The reader may intuitively come to this result by considering the cone $CA$ as a space contractible to a point; here is a rigorous proof of the statement.

In this theorem, we have two pushouts: one for constructing $X \cup CA$ and the other for $X/A$; these are their corresponding diagrams:

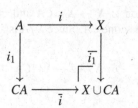

with $i_1: A \rightarrow CA$ given by $x \mapsto [x,1]$ and $i: A \rightarrow X$ the inclusion;

with $c: A \rightarrow *$ the constant map and $\overline{c} = q: X \rightarrow X/A$ the quotient map.

Let $p: CA \rightarrow X/A$ be the constant map to the point $[a_0]$ that is identified with $A$ in $X/A$. Since $qi = pi_1$, there exists a unique map $\ell: X \cup CA \rightarrow X/A$ such that the following diagram is commutative:

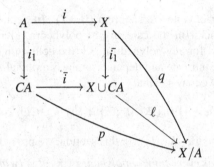

Now we must find a function $\widetilde{\ell}: X/A \rightarrow X \cup CA$ such that $\widetilde{\ell}\ell \sim 1_{X \cup CA}$ and $\ell\widetilde{\ell} \sim 1_{X/A}$. With this in mind, we consider the homotopy

$$H: A \times I \rightarrow X \cup CA , \ (x,t) \mapsto [x, 1-t]$$

and apply the Homotopy Extension Property of $(X,A)$ to get a continuous function

$$G: A \times I \longrightarrow X \cup CA$$

whose restriction to $A \times \{0\}$ coincides with $\overline{i_1}$ and such that $G(i \times 1_I) = H$. The map

$$g: A \longrightarrow X \cup CA , \ g = G(-,1)$$

is homotopic to the inclusion $\overline{i}: X \rightarrow X \cup CA$ and is such that, for every $x \in A$, $g(x) = [x,0]$, the vertex of the cone $CA$. Since $X/A$ is a pushout space, there exists a unique map

$$\widetilde{\ell}: X/A \longrightarrow X \cup CA$$

such that $\widetilde{\ell}q = g$. To prove that $\widetilde{\ell\ell}$ and $\widetilde{\ell\ell}$ are homotopic to their respective identity functions, we construct the following homotopies:

$$H_1 \colon (X \cup CA) \times I \longrightarrow X \cup CA$$

1. $(\forall [x,t] \in CA)$, $H_1([x,t],s) = [x,ts]$,
2. $(\forall x \in X)$, $H_1(x,s) = G(x,1-s)$

and

$$H_2 \colon X/A \times I \longrightarrow X/A$$
$$H_2(q(x),t) = \ell G(x,t) , \ x \in X \smallsetminus A .$$

We ask the reader to verify that these homotopies are well defined and to complete the proof. ∎

We now turn to Theorem (III.4.1). It is sufficient to notice the following facts: (a) $|L|$ is a closed subspace of $|K|$; (b) the pair $(|K|,|L|)$ has the Homotopy Extension Property (see Theorem (III.1.7)), (c) $|K \cup CL| \cong |K| \cup |CL|$ is a pushout space; finally, we apply Theorem (III.4.2).

Theorem (II.4.9) in Sect. II.4 has a corresponding version in the category of polyhedra: let $\{|K_i| \,|\, i = 1,\ldots,p\}$ be a finite set of based polyhedra, each with base point given by a vertex $x_0^i \in K_i$; we then take the *wedge product*

$$\vee_{i=1}^p |K_i| := \cup_{i=1}^n (\{x_0^1\} \times \ldots \times |K_i| \times \ldots \times \{x_0^p\}) .$$

**(III.4.3) Theorem.** *For every $q \geq 1$,*

$$H_q(\vee_{i=1}^p |K_i|, \mathbb{Z}) \cong \oplus_{i=1}^p H_q(|K_i|, \mathbb{Z}) .$$

We consider the topological space

$$X = S^2 \vee (S_1^1 \vee S_2^1)$$

which is the wedge product of a two-dimensional sphere and two circles. The space $X$ is clearly triangulable and therefore, by Theorem (III.4.3), its homology is as follows:

$$H_q(X;\mathbb{Z}) \cong \begin{cases} \mathbb{Z} & q = 0 \\ \mathbb{Z} \oplus \mathbb{Z} & q = 1 \\ \mathbb{Z} & q = 2 \\ 0 & q \neq 0,1,2 . \end{cases}$$

Therefore, $X$ and the torus $T^2$ have the same homology groups. But these spaces are not homeomorphic. To prove that $X$ and $T^2$ are not homeomorphic, we recall Remark (I.1.17). We suppose $f \colon X \to T^2$ to be a homeomorphism and let $x_0$ be the identification point of spheres $S^2$, $S_1^1$, and $S_2^1$; then, $X \smallsetminus \{x_0\}$ and $T^2 \smallsetminus \{f(x_0)\}$ are homeomorphic; but $X \smallsetminus \{x_0\}$ has three connected components while $T^2 \smallsetminus \{f(x_0)\}$ is connected, which leads to a contradiction.

## III.5  Homology of Real Projective Spaces

We have already computed the homology groups of some elementary polyhedra such as the sphere $S^2$ and the torus $T^2$ (see also exercises in Sect. II.4). We have also studied techniques that, in theory, allow us to study the homology of polyhedra (for instance, the Exact Homology Sequence Theorem). In this section, we compute the homology of real projective spaces; we shall see that the methods learned so far are not enough to complete our pre-established task and for this reason we shall provide a new method for computing homology groups.

Those readers who have done the exercises of Sect. II.2 will have found at least one triangulation for $\mathbb{R}P^2$; whatever the case, we consider the triangulation $P$ of $\mathbb{R}P^2$ that has 6 vertices, 15 edges, and 10 faces in Fig. III.3 (the geometric realization

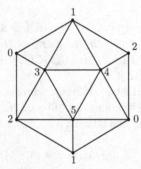

Fig. III.3

of this simplicial complex is homeomorphic to the disk $D_1^2$ with the identification of its boundary antipodal points).

In one of the exercises in Sect. II.4, we have asked the reader to compute the homology groups (with coefficients in $\mathbb{Z}$) of a triangulation of $\mathbb{R}P^2$; the reader should have come to the results that are put together in the following lemma (in any case, this lemma will be proved later on with another method).

**(III.5.1) Lemma.**

$$H_q(\mathbb{R}P^2, \mathbb{Z}) \cong \begin{cases} \mathbb{Z} & , q = 0 \\ \mathbb{Z}_2 & , q = 1 \\ 0 & , q \neq 0, 1 . \end{cases}$$

We now consider the $n$-dimensional real projective spaces $\mathbb{R}P^n$, with $n \geq 3$. As for the two-dimensional case, $\mathbb{R}P^n$ is a quotient space of $S^n$ by identification of antipodal points. Let us rephrase the definition of $\mathbb{R}P^n$ as follows: We define the equivalence relation in $\mathbb{R}^{n+1} \setminus (0, \ldots, 0)$:

$$x = (x_0, \ldots, x_n) \equiv y = (y_0, \ldots, y_n) \iff$$
$$(\exists \lambda \in \mathbb{R}, \lambda \neq 0)(\forall i = 0, \ldots, n) x_i = \lambda y_i .$$

We ask the reader to prove that

$$S^n | \mathbb{Z}_2 = \mathbb{R}P^n \cong (\mathbb{R}^{n+1} \smallsetminus (0, \ldots, 0))/ \equiv \; .$$

In this way, we adopt the definition $\mathbb{R}P^n = (\mathbb{R}^{n+1} \smallsetminus (0, \ldots, 0))/ \equiv$.

We now prove that $\mathbb{R}P^{n+1}$ may be obtained from $\mathbb{R}P^n$ by "adjunction" of an $(n+1)$-dimensional disk $D^{n+1}$; more precisely, $\mathbb{R}P^{n+1}$ is the pushout of the diagram

$$
\begin{array}{ccc}
S^n & \xrightarrow{\;\;q_n\;\;} & \mathbb{R}P^n \\
{\scriptstyle \iota_n}\downarrow & & \\
D^{n+1} & &
\end{array}
$$

where $\iota_n$ is the inclusion of $S^n$ in $D^{n+1}$ and $q_n$ is the map that takes a point $(x_0, \ldots, x_n) \in S^n$ into its equivalence class

$$[(x_0, \ldots, x_n)] \in (\mathbb{R}^{n+1} \smallsetminus (0, \ldots, 0))/ \equiv \; .$$

We start by constructing the pushout

$$
\begin{array}{ccc}
S^n & \xrightarrow{\;\;q_n\;\;} & \mathbb{R}P^n \\
{\scriptstyle \iota_n}\downarrow & & \downarrow {\scriptstyle \bar{\iota}_n} \\
D^{n+1} & \xrightarrow[\;\;\bar{q}_n\;\;]{} & \mathbb{R}P^n \sqcup_{q_n} D^{n+1}
\end{array}
$$

and the following commutative diagram

$$
\begin{array}{ccc}
S^n & \xrightarrow{\;\;q_n\;\;} & \mathbb{R}P^n \\
{\scriptstyle \iota_n}\downarrow & & \downarrow {\scriptstyle j_n} \\
D^{n+1} & \xrightarrow[\;\;g_n\;\;]{} & \mathbb{R}P^{n+1}
\end{array}
$$

where

$$g_n(x_0, \ldots, x_n) = \left[ \left(x_0, \ldots, x_n, \sqrt{1 - \sum_0^n |x_i|^2}\right) \right]$$

and

$$j_n([(x_0, \ldots, x_n)]) = [(x_0, \ldots, x_n, 0)]$$

for every $(x_0, \ldots, x_n) \in S^n$ and $[(x_0, \ldots, x_n)] \in \mathbb{R}P^n$. By the definition of pushout, there is a unique continuous function

$$\ell : \mathbb{R}P^n \sqcup_{q_n} D^{n+1} \to \mathbb{R}P^{n+1}$$

such that

$$\ell\bar{q}_n = g_n \text{ and } \ell\bar{\iota}_n = j_n .$$

We wish to prove that the map $\ell$ is a homeomorphism. We first prove that $\ell$ is bijective; to this end, it is sufficient to prove that the restriction

$$\ell : D^{n+1} \smallsetminus S^n \longrightarrow \mathbb{R}P^{n+1} \smallsetminus \mathbb{R}P^n$$

is bijective. With this in mind, we define the function

$$\tilde{g}_n : \mathbb{R}P^{n+1} \smallsetminus \mathbb{R}P^n \longrightarrow D^{n+1} \smallsetminus S^n$$

such that, for every $[(x_0,\dots,x_{n+1})] \in \mathbb{R}P^{n+1} \smallsetminus \mathbb{R}P^n$,

$$\tilde{g}_n([(x_0,\dots,x_{n+1})]) = \left( \frac{x_0 x_{n+1}}{|x_{n+1}|r}, \dots, \frac{x_n x_{n+1}}{|x_{n+1}|r} \right)$$

where

$$r = \sqrt{\sum_0^{n+1} |x_i|^2} ,$$

and we notice that $g_n\tilde{g}_n = \tilde{g}_n g_n = 1$. The continuity of the inverse function $\ell^{-1}$ derives from the fact that $\ell$ is a bijection from the compact space $\mathbb{R}P^n \sqcup_{q_n} D^{n+1}$ onto the Hausdorff space $\mathbb{R}P^{n+1}$ (see Theorem (I.1.27)).

**(III.5.2) Proposition.**

$$H_q(\mathbb{R}P^3, \mathbb{Z}) \cong \begin{cases} \mathbb{Z} & \text{if } q=0 \\ \mathbb{Z}_2 & \text{if } q=1 \\ \mathbb{Z} & \text{if } q=3 \\ 0 & \text{if } q \neq 0,1,3. \end{cases}$$

*Proof.* The exact homology sequence for the pair $(\mathbb{R}P^3, \mathbb{R}P^2)$ is

$$\cdots \longrightarrow H_n(\mathbb{R}P^2; \mathbb{Z}) \xrightarrow{H_n(i)} H_n(\mathbb{R}P^3; \mathbb{Z}) \xrightarrow{q_*(n)} H_n(\mathbb{R}P^3, \mathbb{R}P^2; \mathbb{Z}) \xrightarrow{\lambda_n}$$

$$\xrightarrow{\lambda_n} H_{n-1}(\mathbb{R}P^2; \mathbb{Z}) \longrightarrow H_{n-1}(\mathbb{R}P^3; \mathbb{Z}) \longrightarrow \cdots$$

(see Theorem (II.4.1)). By Lemma (III.5.1), $H_n(\mathbb{R}P^2; \mathbb{Z})$ is zero if $n \neq 0,1$. By Theorem (II.4.7) and considering that

$$\mathbb{R}P^3/\mathbb{R}P^2 \cong S^3 ,$$

we have that, for every $n \geq 1$, $H_n(\mathbb{R}P^3, \mathbb{R}P^2; \mathbb{Z}) \cong H_n(S^3)$ is zero if $n \neq 3$ and is isomorphic to $\mathbb{Z}$ if $n = 3$. The significant part of the sequence becomes

$$0 \longrightarrow H_3(\mathbb{R}P^3;\mathbb{Z}) \overset{\cong}{\longrightarrow} \mathbb{Z} \longrightarrow 0 \longrightarrow H_2(\mathbb{R}P^3;\mathbb{Z}) \longrightarrow$$

$$\longrightarrow 0 \longrightarrow \mathbb{Z}_2 \underset{\cong}{\longrightarrow} H_1(\mathbb{R}P^3;\mathbb{Z}) \longrightarrow 0,$$

which easily leads to the thesis.                                                                       ∎

At this stage, the reader could be tempted to compute the homology of $\mathbb{R}P^4$ with this method and then apply the Exact Sequence Theorem in Homology to compute the homology of $\mathbb{R}P^n$ by induction; unfortunately, this idea will not work, because the connecting homomorphisms $\lambda_q : H_q(S^n,\mathbb{Z}) \to H_{q-1}(\mathbb{R}P^{n-1},\mathbb{Z})$ are generally not known. Therefore, we must come up with another method for computing the homology of $\mathbb{R}P^n$ when $n \geq 4$; this will be done in the next subsection.

### III.5.1 Block Homology

We begin by reviewing how to compute the homology of the torus $T^2$. To start with, we interpret the torus $T^2$ as deriving from the square with vertices $(0,0)$, $(1,0)$, $(0,1)$, and $(1,1)$ by identifying the two horizontal edges and then the two vertical ones. A possible triangulation $T$ of the torus is represented in Fig. III.4

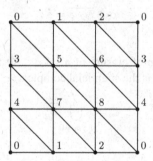

**Fig. III.4**

(see also p. 63). We have 9 vertices, 27 edges, and 18 faces, a total of 54 simplexes in it, too large a number, considering what we started with; in fact, given the identifications made on the Euclidean square, we could take the simplexes in "blocks", consider only one vertex (the four vertices have been identified), two one-dimensional "blocks", namely, one vertical edge and one horizontal edge (without their end-points), and only one two-dimensional "block", namely, the square without its boundary. We then ask ourselves whether it is possible to compute the homology of $T^2$ by means of this block (or "cellular") interpretation of the torus and, in doing so, avoid handling a rather large number of simplexes; because, if the procedure is already a little complicated for the torus, a space that can, after

all, be easily interpreted, imagine what might happen when we try to compute the
homology of a more complex polyhedron. We now explain what a "block" decomposition of a polyhedron is and how the homology of a polyhedron divided into
blocks is computed.

Let $K = (X, \Phi)$ be a given simplicial complex; for every subset $e \subset \Phi$, let $\bar{e}$ be
the smallest subcomplex of $K$ that contains $e$ (we observe that $e$ is not necessarily
a simplicial complex; for instance, take the previous triangulation $T$ of the torus $T^2$
and let $e$ be the set of the simplexes $\{2\}$, $\{0,3,5\}$, and $\{3,4,6\}$; in this example,
$\{0,3\} \subset \{0,3,5\}$ but $\{0,3\} \notin e$). The simplicial complex $\bar{e}$ may be described in
another way:

$$\bar{e} = \bigcup_{\sigma \in e} \bar{\sigma} .$$

The simplicial complex $\bar{e}$ associated with the set $e$ is called *closure* of $e$.

For every $e \subset \Phi$, we define the *boundary* of $e$ as

$$\dot{e} = \overline{\bar{e} \smallsetminus e};$$

this, by definition, is a subcomplex of $K$; furthermore, $\dot{e}$ is a simplicial subcomplex
of $\bar{e}$:

$$\bar{e} \smallsetminus e \subset \bar{e} \Rightarrow \overline{\bar{e} \smallsetminus e} \subset \bar{\bar{e}} = \bar{e} \Rightarrow \dot{e} \subset \bar{e} .$$

Finally, $\bar{e} = e \cup \dot{e}$: in fact, since $e \subset \bar{e}$ and $\dot{e} \subset \bar{e}$, we have the inclusion $e \cup \dot{e} \subset \bar{e}$; on
the other hand, if $\sigma \in \bar{e}$ and $\sigma \notin e$ then $\sigma \in \bar{e} \smallsetminus e$ and thus

$$\sigma \in \overline{\bar{e} \smallsetminus e} = \dot{e} ;$$

we conclude that $e \cup \dot{e} \subset \bar{e}$.

We extend the definition of closure of a set $e \subset \Phi$ to a chain $c_n = \sum_i m_i \sigma_n^i \in$
$C_n(K, \mathbb{Z})$: the *closure* of $c_n$ is the simplicial complex

$$\overline{c_n} = \bigcup_{i \,:\, m_i \neq 0} \overline{\sigma_n^i} .$$

**(III.5.3) Definition.** A *p-block or block of dimension p* of $K$ is a set $e_p \subset \Phi$ such
that:

(i) $e_p$ contains no simplex of dimension $> p$
(ii) $H_p(\overline{e_p}, \dot{e}_p; \mathbb{Z}) \cong \mathbb{Z}$
(iii) $(\forall q \neq p)\ H_q(\overline{e_p}, \dot{e}_p; \mathbb{Z}) \cong 0$

**(III.5.4) Remark.** To compute the homology of the pair of simplicial complexes
$(\overline{e_p}, \dot{e}_p)$, it is necessary to give an orientation to $K$. However, we also could give an
orientation to $\overline{e_p}$ individually by choosing a generator $\beta_p$ of the Abelian group

$$H_p(\overline{e_p}, \dot{e}_p; \mathbb{Z}) \cong Z_p(\overline{e_p}, \dot{e}_p) \cong \mathbb{Z} .$$

Note that a generator $\beta_p$ may be interpreted as a linear combination of $p$-simplexes of $K$ or of $\bar{e}_p$; in any case, $\partial(\beta_p)$ is a chain of $\dot{e}_p$ and therefore

$$\overline{\partial(\beta_p)} \subset \dot{e}_p .$$

**(III.5.5) Definition.** A *block triangulation*[5] of a simplicial complex $K = (X, \Phi)$ (or of a polyhedron $|K|$) is a set $e(K) = \{e_p^i\}$ of blocks with the following conditions:

(1) For every $\sigma \in \Phi$, there is a *unique* block $e(\sigma)$ of the set $e(K)$ such that $\sigma \in e(\sigma)$.
(2) For every $p$-block $e_p^i \in B(K)$, $\dot{e}_p^i$ is a union of blocks with dimension $< p$.

It follows from the preceding definition that

1. For every $i, j, p$, $\dot{e}_p^i \cap e_p^j = \emptyset$
2. If $i \neq j$, then $\overline{e_p^i} \cap e_p^j = \emptyset$

As an example, we give a block triangulation of the torus $T^2$ with the triangulation $T$ previously described. We consider the following sets of simplexes of $T$:

1. $e_0 = \{0\}$
2. $e_1^1 = \{\{3\}, \{4\}, \{0,3\}, \{3,4\}, \{4,0\}\}$
3. $e_1^2 = \{\{1\}, \{2\}, \{0,1\}, \{1,2\}, \{2,0\}\}$
4. $e_2 = T \setminus (e_0 \cup e_1^1 \cup e_1^2)$

We determine that the set

$$e(T^2) = \{e_0, e_1^1, e_1^2, e_2\}$$

is a block triangulation for $T^2$. We notice at once that

$$\overline{e_0} = \{0\} \ , \ \dot{e}_0 = \emptyset \ ,$$
$$\overline{e_1^1} = \overline{\{0,3\}} \cup \overline{\{3,4\}} \cup \overline{\{4,0\}} \ , \ \dot{e}_1^1 = \{0\} \ ,$$
$$\overline{e_1^2} = \overline{\{0,1\}} \cup \overline{\{1,2\}} \cup \overline{\{2,0\}} \ , \ \dot{e}_1^2 = \{0\} \ ,$$
$$\overline{e_2} = T \ , \ \dot{e}_2 = \overline{e_1^1} \cup \overline{e_1^2} \ ;$$

we now recall Theorem (III.4.1) and observe that

$$|\overline{e_1^i}| / |\dot{e}_1^i| \cong S^1 \ , \ i = 1, 2 \ ,$$
$$|\overline{e_2}| / |\dot{e}_2| \cong S^2 \ ;$$

this shows that the elements of $e(T^2)$ are blocks. It is easily proved that they form a block triangulation.

---

[5] In spite of the name, this is not a triangulation. In fact, it is the analog of a cellular decomposition of $|K|$. The block homology can be seen as the cellular homology of such a cellular complex (see e.g. [7, Chap. V]).

For the real projective plane, with the previously described triangulation we have the block triangulation given by:

1. $e_0 = \{0\}$
2. $e_1 = \{\{1\},\{2\},\{0,1\},\{1,2\},\{2,0\}\}$
3. $e_2 = P \smallsetminus (e_0 \cup e_1)$

Let $|K|$ be a polyhedron with a block triangulation $e(K) = \{e_p^i\}$; for every integer $q$ such that $0 \leq q \leq \dim K$, we define

$$e^{(q)} = \{e_p^i \in e(K) | p \leq q\} \, .$$

**(III.5.6) Proposition.** *Let $A$ be any union of $q$-blocks. Then, $e^{(q)} \smallsetminus A$ is a simplicial subcomplex of $K$.*

*Proof.* Let $\sigma$ be any simplex of $e^{(q)} \smallsetminus A$; then $\sigma$ is in a certain block $e_p^i$ of $e^{(q)} \smallsetminus A$, with $p \leq q$. Suppose $\sigma' \subset \sigma$; then either $e(\sigma') = e_p^i$ or $\sigma' \in \overset{\bullet}{e}_p^i$; in the latter case, $e(\sigma')$ is an $r$-block with $r < p$; in any case, $\sigma' \in e^{(q)} \smallsetminus A$. ∎

At this point, we can define the "block homology" of $|K|$ with block triangulation $e(K)$: in a nutshell, it comes from the chain complex $C(e(K)) = \{C_n(e(K))\}$ defined as

$$C_n(e(K)) = H_n(e^{(n)}, e^{(n-1)}; \mathbb{Z}) \cong \bigoplus_{j=1}^{\ell_n} H_n(\overline{e}_n^j, \overset{\bullet}{e}_n^j; \mathbb{Z})$$

where $\{e_n^1, \ldots, e_n^{\ell_n}\}$ is the set of $n$-blocks. The boundary operator $\partial_n^{e(K)}$ will be given by suitable sum of compositions

$$H_n(\overline{e}_n, \overset{\bullet}{e}_n; \mathbb{Z}) \xrightarrow{\lambda_n} H_{n-1}(\overset{\bullet}{e}_n; \mathbb{Z}) \xrightarrow{H_{n-1}(i)} H_{n-1}(\overline{e_{n-1}}; \mathbb{Z}) \xrightarrow{q_*(n-1)} H_{n-1}(\overline{e_{n-1}}, \overset{\bullet}{e}_{n-1}; \mathbb{Z})$$

where $\lambda_n, q_*(n-1)$ are the appropriate homomorphisms of the long exact sequences for $(\overline{e}_n, \overset{\bullet}{e}_n)$, $(\overline{e_{n-1}}, \overset{\bullet}{e}_{n-1})$, respectively, and $H_{n-1}(i)$ is the homomorphism arising from the inclusion $\overset{\bullet}{e}_n \subset \overline{e_{n-1}}$. The problem is also to relate this homology to $H_*(|K|; \mathbb{Z})$. For this, we proceed along the lines of [17, 3.8], and we will define $\partial_*^{e(K)}$ in the process.

We begin with the following key result:

**(III.5.7) Lemma.** *Let $c_p$ be a $p$-chain of $K$ such that $\overline{\partial_p(c_p)} \subset e^{(p-1)}$; then there exists a chain of $p$-blocks $\sum_i m_i \beta_p^i$ homologous to $c_p$.*

*Proof.* Suppose that $\overline{c_p} \subset e^{(q)}$, with $q > p$. Let us first prove the existence of a $p$-chain $c_p' \in C_p(K)$ homologous to $c_p$, where $\overline{c_p'} \subset e^{(p)}$ (this means that we may remove $c_p$ from the blocks with dimension strictly larger than $p$). To this end, it is enough to prove the existence of a $p$-chain $f_p \in C_p(K)$ homologous to $c_p$, with $\overline{f_p} \subset e^{(q-1)}$: in fact, since $c_p$ and $f_p$ are homologous, $\partial_p(c_p) = \partial_p(f_p)$; we may therefore conclude that $\overline{\partial_p(f_p)} \subset e^{(p-1)}$ from the hypothesis $\overline{\partial_p(c_p)} \subset e^{(p-1)}$; if $q - 1 > p$, we apply the preceding argument on $f_p$, and so forth.

Let $e_q^i$ be any $q$-block (recall that $q > p$). We begin by removing $c_p$ from $e_q^i$: we write

$$c_p = c_p^i + f_p^i \quad \text{where } \overline{c_p^i} \subset \overline{e_q^i} \,, \overline{f_p^i} \subset e^{(q)} \smallsetminus e_q^i \,.$$

Hence, $\overline{\partial_p(c_p^i)} \subset \overline{e_q^i}$; in addition,

$$\overline{\partial_p(c_p^i)} \subset \overline{\partial_p(c_p)} \cup \overline{\partial_p(f_p^i)} \subset e^{(p-1)} \cup (e^{(q)} \smallsetminus e_q^i) = e^{(q)} \smallsetminus e_q^i$$

which leads to

$$\overline{\partial_p(c_p^i)} \subset \overline{e_q^i} \cap (e^{(q)} \smallsetminus e_q^i) = \mathring{e}_q^i \,.$$

Therefore, $c_p^i \in Z_p(\overline{e_q^i}, \mathring{e}_p^i)$; since $p \neq q$, we have $H_p(\overline{e_q^i}, \mathring{e}_p^i) = 0$ and so

$$c_p^i = \partial_{p+1}(f_{p+1}^i) + g_p^i \,, \quad \text{where } f_{p+1}^i \in C_{p+1}(\overline{e_q^i}) \text{ and } g_p^i \in C_p(\mathring{e}_q^i) \,.$$

Hence, $c_p$ is homologous to

$$c_p - \partial_{p+1}(f_{p+1}^i) = c_p^i + f_p^i - (c_p^i - g_p^i) = f_p^i + g_p^i \,;$$

note that the closure of the $p$-chain $f_p^i + g_p^i$ is contained in

$$\left( e^{(q)} \smallsetminus e_p^i \right) \cup \mathring{e}_q^i = e^{(q)} \smallsetminus e_q^i \,;$$

since $f_p^i + g_p^i$ is homologous to $c_p$, we may say that we have removed $c_p$ from the block $e_p^i$.

Let us now remember that $i \neq j \implies \overline{e_q^i} \cap e_q^j = \emptyset$; then, because $\overline{\partial_{p+1}(f_{p+1}^i)} \subset \overline{e_q^i}$, the coefficient of each $p$-simplex of $\partial_{p+1}(f_{p+1}^i)$ in $e_q^j$ is zero. Hence for every $i$, we may find a $(p+1)$-chain $f_{p+1}^i$ such that

$$\overline{c_p - \sum_i \partial_{p+1}(f_{p+1}^i)} \subset e^{(q-1)} \,.$$

We define the $p$-chain $f_p = c_p - \sum_i \partial_{p+1}(f_{p+1}^i)$; from the observations made at the beginning of the proof, we conclude that there is a $p$-chain $c_p'$ of $K$ homologous to $c_p$ and such that $\overline{\partial_p(c_p')} \subset e^{(p)}$.

Now the proof proceeds as before: let $e_p^i$ be an arbitrary $p$-block; we write

$$c_p' = k_p^i + h_p^i \quad \text{where } \overline{k_p^i} \subset \overline{e_p^i} \,, \overline{h_p^i} \subset e^{(p)} \smallsetminus e_p^i \text{ and } \overline{h_p^i} \subset e^{(p)} \smallsetminus e_p^i \,.$$

As before, $k_p^i \in Z_p(\overline{e_p^i}, \mathring{e}_p^i)$. However, $Z_p(\overline{e_p^i}, \mathring{e}_p^i) \cong \mathbb{Z}$; let $\beta_p^i$ be one of its generators; then, $k_p^i = m_i \beta_p^i$ for a certain nonzero integer $m_i$. Repeating this procedure for each $i$ we obtain a chain of $p$-blocks with

$$\overline{c_p' - \sum_i m_i \beta_p^i} \subset e^{(p-1)} \,;$$

since $e^{(p-1)}$ contains no $p$-simplex, we conclude that

$$c'_p = \sum_i m_i \beta_p^i$$

and so $c_p$ is homologous to $\sum_i m_i \beta_p^i$.  ▪

This lemma has an important consequence, essential to the definition of block homology:

**(III.5.8) Theorem.** *The following results hold for any polyhedron $|K|$ with a block triangulation:*

*(1) Every $p$-cycle $z_p \in Z_p(K)$ is homologous to a cycle of $p$-blocks $\sum_i m_i \beta_p^i$.*

*(2) If $\sum_i m_i \beta_p^i$ is a boundary, there exists a chain of $(p+1)$-blocks $\sum_j n_j \beta_{p+1}^j$ such that*

$$\partial_{p+1}\left(\sum_j n_j \beta_{p+1}^j\right) = \sum_i m_i \beta_p^i ;$$

*(3) $\partial_p(\beta_p^i)$ is a chain of $(p-1)$-blocks like $\sum_j m_j^i \beta_{p-1}^j$.*

*Proof.* (1) Let $z_p$ be a $p$-cycle of $K$; then $\partial_p(z_p) = 0$ and so $\overline{\partial_p(z_p)} \subset e^{(p-1)}$; we conclude from Lemma (III.5.7) that there is a chain of $p$- blocks $c'_p = \sum_i m_i \beta_p^i$, which is homologous to $z_p$; then, $c'_p$ is a cycle because $\partial_p(c'_p) = \partial_p(z_p) = 0$.

(2) Let us suppose that $\sum_i m_i \beta_p^i = \partial_{p+1}(c_{p+1})$, where $c_{p+1} \in C_{p+1}(K)$; since $\overline{\partial_{p+1}(c_{p+1})} \subset e^{(p)}$, the $(p+1)$-chain $c_{p+1}$ is homologous to a chain of $(p+1)$-blocks $\sum_j n^j \beta_{p+1}^j$; hence

$$\sum_i m_i \beta_p^i = \partial_{p+1}(c_{p+1}) = \partial_{p+1}\left(\sum_j n^j \beta_{p+1}^j\right).$$

(3) By definition, $\beta_p^i$ is a generator of $H_p(\overline{e_p^i}, \mathring{e}_p^i) \cong \mathbb{Z}$; hence

$$\overline{\partial_p(\beta_p^i)} \subset \mathring{e}_p^i \subset e^{(p-1)} ;$$

we now proceed as we did for the second part of Lemma (III.5.7); we substitute $\partial_p(\beta_p^i)$ for $c'_p$ and $p-1$ for $p$.  ▪

Let $|K|$ be a polyhedron with a block triangulation $e(K)$. We now construct the chain complex

$$C(e(K)) = \left\{ \left(C_n(e(K)), \partial_n^{e(K)}\right) \mid n \geq 0 \right\}$$

as follows: for each $n \geq 0$, $C_n(e(K))$ is the free Abelian group generated by $\beta_n^i$; the boundary operators $\partial_n^{e(K)}$ are defined on generators according to part (3) of Theorem (III.5.8):

$$\partial_p^{e(K)}(\beta_p^i) = \partial_p(\beta_p^i) = \sum_j m_j^i \beta_{p-1}^j ;$$

hence, $\partial_p^{e(K)} \partial_{p+1}^{e(K)} = 0$. As for the simplicial homology, we define

$$Z_n(e(K);\mathbb{Z}) = \ker \partial_n^{e(K)} \ , \ B_n(e(K);\mathbb{Z}) = \operatorname{im} \partial_{n+1}^{e(K))} \ ,$$

$$H_n(e(K),\mathbb{Z}) := Z_n(e(K);\mathbb{Z})/B_n(e(K);\mathbb{Z})$$

if $n \geq 0$, and $H_n(e(K),\mathbb{Z}) = 0$ when $n < 0$.

**(III.5.9) Theorem.** *Let $|K|$ be a polyhedron with a block triangulation $e(K)$. Then for every $n \in \mathbb{Z}$, there is a group isomorphism*

$$H_n(e(K),\mathbb{Z}) \cong H_n(|K|,\mathbb{Z}) \ .$$

*Proof.* Every chain of $n$-blocks may be interpreted as a normal simplicial $n$-chain of $K$; hence, there exists a homomorphism

$$i_n : C_n(e(K)) \to C_n(K)$$

which is easily seen to commute with the boundary operators and induces therefore a homomorphism $i_n : H_n(e(K),\mathbb{Z}) \to H_n(|K|,\mathbb{Z})$ for every $n$. By Theorem (III.5.8), parts (1) and (2), $i_n$ is bijective. ∎

We now compute the homology of $T^2$ and $\mathbb{R}P^2$ through their block triangulations that we have mentioned before. Since both polyhedra are connected,

$$H_0(e(T^2),\mathbb{Z}) \cong H_0(T^2,\mathbb{Z}) \cong \mathbb{Z}$$
$$H_0(e(\mathbb{R}P^2),\mathbb{Z}) \cong H_0(\mathbb{R}P^2,\mathbb{Z}) \cong \mathbb{Z}$$

(see Lemma (II.4.5)). For the torus $T^2$, we have the following results: (a) $C_1(e(T^2))$ has two generators, $\beta_1^1$ and $\beta_1^2$, corresponding to the 1-blocks $e_1^1$ and $e_1^2$, respectively; these generators are the only 1-cycles of the block triangulation $e(T^2)$ and so

$$H_1(e(T^2),\mathbb{Z}) \cong H_1(T^2,\mathbb{Z}) \cong \mathbb{Z} \oplus \mathbb{Z} \ ;$$

(b) $C_2(e(T^2))$ has only one generator $\beta_2$ corresponding to $e_2$; since $\beta_2$ is a cycle,

$$H_2(e(T^2),\mathbb{Z}) \cong H_2(T^2,\mathbb{Z}) \cong \mathbb{Z} \ .$$

For $\mathbb{R}P^2$, we have that $C_1(e(\mathbb{R}P^2))$ has only one generator $\beta_1$ corresponding to $e_1$; but $2\beta_1$ is a boundary (of $\overline{e_2}$) and therefore

$$H_1(e(\mathbb{R}P^2),\mathbb{Z}) \cong H_1(\mathbb{R}P^2,\mathbb{Z}) \cong \mathbb{Z}_2 \ ;$$

on the other hand, $C_2(e(\mathbb{R}P^2))$ has a generator (corresponding to $e_2$) but has no cycles and so

$$H_2(e(\mathbb{R}P^2),\mathbb{Z}) \cong H_2(\mathbb{R}P^2,\mathbb{Z}) \cong 0$$

(this is the proof of Lemma (III.5.1)).

## III.5.2  Homology of $\mathbb{R}P^n$, with $n \geq 4$

We first wish to find a convenient triangulation for $\mathbb{R}P^n$. We remind the reader that $\mathbb{R}P^n$ is also interpreted as the pushout space of the diagram

$$
\begin{array}{ccc}
S^{n-1} & \xrightarrow{\;q_{n-1}\;} & \mathbb{R}P^{n-1} \\[2pt]
\Big\downarrow{\scriptstyle \iota_{n-1}} & & \\[6pt]
D^n & &
\end{array}
$$

where $q_{n-1}$ is the map that identifies antipodal points of $S^{n-1}$ and $\iota_{n-1}$ is the inclusion. In fact, in this way we have a sequence of real projective spaces

$$ \mathbb{R}P^1 \subset \ldots \subset \mathbb{R}P^{n-1} \subset \mathbb{R}P^n \, . $$

Let $S^n_+$ be the northern hemisphere of $S^n$, that is to say, the set of all points $(x_1,\ldots,x_{n+1}) \in \mathbb{R}^{n+1}$, where $r = \sqrt{\sum_{i=1}^{n+1} x_i^2} = 1$ and $x_{n+1} \geq 0$. We define the function $f : D^n \to S^n_+$ as follows:

$$
f(x_1,\ldots,x_n) = \begin{cases} (\frac{x_1}{r}\sin\frac{\pi r}{2},\ldots,\frac{x_n}{r}\sin\frac{\pi r}{2},\cos\frac{\pi r}{2}) & \text{if } r = \sqrt{\sum_{i=1}^n x_i^2} \neq 0 \\ (0,0,\ldots,1) & \text{if } r = 0 \, . \end{cases}
$$

This function is a homeomorphism whose restriction to the boundary of $S^n_+$ is the identity. With this in mind, we may say that $\mathbb{R}P^n$ is obtained from $S^n_+$ by identifying the antipodal points of $\partial S^n_+ \cong S^{n-1}$.

Let $K^{n-1}$ be the standard triangulation of $S^{n-1}$ (see p. 55). We define a triangulation $K^n_+$ of $S^n_+$ as the join of $K^{n-1}$ and $\{a_{n+1}\}$, where $a_{n+1} = (0,\ldots,0,1) \in \mathbb{R}^{n+1}$. We are tempted to define a triangulation for $\mathbb{R}P^n$ by identifying the antipodal vertices of $S^{n-1}$; it does not work, as we could have different simplexes defined by the same vertices in $\mathbb{R}P^n$. The trick is to work with barycentric subdivisions. Hence, we define the triangulation $M^n$ of $\mathbb{R}P^n$, identifying the antipodal vertices of $(K^{n-1})^{(1)}$ in $(K^n_+)^{(1)}$. The set $e_n$ of $n$-simplexes of $M^n$ is an $n$-block of $M^n$: in fact, the simplicial function $((K^n_+)^{(1)},(K^{n-1})^{(1)}) \to (e_n,\overset{\bullet}{e}_n)$ is injective into $(K^n_+)^{(1)} \smallsetminus (K^{n-1})^{(1)}$ and induces therefore an isomorphism among the chain complexes $C((K^n_+)^{(1)},(K^{n-1})^{(1)})$, and $C(e_n,\overset{\bullet}{e}_n)$; on the other hand, the relative version of the chain complexes homomorphism

$$ \aleph : C((K^n_+),(K^{n-1});\mathbb{Z}) \to C((K^n_+)^{(1)},(K^{n-1})^{(1)};\mathbb{Z}) \, , $$

defined in Theorem (III.2.2), induces an isomorphism between the relative homology of pairs $((K^n_+)^{(1)},(K^{n-1})^{(1)}))$ and $(e_n,\overset{\bullet}{e}_n)$; hence,

$$
H_r(e_n,\overset{\bullet}{e}_n;\mathbb{Z}) \cong \begin{cases} \mathbb{Z} & \text{if } r = n \\ 0 & \text{if } r \neq n. \end{cases}
$$

We find a triangulation $M^{n-1}$ of $\mathbb{R}P^{n-1}$ in a similar way and so on. Note that $\mathring{e}_n$ coincides with the set of $(n-1)$-simplexes $e_{n-1}$ of $M^{n-1}$, which in turn is an $(n-1)$-block, etc. In this way, we obtain a block triangulation of $\mathbb{R}P^n$.

All that is left to do is to look into the generators of $H_r(e_r, \mathring{e}_r; \mathbb{Z})$ and their boundaries. We denote a generator of $H_{n-1}(S^{n-1}, \mathbb{Z})$ with $z_{n-1} \in Z_{n-1}(S^{n-1}; \mathbb{Z})$.[6] The exact sequence of the homology groups of $(K_+^n, K^{n-1})$ shows that $z_{n-1} * a_{n+1}$ is a generating cycle of $H_n(K_+^n, K^{n-1}; \mathbb{Z})$. It follows that $\aleph(z_{n-1} * a_{n+1})$ is a generator for $Z_n((K_+^n)^{(1)}, (K^{n-1})^{(1)}; \mathbb{Z})$ and if $q : |K_+^n| \to |M^n|$ is the quotient map, then $q\aleph(z_{n-1} * a_{n+1})$ is a generator of $Z_n(|M^n|, |M^{n-1}|; \mathbb{Z})$. Now,

$$\partial q \aleph(z_{n-1} * a_{n+1}) = q\aleph \partial(z_{n-1} * a_{n+1}) =$$

$$(-1)^n q\aleph(z_{n-1}) = (-1)^n q\aleph(z_{n-2} * a_n - z_{n-2} * a_n') .$$

We interchange in $z_{n-2}$ all points $a_r$ with $a_r'$; this gives rise to an element which we denote with $z_{n-2}'$; indeed, $z_{n-2} = (-1)^{n-1} z_{n-2}'$. Hence,

$$\partial q \aleph(z_{n-1} * a_{n+1}) = (-1)^n q\aleph(z_{n-2} * a_n - (-1)^{n-1} z_{n-2}' * a_n')$$
$$= (-1)^n q\aleph(z_{n-2} * a_n) - (-1)^n (-1)^{n-1} q\aleph(z_{n-2}' * a_n') .$$

Finally, since $q\aleph(z_{n-2} * a_{n+1}) = q\aleph(z_{n-2}' * a_n')$, we conclude that

$$\partial q \aleph(z_{n-1} * a_{n+1}) = (1 + (-1)^n) q\aleph(z_{n-2} * a_n) .$$

To simplify the notation, we write $\beta_r = q\aleph(z_{r-1} * a_{r+1})$; this is a generator of $C_r(e(M^n); \mathbb{Z})$ and has the property

$$\partial(\beta_r) = (1 + (-1)^r)\beta_{r-1} .$$

Specifically for $0 < 2r \le n$, the group $C_{2r}(e(M^n); \mathbb{Z})$ is generated by the elements $\beta_{2r}$ and $\partial(\beta_{2r}) = 2\beta_{2r-1}$; therefore, $Z_{2r}(e(M^n); \mathbb{Z}) = 0$ and $H_{2r}(\mathbb{R}P^n; \mathbb{Z}) = 0$. If $2r-1 < n$, then the group $C_{2r-1}(e(M^n); \mathbb{Z})$ is generated by $\beta_{2r-1}$ and $\partial(\beta_{2r-1}) = 0$; hence, the group $Z_{2r-1}(e(M^n); \mathbb{Z}) \cong \mathbb{Z}$ is generated by $\beta_{2r-1}$, the group $B_{2r-1}(e(M^n); \mathbb{Z})$ is generated by $2\beta_{2r-1}$, and there is an isomorphism $H_{2r-1}(\mathbb{R}P^n; \mathbb{Z}) \cong \mathbb{Z}_2$. We have proved the following result:

**(III.5.10) Theorem.**

$$H_p(\mathbb{R}P^n, \mathbb{Z}) \cong \begin{cases} \mathbb{Z} & \text{if } p = 0 \\ 0 & \text{if } 0 < p = 2q \le n \\ \mathbb{Z}_2 & \text{if } 0 < p = 2q - 1 < n \\ \mathbb{Z} & \text{if } n \text{ is odd and } p = n . \end{cases}$$

---

[6] We start with $S^0$ and select a generator $z^0 = \{a_1\} - \{a_1'\}$; next, for $S^1$, we define $z_1$ as the join of $z_0$ and $\{a_1, a_1'\}$, etc.

## III.6  Homology of the Product of Two Polyhedra

In this section, we study the homology of the product of two polyhedra. The main result is given by the Acyclic Models Theorem.

### III.6.1  Acyclic Models Theorem

This theorem is due to Samuel Eilenberg and Saunders MacLane (see [12]). The reader who wishes to go considerably deeply in this subject is advised to consult the book [4] by Michael Barr.

In former sections, we have noticed that we can associate a chain complex $C(K)$ with an augmentation

$$\varepsilon\colon C_0(K) \to \mathbb{Z} \;,\quad \Sigma_{i=1}^n a_i\{x_i\} \mapsto \Sigma_{i=1}^n a_i$$

to every oriented simplicial complex $K$. We have also proved the Acyclic Carrier Theorem (II.3.9) which makes it possible to compare two augmented chain complexes under certain conditions (broadly speaking, these conditions require that some of the local homology groups be trivial). Unfortunately, the Acyclic Carrier Theorem is not powerful enough for studying the homology of a product of two polyhedra; to this end, we need the Acyclic Models Theorem.

We have defined the category $\mathfrak{C}$ of chain complexes in Sect. II.3; in this section, we work with a subcategory of $\mathfrak{C}$, namely, the category **Clp** whose objects are chain complexes of free Abelian groups with an augmentation homomorphism; as we have done in $\mathfrak{C}$, we indicate the objects of **Clp** with sequences of free Abelian groups

$$\cdots \longrightarrow C_n \xrightarrow{\;\partial_n\;} C_{n-1} \xrightarrow{\;\partial_{n-1}\;} \cdots \xrightarrow{\;\partial_1\;} C_0 \xrightarrow{\;\varepsilon\;} \mathbb{Z}$$

and a morphism between objects $C$ and $C'$ with a commutative diagram

$$
\begin{array}{ccccccccc}
\cdots \longrightarrow & C_n & \xrightarrow{\;\partial_n\;} & C_{n-1} & \xrightarrow{\;\partial_{n-1}\;} & \cdots \xrightarrow{\;\partial_1\;} & C_0 & \xrightarrow{\;\varepsilon\;} & \mathbb{Z} \\
& \downarrow{\scriptstyle f_n} & & \downarrow{\scriptstyle f_{n-1}} & & & \downarrow{\scriptstyle f_0} & & \downarrow{\scriptstyle \bar f} \\
\cdots \longrightarrow & C'_n & \xrightarrow{\;\partial'_n\;} & C'_{n-1} & \xrightarrow{\;\partial_{n-1}\;} & \cdots \xrightarrow{\;\partial_1\;} & C'_0 & \xrightarrow{\;\varepsilon'\;} & \mathbb{Z}
\end{array}
$$

Let a category $\mathfrak{C}$ and a (covariant) functor

$$F\colon \mathfrak{C} \longrightarrow \mathbf{Clp}$$

be such that:

1. For every $n \geq 0$, there is a set $\mathfrak{M}_n$ of objects $M_n^i \in \mathfrak{C}$, $i \in J_n$ ($J_n$ is a set of indexes).
2. For each $M_n^i$ there is an element $x_n^i \in F(M_n^i)_n$ such that, for every $X \in \mathfrak{C}$, the set

$$\{F(f)_n(x_n^i) | f \in \mathfrak{C}(M_n^i, X), i \in J_n\}$$

is a basis for the free Abelian group $F(X)_n$ in the chain $F(X)$.

Such a functor (if any exists!) is a *free* functor with *models* $\mathfrak{M} = \{\mathfrak{M}_n | n \geq 0\}$ and *universal elements* $\{x_n^i | i \in J_n, n \geq 0\}$.

We show the existence of free functors with models and universal elements by means of an example.

**Example.** Starting from the category **Csim** of simplicial complexes, we consider the functor

$$C \colon \textbf{Csim} \longrightarrow \textbf{Clp}$$

that takes each simplicial complex $K$ to the augmented chain complex

$$C(K) = \{C_n(K; \mathbb{Z}), \partial_n\}.$$

For every $n \geq 0$, we choose an $n$-simplex $\sigma_n$ (fixed) and the corresponding oriented simplicial complex $\overline{\sigma_n}$; we then set

$$\mathfrak{M}_n = \{\overline{\sigma_n}\}$$

(the set $J_n$ has therefore only one element); these are the sets of models. We now look for the universal elements. For each $n \geq 0$, let $x_n \in C_n(\overline{\sigma_n}, \mathbb{Z})$ be the generator represented by the oriented simplex $\sigma_n$ (the only $n$-simplex of the simplicial complex $\overline{\sigma_n}$). For every model $\overline{\sigma_n}$, the only simplicial functions $f \colon \overline{\sigma_n} \to K$, which produce nontrivial homomorphisms are precisely the bijective simplicial functions that take $\sigma_n$ into the various oriented $n$-simplexes of $K$; hence, the set $\{C(f)(x_n)\}$ contains all oriented $n$-simplexes of $K$ and is therefore a basis of $C_n(K; \mathbb{Z})$.

Before we go over other examples, let us define the tensor product of two chain complexes. Given $C, C' \in \mathfrak{C}$ arbitrarily, we define the Abelian group

$$(C \otimes C')_n = \bigoplus_{i+j=n} C_i \otimes C'_j,$$

for every $n \geq 0$, and the homomorphism

$$d_n^{\otimes} \colon (C \otimes C')_n \longrightarrow (C \otimes C')_{n-1}$$

for every $n \geq 1$ such that, for every $x_i \in C_i$ and $y_j \in C'_j$,

$$d_{i+j}^{\otimes}(x_i \otimes y_j) = \partial_i(x) \otimes y_j + (-1)^i x_i \otimes \partial'_j(y_j).$$

Note that $d^{\otimes} = \{d_n^{\otimes} | n \geq 1\}$ is really a differential:

$$d_{i+j-1}^{\otimes} d_{i+j}^{\otimes}(x_i \otimes y_j) = \partial_{i-1}\partial_i(x) \otimes y_j + (-1)^{i-1}\partial_i(x_i) \otimes \partial_j'(y_j) +$$
$$(-1)^i \partial_i(x_i) \otimes \partial_j'(y_j) + x_i \otimes \partial_{j-1}'\partial_j'(y_j) = 0.$$

It is sometimes convenient to leave the chain morphisms indexes out, to simplify the definition and, therefore, also the proofs; for instance, we define the boundary morphism

$$d^{\otimes} : C \otimes C' \to C \otimes C'$$

on generators $x \otimes y \in C \otimes C'$ by the formula

$$d^{\otimes}(x \otimes y) = d(x) \otimes y + (-1)^{|x|} x \otimes d'(y)$$

(remember that $|x|$ indicates the degree of $x$, that is to say, $|x| = n \iff x \in C_n$ ). As an exercise, we leave to the reader the task of proving that the tensor product of two free chain complexes with an augmentation homomorphism is a free chain complex with augmentation. The reader is advised to do the exercises on tensor products of chain complexes, at the end of this section. Let $K$ and $L$ be two simplicial complexes. In Sect. III.1, we have proved that the product of two polyhedra is a polyhedron. Actually, we have proved that, given the polyhedra $|K|$ and $|L|$, there exists a simplicial complex $K \times L$ such that $|K \times L| \cong |K| \times |L|$ (see Theorem (III.1.1)); the reader is also advised to review the construction of the simplicial complex $K \times L$ in the proof of Theorem (III.1.1).

Therefore, given two simplicial complexes $K$ and $L$, we can define two new functors from the product category $\mathbf{Csim} \times \mathbf{Csim}$ (see examples of categories in Sect. I.2) to the category $\mathbf{Clp}$:

$$C_{\times} : \mathbf{Csim} \times \mathbf{Csim} \to \mathbf{Clp} , \ (K, L) \mapsto C(K \times L) ,$$
$$C \otimes C : \mathbf{Csim} \times \mathbf{Csim} \to \mathbf{Clp} , \ (K, L) \mapsto C(K) \otimes C(L) .$$

The functors $C_{\times}$ and $C \otimes C$ are examples of free functors with models and universal elements. In fact, as we did for the functor $C : \mathbf{Csim} \to \mathbf{Clp}$, we fixate an $n$-simplex $\sigma_n$ for each $n \geq 0$ and we take as models the pairs $(\overline{\sigma}_j, \overline{\sigma}_{n-j})$ for $C_{\times}$, and $(\overline{\sigma}_n, \overline{\sigma}_n)$ for $C \otimes C$. The universal elements are easily described.

We note that on each model $\overline{\sigma}_n$ the functor $C : \mathbf{Csim} \to \mathbf{Clp}$ produces a positive, acyclic free chain complex, that is to say,

$$H_i(\overline{\sigma}_n, \mathbb{Z}) \cong \begin{cases} \mathbb{Z} & \text{if } i = 0 \\ 0 & \text{if } i > 0 \end{cases}$$

(see the definition of acyclic chain complexes in Sect. II.3 and how to compute the homology of $\overline{\sigma}_n$ in Sect. II.4). For this reason, we say that the models $\overline{\sigma}_n$ are *acyclic* for the functor $C$.

**(III.6.1) Lemma.** *The models* $(\overline{\sigma}_j, \overline{\sigma}_{n-j})$ *are acyclic for the functor* $C_{\times}$ *and the models* $(\overline{\sigma}_n, \overline{\sigma}_n)$ *are acyclic for* $C \otimes C$.

*Proof.* In regard to the first part of the statement, we need only to note that

$$H_i(\overline{\sigma_j} \times \overline{\sigma_{n-j}}, \mathbb{Z}) \cong H_i(|\overline{\sigma_j}| \times |\overline{\sigma_{n-j}}|, \mathbb{Z}) \cong H_i(|\overline{\sigma^n}|, \mathbb{Z})$$

$$\cong \begin{cases} \mathbb{Z} & \text{if } i = 0 \\ 0 & \text{if } i > 0. \end{cases}$$

For the proof of the second part, we recall Lemma (II.3.8), which tells us that

**I.** There exists a function

$$\eta \colon \mathbb{Z} \to C_0(\overline{\sigma_n})$$

such that $\varepsilon \eta = 1$,

**II.** There exists a homotopy

$$s \colon C(\overline{\sigma_n}) \to C(\overline{\sigma_n})$$

such that

1. $\partial_1 s_0 = 1 - \eta \varepsilon$
2. $\partial_{n+1} s_n + s_{n-1} \partial_n = 1$ for every $n \geq 1$

. We define the morphism

$$S \colon C(\overline{\sigma_n}) \otimes C(\overline{\sigma_n}) \to C(\overline{\sigma_n}) \otimes C(\overline{\sigma_n})$$

on generators by the formula

$$S(x \otimes y) = s(x) \otimes y + \eta \varepsilon(x) \otimes s(y)$$

provided that $\eta \varepsilon(x) = 0$ if $|x| \neq 0$. Since $\varepsilon \partial_1 = 0$, we conclude that

$$d^\otimes S(x \otimes y) = d^\otimes [s(x) \otimes y) + \eta \varepsilon(x) \otimes s(y)]$$
$$= d(s(x)) \otimes y + (-1)^{|x|+1} s(x) \otimes d(y) + \eta \varepsilon(x) \otimes d(s(y)).$$

On the other hand,

$$S(d^\otimes(x \otimes y)) = S(d(x) \otimes y + (-1)^{|x|} x \otimes d(y))$$
$$= s(d(x)) \otimes y + (-1)^{|x|} s(x) \otimes d(y) + \eta \varepsilon(x) \otimes s(d(y))$$

and then,

$$(\forall x \otimes y \in (C(\overline{\sigma_n}) \otimes C(\overline{\sigma_n}))_i \text{ with } i > 0) \ (d^\otimes S + S d^\otimes)(x \otimes y) = x \otimes y$$

while

$$(\forall x \otimes y \in (C(\overline{\sigma_n}) \otimes C(\overline{\sigma_n}))_0) \ d^\otimes S(x \otimes y) = (1 - \eta \varepsilon \otimes \eta \varepsilon)(x \otimes y).$$

We conclude that properties II.1 and 2 of Lemma (II.3.8) hold for $\eta \otimes \eta$ and for $S$. ∎

We may now concern ourselves with the **Acyclic Models Theorem**:

**(III.6.2) Theorem.**  *Let $\mathfrak{C}$ be any category and let two (covariant) functors*

$$F, G\colon \mathfrak{C} \longrightarrow \mathbf{Clp};$$

*be given; suppose that $F$ has models $\mathfrak{M} = \{\mathfrak{M}_n | n \geq 0\}$ that are acyclic for the functor $G$. Then, for every homomorphism $\bar{f}\colon \mathbb{Z} \to \mathbb{Z}$, there exists a natural transformation*

$$\eta_{\bar{f}}\colon F \longrightarrow G.$$

*such that, for each object $X \in \mathfrak{C}$, $\eta_{\bar{f}}(X)$ is an extension of $\bar{f}$. Moreover, if $\tau$ is another natural transformation from $F$ to $G$ with the same property held by $\eta$, there exists a natural homotopy of chain complexes $E$ such that*

$$d^G E + E d^F = \eta - \tau.$$

Before we begin the proof of the Acyclic Models Theorem, we observe that the assertion *for each $X \in \mathfrak{C}$ the morphism*

$$\eta_{\bar{f}}(X)\colon F(X) \longrightarrow G(X)$$

*is an extension of $\bar{f}$* means that there is a commutative diagram

$$
\begin{array}{ccccccccc}
\cdots F(X)_n & \xrightarrow{d_n^F} & F(X)_{n-1} & \xrightarrow{d_{n-1}^F} & \cdots & \xrightarrow{d_1^F} & F(X)_0 & \xrightarrow{\varepsilon} & \mathbb{Z} \\
\downarrow{\scriptstyle \eta_{\bar{f}}(X)_n} & & \downarrow{\scriptstyle \eta_{\bar{f}}(X)_{n-1}} & & & & \downarrow{\scriptstyle \eta_{\bar{f}}(X)_0} & & \downarrow{\scriptstyle \bar{f}} \\
\cdots G(X)_n & \xrightarrow{d_n^G} & G(X)_{n-1} & \xrightarrow{d_{n-1}^G} & \cdots & \xrightarrow{d_1^G} & G(X)_0 & \xrightarrow{\varepsilon'} & \mathbb{Z}
\end{array}
$$

Note that the boundary homomorphisms depend on $X$; and from now on, for the sake of a simpler notation, we shall write $\eta_n$ instead of $\eta_{\bar{f}}(X)_n$.

We advise the reader to go back to Sect. II.3 to review the definition of homotopy among chain complexes and for properly interpreting the last part of the statement on each given object $X \in \mathfrak{C}$.

*Proof.* We begin by building $\eta_0$. For each $M_0^i \in \mathfrak{M}_0$, there is an element $x_0^i \in F(M_0^i)_0$ such that, for every $X \in \mathfrak{C}$, the set

$$\{F(f)_0(x_0^i) | f \in \mathfrak{C}(M_0^i, X), i \in J_0\}$$

is a basis for $F(X)_0$. Since $\varepsilon'\colon G(M_0^i)_0 \to \mathbb{Z}$ is surjective, for each element $\bar{f}\varepsilon(x_0^i) \in \mathbb{Z}$ there exists $y_0^i \in G(M_0^i)_0$ such that

$$\varepsilon'(y_0^i) = \bar{f}\varepsilon(x_0^i);$$

we define

$$\eta_0(F(f)_0(x_0^i)) := G(f)_0(y_0^i)$$

on the generators $F(f)_0(x_0^i)$ and linearly extend $\eta_0$ to the entire free group $F(X)_0$.

The diagram

$$
\begin{array}{ccc}
F(X)_0 & \xrightarrow{\;\varepsilon\;} & \mathbb{Z} \\
\downarrow{\scriptstyle \eta_0} & & \downarrow{\scriptstyle \bar{f}} \\
G(X)_0 & \xrightarrow[\;\varepsilon'\;]{} & \mathbb{Z}
\end{array}
$$

is commutative. In fact,

$$\bar{f}\varepsilon(F(f)_0(x_0^i)) = \varepsilon'G(f)_0(y_0^i) = \varepsilon'\eta_0(F(f)_0(x_0^i))$$

and therefore $\bar{f}\varepsilon = \varepsilon'\eta_0$.

We now prove that $\eta_0$ is natural. For any given $g \in \mathfrak{C}(X,Y)$,

$$\eta_0 F(g)_0(F(f)_0(x_0^i)) = \eta_0 F(gf)_0(x_0^i)$$
$$G(gf)_0(y_0^i) = G(g)_0 G(f)_0(y_0^i) = G(g)_0 \eta_0(F(f)_0(x_0^i)),$$

that is to say, the diagram

$$
\begin{array}{ccc}
F(X)_0 & \xrightarrow{\;F(g)_0\;} & F(Y)_0 \\
\downarrow{\scriptstyle \eta_0} & & \downarrow{\scriptstyle \eta_0} \\
G(X)_0 & \xrightarrow[\;G(g)_0\;]{} & G(Y)_0
\end{array}
$$

is commutative.

We now construct $\eta_1$. Let us consider the diagram

$$
\begin{array}{ccccc}
F(M_1^i)_1 & \xrightarrow{\;d\;} & F(M_1^i)_0 & \xrightarrow{\;\varepsilon\;} & \mathbb{Z} \\
& & \downarrow{\scriptstyle \eta_0} & & \downarrow{\scriptstyle \bar{f}} \\
G(M_1^i)_1 & \xrightarrow[\;d'\;]{} & G(M_1^i)_0 & \xrightarrow[\;\varepsilon'\;]{} & \mathbb{Z}
\end{array}
$$

for a model $M_1^i \in \mathfrak{M}_1$ and let us take

$$\eta_0 d(x_1^i) \in G(M_1^i)_1$$

for every universal element $x_1^i \in F(M_1^i)_1$.

Here is where the hypothesis on the acyclicity of $G$ on the models of the set $\mathfrak{M}$ is needed. Indeed, since $\varepsilon d = 0$, we have

$$\eta_0 d(x_1^i) \in \ker \varepsilon' = \operatorname{im} d'$$

and, consequently, there is $y_1^i \in G(M_1^i)_1$ such that $d'(y_1^i) = \eta_0 d(x_1^i)$; we then define

$$\eta_1(F(f)_1(x_1^i)) := G(f)_1(y_1^i)$$

for every $f: M_1^i \to X$. We ask the reader to prove that $\eta_0 d = d'\eta_1$ and that $\eta_1$ is natural. The functor $\eta_n$ is obtained by induction.

We now suppose $\theta: F \to G$ to be a natural transformation extending $\bar{f}$. We wish to construct the homotopy $E: F \to G$ that connects $\eta$ to $\theta$. We begin with $E_0: F(X)_0 \to G(X)_1$ for an arbitrary $X$. Let us take a model $M_0^i \in \mathfrak{M}_0$ and a universal element $x_0^i \in F(M_0^i)_0$. From

$$\varepsilon'[\eta(x_0^i) - \theta(x_0^i)] = 0$$

and since $G$ is acyclic for the model $M_0^i$, it follows that there exists $z_0^i \in G(M_0^i)_1$ such that

$$d'(z_0^i) = \eta(x_0^i) - \theta(x_0^i);$$

we now define $E_0: F(X)_0 \to G(X)_1$ on the generators of $F(X)_0$ by the formula

$$E_0(F(f)_0(x_0^i)) := G(f)_1(z_0^i)$$

and then extend it on $F(X)_0$ by linearity. We now suppose that the morphisms $E_1, \ldots, E_{n-1}$ have been defined; our aim is to define $E_n$.

As usual, we take a model $M_n^i$ and a universal element $x_n^i \in F(M_n^i)_n$. We note that

$$d'(\eta_{n-1}d(x_n^i) - \theta_{n-1}d(x_n^i) - E_{n-1}d(x_n^i)) = 0$$

because $\eta_{n-1}d(x_n^i) - \theta_{n-1}d(x_n^i) - E_{n-1}d(x_n^i)) = E_{n-2}d(d(x_n^i))$. Therefore, since $G$ is acyclic on the model,

$$(\exists t_n^i \in G(M_n^i)_{n+1})d't_n^i = \eta_{n-1}d(x_n^i) - \theta_{n-1}d(x_n^i) - E_{n-1}d(x_n^i).$$

We define $E_n$ on the generators $F(f)_n(x_n^i)$ of $F(X)_n$ by

$$E_n(F(f)_n(x_n^i)) := G(f)_{n+1}(t_n^i)$$

and linearly extend it. We leave to the reader the final details of the proof.  ■

The next result, known as the **Eilenberg–Zilber Theorem**, is a direct consequence of the Acyclic Models Theorem.

**(III.6.3) Theorem.** *Let*

$$C: \mathbf{Csim} \longrightarrow \mathbf{Clp}$$

*be the functor that takes each simplicial complex $K$ to the augmented chain complex*

$$C(K) = \{C_n(K), \partial_n\}.$$

*There are natural transformations*

$$\eta: C_\times \longrightarrow C \otimes C$$

$$\tau: C \otimes C \longrightarrow C_\times$$

*which extend the identity homomorphism $1_{\mathbb{Z}}: \mathbb{Z} \to \mathbb{Z}$. Besides, $\eta\tau$ and $\tau\eta$ are homotopic to the natural transformations given by the identities.*

## III.6.2 Homology of the Product of Two Polyhedra

Let two polyhedra $|K|, |L| \in \mathbf{Csim}$ be given. Our aim here is to compute the homology groups of the polyhedron $|K| \times |L|$ in terms of those of $|K|$ and $|L|$.

By the Eilenberg–Zilber Theorem (III.6.3), the chain complexes $C(K) \otimes C(L)$ and $C(K \times L)$ are chain equivalent; therefore, we have a graded group isomorphism

$$H_*(|K| \times |L|; \mathbb{Z}) \cong H_*(C(K) \otimes C(L)).$$

Following the steps taken in Sect. II.5 when studying the relationship between $H_*(K; G)$ and $H_*(K, \mathbb{Z})$, we interpret the graded Abelian groups $Z(K) = \{Z_n(K) | n \geq 0\}$ and $B(K) = \{B_n(K) | n \geq 0\}$ as chain complexes with trivial boundary operator 0; we then construct the chain complexes

1. $(Z(K) \otimes C(L), 0 \otimes d^L)$
2. $(C(K) \otimes C(L), \partial^K \otimes \partial^L)$
3. $(\widetilde{B(K)} \otimes C(L), 0 \otimes \partial^L)$, where $\widetilde{B(K)}_n = B_{n-1}(K)$

since the chain complex sequence

$$Z(K) \xrightarrowtail{\ i\ } C(K) \xrightarrow{\ d^K\ } \widetilde{B(K)}$$

is exact and short, with an argument similar to that used in Lemma (II.5.1), we conclude that

$$(Z(K) \otimes C(L), 0 \otimes \partial^L) \rightarrowtail (C(K) \otimes C(L), d^K \otimes \partial^L)$$

$$\longrightarrow (\widetilde{B(C)} \otimes C(L), 0 \otimes \partial^L)$$

is a short exact sequence.

By the Long Exact Sequence Theorem (II.3.1), we obtain the exact sequence

$$\ldots \to H_n(Z(K) \otimes C(L)) \xrightarrow{i_*} H_n(C(K) \otimes C(L)) \xrightarrow{\partial_*} H_n(\widetilde{B(K)} \otimes C(L))$$

$$\xrightarrow{\lambda_n} H_{n-1}(Z(K) \otimes C(L)) \to \ldots .$$

We now have to understand the nature of the groups

$$H_n(Z(K) \otimes C(L)) \text{ and } H_n(\widetilde{B(K)} \otimes C(L)) .$$

We begin by observing that

$$Z_n(Z(K) \otimes C(L)) = \ker(0 \otimes \partial^L) \cong \sum_{i+j=n} (Z_i(K) \otimes Z_j(L)) ,$$

$$Z_n(\widetilde{B(K)} \otimes C(L)) = \ker(0 \otimes \partial^L) \cong \sum_{i+j=n} (\widetilde{B(K)}_i \otimes Z_j(L))$$

$$\cong \sum_{i+j=n-1} (B_{i-1}(K) \otimes Z_j(L)) ,$$

$$B_n(Z(K) \otimes C(L)) = \operatorname{im}(0 \otimes \partial^L) \cong \sum_{i+j=n} (Z_i(K) \otimes B_j(L)) ,$$

$$B_n(\widetilde{B(K)} \otimes C(L)) = \operatorname{im}(0 \otimes \partial^L) \cong \sum_{i+j=n} (\widetilde{B(K)}_i \otimes B_j(L))$$

$$\cong \sum_{i+j=n-1} (B_{i-1}(K) \otimes B_j(L)).$$

Therefore,

$$H_n(Z(K) \otimes C(L)) \cong \sum_{i+j=n} (Z_i(K) \otimes H_j(L; \mathbb{Z}))$$

$$H_n(\widetilde{B(K)} \otimes C(L)) \cong \sum_{i+j=n-1} (B_i(K) \otimes H_j(L; \mathbb{Z}))$$

and the long exact homology sequence takes the form

$$\ldots \to \sum_{i+j=n} (Z_i(K) \otimes H_j(L; \mathbb{Z})) \xrightarrow{i_*} H_n(K \times L; \mathbb{Z}) \xrightarrow{\partial_*} \sum_{i+j=n-1} (B_i(K) \otimes H_j(L; \mathbb{Z}))$$

$$\xrightarrow{\lambda_n} \sum_{i+j=n-1} (Z_i(K) \otimes H_j(L; \mathbb{Z})) \to \ldots .$$

We consider the homomorphism

$$\partial_n \colon H_n(K \times L; \mathbb{Z}) \to \sum_{i+j=n-1} (B_i(K) \otimes H_j(L; \mathbb{Z}))$$

and the short exact sequence

$$\ker \partial_n \rightarrowtail H_n(K \times L; \mathbb{Z}) \twoheadrightarrow \operatorname{im} \partial_n;$$

note that

$$\operatorname{im} \partial_n \cong \sum_{i+j=n} (Z_i(K) \otimes H_j(L; \mathbb{Z}))/\ker i_n$$

$$\cong \sum_{i+j=n} (Z_i(K) \otimes H_j(L; \mathbb{Z}))/\operatorname{im} \lambda_{n+1} \cong \operatorname{coker} \lambda_{n+1}$$

and so, for every $n \geq 1$, we obtain the short exact sequence

$$\operatorname{coker} \lambda_{n+1} \rightarrowtail H_n(K \times L; \mathbb{Z}) \twoheadrightarrow \ker \lambda_n.$$

As for changing the group of coefficients in homology (Sect. II.5), we prove that

$$\operatorname{coker} \lambda_{n+1} \cong \sum_{i+j=n} (H_i(K; \mathbb{Z}) \otimes H_j(L; \mathbb{Z}))$$

$$\ker \lambda_n \cong \sum_{i+j=n-1} \operatorname{Tor}(H_i(K; \mathbb{Z}), H_j(L; \mathbb{Z}))$$

and so we have the next result, known as the **Künneth Theorem**:

**(III.6.4) Theorem.** *For every pair of polyhedra $|K|$ and $|L|$ and every $n \geq 1$, the following short sequence of Abelian groups*

$$\sum_{i+j=n} (H_i(K; \mathbb{Z}) \otimes H_j(L; \mathbb{Z})) \rightarrowtail H_n(K \times L; \mathbb{Z})$$

$$\twoheadrightarrow \sum_{i+j=n-1} \operatorname{Tor}(H_i(K; \mathbb{Z}), H_j(L; \mathbb{Z}))$$

*is exact.*


## Exercises

**1.** Prove that, for every $C, C' \in \mathbf{Clp}$, the tensor product $C \otimes C'$ belongs to **Clp**.

**2.** Let $f \in \mathfrak{C}(C, D)$ and $g \in \mathfrak{C}(C', D')$ be two morphisms of given chain complexes; prove that

$$f \otimes g \colon C \otimes C' \to D \otimes D'$$

defined on the generators by

$$(f \otimes g)(x \otimes y) = f(x) \otimes g(y)$$

is a morphism of chain complexes.

**3.** Let $C, C' \in \mathfrak{C}$ be two chain complexes. Prove that

$$\mu \colon C \otimes C' \to C' \otimes C$$

defined on generators by the formula

$$\mu(x \otimes y) = (-1)^{|x||y|} y \otimes x$$

is an isomorphism of chain complexes.

# Chapter IV
# Cohomology

## IV.1 Cohomology with Coefficients in $G$

In Sect. II.5, we have seen that the homology groups $H_n(K;\mathbb{Q})$ of an oriented simplicial complex $K$, with rational coefficients, have the structure of vector spaces and may therefore be dualized. The possibility of dualizing such vector spaces led mathematicians to ask whether it was also possible to "dualize" the homology groups with coefficients in a different Abelian group $G$. Let us remember that we used the tensor product to change the coefficients of the homology groups; to "dualize" homology or change the coefficients of the dualized theory, we use the functor

$$\mathrm{Hom}(-,G)\colon \mathbf{Ab} \to \mathbf{Ab}$$

where $G$ is a fixed Abelian group. More precisely, given an Abelian group $A$, we define $\mathrm{Hom}(A,G)$ to be the Abelian group of all homomorphisms from $A$ to $G$ with the addition

$$\mathrm{Hom}(A,G) \times \mathrm{Hom}(A,G)) \to \mathrm{Hom}(A,G)$$

defined, for each pair $(\phi,\psi)$ and for each $a \in A$, by

$$(\phi + \psi)(a) = \phi(a) + \psi(a)\,.$$

On morphisms, $\mathrm{Hom}(-,G)$ acts as follows: given $f\colon A \to A'$,

$$\widetilde{f} = \mathrm{Hom}(f,G)\colon\ \mathrm{Hom}(A',G) \to \mathrm{Hom}(A,G)\,,\ \phi \mapsto \phi\, f\,.$$

The homomorphism $\widetilde{f} = \mathrm{Hom}(f,G)$ is called *adjoint* of $f$. Note that the functor

$$\mathrm{Hom}(-,G)\colon \mathbf{Ab} \to \mathbf{Ab}$$

is *contravariant*.

D.L. Ferrario and R.A. Piccinini, *Simplicial Structures in Topology*,
CMS Books in Mathematics, DOI 10.1007/978-1-4419-7236-1_IV,
© Springer Science+Business Media, LLC 2011

**(IV.1.1) Theorem.** *Suppose that the sequence of Abelian groups*

$$A \xrightarrow{\ f\ } B \xrightarrow{\ g\ } C \xrightarrow{\quad} 0$$

*is exact. Then, for every Abelian group $G$, the sequence of Abelian groups*

$$0 \xrightarrow{\quad} \operatorname{Hom}(C,G) \xrightarrow{\ \widetilde{g}\ } \operatorname{Hom}(B,G) \xrightarrow{\ \widetilde{f}\ } \operatorname{Hom}(A,G)$$

*is exact.*

*Proof.* Let us prove that $\widetilde{g}$ is injective. Take $\phi \in \operatorname{Hom}(C,G)$ such that $\widetilde{g}(\phi) = 0$. Then, for every $b \in B$, $\phi(g(b)) = 0$. Since $g$ is surjective, for every $c \in C$, there exists $b \in B$ such that $c = g(b)$. Hence, for every $c \in C$, $\phi(c) = \phi(g(b)) = 0$; that is to say, $\phi = 0$.

We now prove the exactness at $\operatorname{Hom}(B,G)$. For every $\phi \in \operatorname{Hom}(C,G)$,

$$\widetilde{f}\widetilde{g}(\phi) = \phi(gf) = 0$$

since $gf = 0$; hence, $\operatorname{im}\widetilde{g} \subset \ker\widetilde{f}$. Let $\psi \in \operatorname{Hom}(B,G)$ be such that $\widetilde{f}(\psi) = \psi f = 0$. Then the restriction of $\psi$ to $f(A)$ is null and so there exists a homomorphism $\psi' \colon B/f(A) \to G$ such that the composite function

$$B \xrightarrow{\ q\ } B/f(A) \xrightarrow{\ \psi'\ } G$$

(where $q$ is the quotient homomorphism) coincides with $\psi$. On the other hand, since $\operatorname{im} f = \ker g$, there exists an isomorphism $g' \colon B/f(A) \cong C$ for which $g'q = g$. The homomorphism $\phi = \psi'(g')^{-1} \colon C \to G$ is such that $\widetilde{g}(\phi) = \psi$ and therefore, $\ker\widetilde{f} \subset \operatorname{im}\widetilde{g}$. ∎

We note that if $A$ is the direct sum of the Abelian group of the integers $\mathbb{Z}$ with itself $n$ times ($A = \mathbb{Z} \times \ldots \times \mathbb{Z} = \mathbb{Z}^n$), then

$$\operatorname{Hom}(\underbrace{\mathbb{Z} \times \ldots \times \mathbb{Z}}_{n \text{ times}}, G) \cong \underbrace{\operatorname{Hom}(\mathbb{Z},G) \times \ldots \times \operatorname{Hom}(\mathbb{Z},G)}_{n \text{ times}} \cong \underbrace{G \times \ldots \times G}_{n \text{ times}};$$

indeed, the function

$$\phi \colon \operatorname{Hom}(\mathbb{Z} \times \ldots \times \mathbb{Z}, G) \longrightarrow G \times \ldots \times G$$

(defined by $\phi(f) = (f(1,0,\ldots,0),\ldots,f(0,\ldots,1))$ for every $f \in \operatorname{Hom}(\mathbb{Z} \times \ldots \times \mathbb{Z}, G)$) is an isomorphism.

We use Theorem (IV.1.1) for computing $\operatorname{Hom}(\mathbb{Z}_2, G)$. In view of the exact sequence

$$0 \xrightarrow{\quad} \mathbb{Z} \xrightarrow{\ 2\ } \mathbb{Z} \xrightarrow{\quad} \mathbb{Z}_2 \xrightarrow{\quad} 0,$$

we conclude that the sequence

$$0 \longrightarrow \mathrm{Hom}(\mathbb{Z}, G) \longrightarrow \mathrm{Hom}(\mathbb{Z}, G) \xrightarrow{\widetilde{2}} \mathrm{Hom}(\mathbb{Z}_2, G)$$

is exact. However, $\mathrm{Hom}(\mathbb{Z}, G) \cong G$ and since $\widetilde{2}$ is precisely the multiplication by 2 in $G$, we conclude from the last exact sequence that $\mathrm{Hom}(\mathbb{Z}_2, G) \cong \ker(2 \colon G \to G)$. In particular, $\mathrm{Hom}(\mathbb{Z}_2, \mathbb{Z}) = 0$.

As in the case for changing the coefficients in homology, where we have applied the functor $- \otimes G$ to a chain complex $(C, \partial)$ – in particular, the complex $(C(K), \partial)$ – we can apply the functor $\mathrm{Hom}(-, G)$ to $(C, \partial)$.

We now construct the *cohomology* (with coefficients in an Abelian group $G$) of a chain complex $(C, \partial)$. The image of $(C, \partial)$ by the functor $\mathrm{Hom}(-, G)$ is the graded Abelian group

$$\mathrm{Hom}(C, G) = \{C^n(C, G)\} := \{\mathrm{Hom}(C_n, G)\}.$$

We denote the adjoint homomorphism of the boundary homomorphism

$$\partial_n \colon C_n \to C_{n-1}$$

with $\partial^{n-1}$, that is to say,

$$\partial^{n-1} := \mathrm{Hom}(\partial_n, G) \colon C^{n-1}(C, G) \to C^n(C, G).$$

Since $\partial_n \partial_{n+1} = 0$, we immediately conclude that $\partial^n \partial^{n-1} = 0$ and so that

$$\mathrm{im}\, \partial^{n-1} := B^n(C, G) \subset \ker \partial^n := Z^n(C, G).$$

The quotient group

$$H^n(C; G) = Z^n(C, G)/B^n(C, G)$$

is the *nth-cohomology group* of the chain complex $(C, \partial)$ with coefficients in $G$. In particular, if $(C, \partial) = (C(K), \partial)$ is the positive free chain complex associated with the oriented chain simplex $K = (X, \Phi)$, the graded Abelian group $H^*(K; \mathbb{Z})$ is the *simplicial cohomology* of $K$ with integral coefficients. In this case, since the Abelian group $C_n(K)$ is generated by the oriented $n$-simplexes $\sigma$ of $K$, we define the homomorphisms

$$c_\sigma \colon C_n(K) \to \mathbb{Z}, \ (\forall \tau \in \Phi, \dim \tau = n)\ c_\sigma(\tau) = \begin{cases} 1, & \tau = \sigma \\ 0, & \tau \neq \sigma \end{cases}$$

Consequently,

$$C^n(K) = \mathrm{Hom}(C_n(K), \mathbb{Z})$$

is the free Abelian group generated by the homomorphisms $c_\sigma$, where $\sigma$ runs over the set of all oriented $n$-simplexes of $K$.

**(IV.1.2) Remark.** It is easily proved that, for every $c_{n+1} \in C_{n+1}(K)$ and for every $c^n \in C^n(K)$,

$$(\partial^n(c^n))(c_{n+1}) = c^n \partial_{n+1}(c_{n+1}) \; ;$$

in particular, if $c^n = \sum\limits_{\substack{\sigma \in \Phi \\ \dim \sigma = n}} n_\sigma c_\sigma$, we have

$$\partial^n(c^n) = \sum\limits_{\substack{\sigma \in \Phi \\ \dim \sigma = n}} n_\sigma \partial^n(c_\sigma)$$

and therefore, for every oriented $(n+1)$-simplex $\tau$ of $K$,

$$\partial^n(c^n)(\tau) = \sum\limits_{\substack{\sigma \in \Phi \\ \dim \sigma = n}} n_\sigma c_\sigma(\partial_{n+1}(\tau)) \; .$$

We now prove that
$$H^*(-;\mathbb{Z}) \colon \mathbf{Csim} \longrightarrow \mathbf{Ab}^{\mathbb{Z}}$$

is a contravariant functor. Let $K$, $L$ be two simplicial complexes and $f \colon K \to L$ be a simplicial function. We know that, for each $n \geq 0$, $f$ defines a homomorphism $C_n(f) \colon C_n(K) \to C_n(L)$; therefore, for every $n \geq 0$, we have a homomorphism of Abelian groups

$$C^n(f) \colon C^n(L) \longrightarrow C^n(K)$$

such that $C^n(f)(c^n) = c^n C_n(f)$ for every $c^n \in C^n(L)$. It is easily proved that, for every $n \geq 0$,

$$C^{n+1}(f)\partial_L^n = \partial_K^n C^n(f) \; ;$$

consequently, $C^n(f)$ induces a homomorphism of Abelian groups

$$H^n(f) \colon H^n(L;\mathbb{Z}) \to H^n(K;\mathbb{Z}).$$

Here, the situation is completely analogous to the one for homology.

As in the case of simplicial homology, we define the cohomology $H^*(K,L;\mathbb{Z})$ of a pair of simplicial complexes $(K,L)$: for that, it is enough to apply the functor $\mathrm{Hom}(-,\mathbb{Z})$ to the chain complex

$$(C(K,L),\partial^{K,L}) := \{C_n(K)/C_n(L), \partial_n^{K,L}\} \; .$$

What is more, $(K,L)$ produces a long exact sequence in simplicial cohomology, that is to say, the following **Cohomology Long Exact Sequence Theorem** holds:

**(IV.1.3) Theorem.** *Let $(K,L)$ be a pair of simplicial complexes. For every $n > 0$, there exists a homomorphism*

$$\tilde{\lambda}^n \colon H^n(L;\mathbb{Z}) \to H^{n+1}(K,L;\mathbb{Z})$$

*for which the following sequence of cohomology groups*

$$\ldots \to H^n(K;\mathbb{Z}) \xrightarrow{H^n(i)} H^n(L;\mathbb{Z}) \xrightarrow{\tilde{\lambda}^n} H^{n+1}(K,L;\mathbb{Z}) \xrightarrow{q^*(n+1)} H^{n+1}(K;\mathbb{Z}) \to \ldots$$

*is exact.*

The proof of this theorem follows the steps taken in proving the corresponding theorem in homology Theorem (II.3.1); we only wish to point out that the main result needed for this proof is the following lemma:

**(IV.1.4) Lemma.** *If*

$$A \xrightarrowtail{f} B \xrightarrow{g} \twoheadrightarrow C$$

*is a short exact sequence of free groups and $G$ is an Abelian group, then also the sequence*

$$\mathrm{Hom}(C,G) \xrightarrowtail{\tilde{g}} \mathrm{Hom}(B,G) \xrightarrow{\tilde{f}} \twoheadrightarrow \mathrm{Hom}(A,G)$$

*is exact.*

The reader may take the proof of Lemma (II.5.1) as a basis for proving this one; actually, the only tensor product property used in Lemma (II.5.1) is that $-\otimes G$ is a functor which takes sums of morphisms into sums of morphisms; this very same property holds also for the (contravariant) functor $\mathrm{Hom}(-,G)$.

The cohomology determined by an oriented chain simplex $K$ with coefficients in an Abelian group $G$ is defined in the same manner as the one with coefficients in $\mathbb{Z}$, except that we apply to the chain complex $(C(K),\partial)$ the contravariant functor $\mathrm{Hom}(-,G)$ instead of $\mathrm{Hom}(-,\mathbb{Z})$.

We wish to determine the cohomology with coefficients in $G$ of a simplicial complex, based on its homology with coefficients in $\mathbb{Z}$ and the Abelian group $G$. We seek to use the same previous ideas to our benefit, bearing in mind that the functor $A \mapsto A \otimes G$ is covariant, whereas $A \mapsto \mathrm{Hom}(A,G)$ is contravariant. We begin by reviewing what has been done so far in general terms. A *cochain complex* (or *cocomplex*) is a graded Abelian group $\{C^n\}$ together with an endomorphism of degree $+1$, called *coboundary homomorphism* $\partial^* = \{\partial^n \colon C^n \to C^{n+1},\}$, for which $\partial^{n+1}\partial^n = 0$. A *cochain homomorphism* between two cochain complexes $(C,\partial^*)$ and $(C',\partial'^*)$ is a homomorphism between the graded Abelian groups $C$ and $C'$ that commutes with the coboundary homomorphisms.

Also the other concepts defined for chain complexes have their dual correspondents: the elements of $C^n$ are called *n-cochains*; the elements of

$$Z^n(C) := \ker(\partial^n \colon C^n \to C^{n+1})$$

are *n-cocycles*, and those of

$$B^n(C) := \mathrm{im}(\partial^{n-1} \colon C^{n-1} \to C^n)$$

are *n-coboundaries*; finally, the quotient $H^n(C) := Z^n(C)/B^n(C)$ is the *cohomology group* of the cocomplex $(C,\partial^*)$ in dimension $n$.

If, starting from a cocomplex $(C^*, \partial^*)$, we define $C_n := C^{-n}$ and $\partial_n := \partial^{-n}$, it becomes clear that $(C_*, \partial)$ is a chain complex and all concepts have their respective equivalents. Specifically, the cohomology of $(C^*, \partial^*)$ is exactly the homology of the complex $(C_*, \partial)$. This gives us an indication of how to write all results in cohomological terms (and we shall do so from now on, without further comments).

For every complex $(C, \partial)$ and for every Abelian group $G$, we may define a cocomplex $(C^*, \partial^*)$ where $C^n := \text{Hom}(C_n, G)$ is the homomorphism group from $C_n$ into $G$, and $\partial^n: C^n \to C^{n+1}$ is the map adjoint to $\partial_{n+1}: C_{n+1} \to C_n$. This cocomplex cohomology is called *cohomology of the complex* $(C, \partial)$ *with coefficients in* $G$.

We therefore assume that the complex $(C, \partial)$ is free and, so, both $Z_n C$ and $B_n C$ are also free; for every $n$, the sequence of free Abelian groups

$$Z_n(C) \rightarrowtail C_n \xrightarrow{\partial_n} B_{n-1}(C)$$

is short exact; hence, by Lemma (IV.1.4), the sequence

$$\text{Hom}(B_{n-1}(C), G) \rightarrowtail \text{Hom}(C_n, G) \twoheadrightarrow \text{Hom}(Z_n(C), G)$$

is exact. This enables us to construct the short exact complex sequence

$$(\text{Hom}(\widetilde{B(C)}, G), 0) \rightarrowtail (\text{Hom}(C, G), \partial) \twoheadrightarrow (\text{Hom}(Z(C), G), 0)$$

where

$$\widetilde{B(C)} = \{\widetilde{B(C)}^n\} := \{B_{n-1}(C)\} .$$

The Cohomology Long Exact Sequence Theorem, in its general form (in terms of chain complexes), allows us to write the long exact sequence

$$\cdots \longrightarrow \text{Hom}(Z_{n-1}(C), G) \xrightarrow{\widetilde{i}_{n-1}} \text{Hom}(B_{n-1}(C), G) \xrightarrow{h_n} H^n(C; G)$$

$$\xrightarrow{j_n} \text{Hom}(Z_n(C), G) \xrightarrow{\widetilde{i}_n} \text{Hom}(B_n(C), G) \longrightarrow \cdots$$

(where $\widetilde{i}_n$ is the adjoint of the inclusion $i_n: B_n(C) \to Z_n(C)$) that breaks down into short exact sequences

$$\text{im}\, h_n \rightarrowtail H^n(C; G) \twoheadrightarrow \text{im}\, \widetilde{j}_n .$$

Since we have the isomorphisms $\text{im}\, h_n \cong \text{Hom}(B_{n-1}(C), G)/\ker h_n$ and $\ker h_n = \text{im}\, \widetilde{i}_{n-1}$, we have $\text{im}\, h_n = \text{coker}(\widetilde{i}_{n-1})$; besides, $\text{im}\, \widetilde{j}_n = \ker \widetilde{i}_n$ and so these last short exact sequences become the short exact sequences

$$\text{coker}(\widetilde{i}_{n-1}) \rightarrowtail H^n(C; G) \twoheadrightarrow \ker \widetilde{i}_n .$$

**(IV.1.5) Lemma.** $\ker(\widetilde{i_n}) \cong \operatorname{Hom}(H_n(C), G)$.

*Proof.* We define

$$\phi: \ker(\widetilde{i_n}) \to \operatorname{Hom}(H_n(C), G)$$

so that, for every $x + B_n(C) \in H_n(C)$, $\phi(f)(x + B_n(C)) = f(x)$. This function is well defined and is a homomorphism. On the other hand, we define

$$\psi: \operatorname{Hom}(H_n(C), G) \to \ker(\widetilde{i_n})$$

by setting $\psi(g) = gp$, where $p: Z_n(C) \to H_n(C)$ is the quotient homomorphism. Also this function is a homomorphism; in addition, the compositions satisfy $\psi\phi = 1_{\ker(\widetilde{i_n})}$ and $\phi\psi = 1_{\operatorname{Hom}(H_n(C), G)}$.    ∎

This result shows that $\ker(\widetilde{i_n})$ is independent from the groups $\operatorname{Hom}(B_n(C), G)$ and $\operatorname{Hom}(Z_n(C), G)$, and depends only on $H_n(C)$, the cokernel of the monomorphism $i_n: B_n(C) \to Z_n(C)$, and on $G$. As in the case of the functor $- \otimes G$, we wonder whether the same is also true for $\operatorname{coker}(\widetilde{i_{n-1}})$. In fact, we have the following result which is dual to Proposition (II.5.4):

**(IV.1.6) Proposition.** *Let $H$ be the cokernel of the monomorphism $i: B \to Z$ between free Abelian groups and any Abelian group $G$. Then, both the kernel and the cokernel of the homomorphism $\widetilde{i}: \operatorname{Hom}(Z, G) \to \operatorname{Hom}(B, G)$ depend only on $H$ and $G$. Moreover, $\ker(\widetilde{i}) \cong \operatorname{Hom}(H, G)$, while $\operatorname{coker}(\widetilde{i})$ gives rise to a new contravariant functor*

$$\operatorname{Ext}(-, G): \mathbf{Ab} \longrightarrow \mathbf{Ab},$$

*called* extension product.

*Proof.* The method used in this proof is analogous to the one for proving Proposition (II.5.4), that is to say, merely replacing the functor $- \otimes G$ with the functor $\operatorname{Hom}(-, G)$ and keeping in mind the contravariance of the latter. We leave the details to the reader. Nevertheless, we note that $\operatorname{Ext}(-, G)$ does not depend on the free presentation of $H$.    ∎

Let us compute $\operatorname{Ext}(\mathbb{Z}_n, G)$ for any Abelian group $G$. Since $\operatorname{Ext}(-, G)$ does not depend on any particular free presentation of $\mathbb{Z}_n$, we choose the presentation

$$\mathbb{Z} \xrightarrowtail{\quad n \quad} \mathbb{Z} \longtwoheadrightarrow \mathbb{Z}/n$$

where $n$ is the multiplication by $n$. Then, we have the exact sequence

$$\operatorname{Hom}(\mathbb{Z}_n, G) \longrightarrow \operatorname{Hom}(\mathbb{Z}, G) \xrightarrow{\widetilde{n}} \operatorname{Hom}(\mathbb{Z}, G) \longrightarrow \operatorname{Ext}(\mathbb{Z}_n, G) \longrightarrow 0$$

and since $\operatorname{Hom}(\mathbb{Z}, G) \cong G$, we conclude that $\operatorname{Ext}(\mathbb{Z}_n, G) \cong G/nG$. In particular, $\operatorname{Ext}(\mathbb{Z}_n, \mathbb{Z}) \cong \mathbb{Z}_n$.

Due to Lemma (IV.1.4), if $H$ is free, $\operatorname{Ext}(H, G) = 0$.

When we apply this result to these exact sequences, we get the **Universal Coefficients Theorem in Cohomology**:

**(IV.1.7) Theorem.** *The cohomology of a free positive complex* $(C, \partial)$ *with coefficients in an Abelian group G is determined by the following short exact sequences:*

$$\text{Ext}(H_{n-1}(C), G) \rightarrowtail H^n(C; G) \twoheadrightarrow \text{Hom}(H_n(C), G) .$$

We note that for an oriented simplicial complex $K$, we apply the functor $\text{Hom}(-, G)$ to the positive free chain complex $(C(K), \partial)$ to obtain, for every $n \geq 0$, the following short exact sequences:

$$\text{Ext}(H_{n-1}(K; \mathbb{Z}), G) \rightarrowtail H^n(K; G) \twoheadrightarrow \text{Hom}(H_n(K; \mathbb{Z}), G) .$$

### IV.1.1  Cohomology of Polyhedra

We now wish to study the cohomology (with coefficients in $\mathbb{Z}$, to make it simple) as a contravariant functor from the category **P** of polyhedra to the category $\mathbf{Ab}^{\mathbb{Z}}$,

$$H^*(-; \mathbb{Z}): \mathbf{P} \longrightarrow \mathbf{Ab}^{\mathbb{Z}} .$$

The definition of the functor on the objects is obvious:

$$(\forall |K| \in \mathbf{P}) \, H^*(|K|; \mathbb{Z}) = H^*(K; \mathbb{Z}).$$

As for the morphisms, let $f: |K| \to |L|$ be a continuous function; by the Simplicial Approximation Theorem, there exists a simplicial function $g: K^{(r)} \to L$ that approximates $f$ simplicially. It produces a homomorphism

$$H^n(g; \mathbb{Z}): H^n(L; \mathbb{Z}) \to H^n(K^{(r)}; \mathbb{Z})$$

for every $n \in \mathbb{Z}$. All results in Sect. III.2, needed to prove that a map $f: |K| \to |L|$ defines a homomorphism $H_*(f; \mathbb{Z})$ between the homology of the polyhedron $|K|$ and the homology of $|L|$, hold in cohomology; in particular, we point out that if a chain complex $C$ is acyclic, also the chain complex $\text{Hom}(C, \mathbb{Z})$ is acyclic (use the Universal Coefficients Theorem in Cohomology). In addition, if we return to the proof of Theorem (III.2.2), we see that, for any projection $\pi^r: K^{(r)} \to K$, we may find a homomorphism of chain complexes

$$\aleph^r: C(K) \to C(K^{(r)})$$

such that $\aleph^r C(\pi^r)$ is chain homotopic to $1_{C(K^{(r)})}$, and $C(\pi^r) \aleph^r$ is chain homotopic to $1_{C(K)}$. This means that, homologically speaking, the homomorphisms induced by $\aleph^r$ and $C(\pi^r)$ are isomorphisms and the inverse of each other. We now consider the cochain complexes

$$\text{Hom}(C(K);\mathbb{Z}) = \{\text{Hom}(C_n(K);\mathbb{Z})\}\,,$$

$$\text{Hom}(C(K^{(r)});\mathbb{Z}) = \{\text{Hom}(C_n(K^{(r)});\mathbb{Z})\}$$

and the morphisms

$$\text{Hom}(C(\pi^r);\mathbb{Z})\colon\ \text{Hom}(C(K);\mathbb{Z}) \to \text{Hom}(C(K^{(r)});\mathbb{Z})\,,$$

$$\text{Hom}(\aleph^r;\mathbb{Z})\colon\ \text{Hom}(C(K^{(r)});\mathbb{Z}) \to \text{Hom}(C(K);\mathbb{Z})\,.$$

The chain homotopies mentioned before become homotopies of the functor $\text{Hom}(-;\mathbb{Z})$ and so we obtain homomorphisms

$$H^*(\aleph^r;\mathbb{Z})\colon\ H^*(K^{(r)};\mathbb{Z}) \to H^*(K;\mathbb{Z})\,,$$

$$H^*(\pi^r;\mathbb{Z})\colon\ H^*(K;\mathbb{Z}) \to H^*(K^{(r)};\mathbb{Z})$$

that are isomorphisms and the inverse of each other. We then define

$$H^*(f;\mathbb{Z})\colon\ H^*(|L|;\mathbb{Z}) \to H^*(|K|;\mathbb{Z})$$

as

$$H^*(f;\mathbb{Z}) = H^*(\aleph^r;\mathbb{Z})H^*(g;\mathbb{Z})\,.$$

Considerations similar to the ones in Corollary (III.2.7) allow us to conclude that $H^*(f;\mathbb{Z})$ is well defined.

Finally, we use the short exact sequences

$$\text{Ext}(H_{n-1}(|K|;\mathbb{Z}),G) \rightarrowtail H^n(|K|;G) \twoheadrightarrow \text{Hom}(H_n(|K|;\mathbb{Z}),G)\,,$$

associated with the polyhedron $|K|$, for computing the cohomology groups, with coefficients in $\mathbb{Z}$, of the torus $T^2$ and of the real projective plane $\mathbb{R}P^2$:

$$H^i(T^2;\mathbb{Z}) \cong \begin{cases} \mathbb{Z} & \text{if } i=0 \\ \mathbb{Z}\times\mathbb{Z} & \text{if } i=1 \\ \mathbb{Z} & \text{if } i=2 \\ 0 & \text{if } i\neq 0,1,2. \end{cases}$$

$$H^i(\mathbb{R}P^2;\mathbb{Z}) \cong \begin{cases} \mathbb{Z} & \text{if } i=0 \\ 0 & \text{if } i=1 \\ \mathbb{Z}_2 & \text{if } i=2 \\ 0 & \text{if } i\neq 0,1,2. \end{cases}$$

## Exercises

**1.** An Abelian group $G$ is said to be *divisible* if, for every positive integer $n$ and every $g \in G$, there exists a $g' \in G$ (not necessarily unique) such that $ng' = g$. The Abelian groups $\mathbb{Q}$ and $\mathbb{R}$ are examples of divisible groups. Prove that, if $G$ is a finitely generated Abelian group without a free part (see the decomposition theorem of finitely generated Abelian groups) and $H$ is a divisible group, then $G \otimes H \cong 0$.

**2.** Prove that if $G$ is free or if $H$ is divisible, then $\text{Ext}(G, H) \cong 0$.

**3.** Prove that $\text{Ext}(\mathbb{Z}_p, \mathbb{Z}_q) \cong \mathbb{Z}_{(p,q)}$.

**4.** Prove that $\text{Ext}(\mathbb{Z}, \mathbb{Z}_q) \cong 0$.

**5.** From the homology of real projective spaces (see Sect. III.5), prove that

$$H^p(\mathbb{R}P^n, \mathbb{Z}) \cong \begin{cases} \mathbb{Z} & \text{if } p = 0 \\ \mathbb{Z}_2 & \text{if } 0 < p = 2q \leq n \\ 0 & \text{if } 0 < p = 2q - 1 < n \\ \mathbb{Z} & \text{if } n \text{ odd and } p = n . \end{cases}$$

## IV.2 The Cohomology Ring

After studying the section on the cohomology of simplicial complexes, the reader could arrive to the conclusion that the graded (homology and cohomology) Abelian groups $H_*(K; \mathbb{Z})$ and $H^*(K; \mathbb{Z})$ are simply similar algebraic structures connected to $K$ and, therefore, that all the work done to construct the contravariant functor $H^*(-; \mathbb{Z})$ will produce no information on $K$, which is not already given by $H_*(-; \mathbb{Z})$; this is not the case, because cohomology has a richer structure than homology and, for this reason, produces more results than homology.

This section is written in the context of cohomology with coefficients in $\mathbb{Z}$ but all results could be written for cohomology with coefficients in a commutative ring $A$ with unity $1_A$. We shall first of all see that $H^*(K; \mathbb{Z})$ has the structure of a ring.

Let $K = (X, \Phi)$ be a finite oriented simplicial complex; let

$$c^p \in C^p(K; \mathbb{Z}) = \text{Hom}(C_p(K), \mathbb{Z})$$

and $c^q \in C^q(K; \mathbb{Z})$ be any two cochains. We define the $(p + q)$-cochain $c^p \cup c^q \in C^{p+q}(K, \mathbb{Z})$ as follows: for any generator $\{x_0, x_1, \ldots, x_{p+q}\}$ of $C_{p+q}(K)$, we set

$$c^p \cup c^q(\{x_0, x_1, \ldots, x_{p+q}\}) := c^p(\{x_0, x_1, \ldots, x_p\}) \times c^q(\{x_p, x_{p+1}, \ldots, x_{p+q}\})$$

where $\times$ is the product in the ring $\mathbb{Z}$; we linearly extend the action of $c^p \cup c^q$ to every element of $C_{p+q}(K)$. In this manner, we have defined a function

$$\cup : \ C^p(K; \mathbb{Z}) \times C^q(K; \mathbb{Z}) \longrightarrow C^{p+q}(K; \mathbb{Z})$$

known as *cup product*. Proving that the cup product is associative and distributive relatively to addition is not difficult; besides, the element $c^0 \in C^0(K; \mathbb{Z})$ defined by

$$(\forall x \in X) \ c^0(\{x\}) = 1$$

is such that
$$(\forall c^p \in C^p(K;\mathbb{Z}))\ c^0 \cup c^p = c^p \cup c^0 = c^p.$$

We conclude from these remarks that the cup product provides the graded Abelian group $C^*(K;\mathbb{Z})$ with a structure of graded ring with identity $c^0$.

**(IV.2.1) Theorem.** *For every $c^p \in C^p(K;\mathbb{Z})$ and $c^q \in C^q(K;\mathbb{Z})$,*

$$\partial^{p+q}(c^p \cup c^q) = \partial^p(c^p) \cup c^q + (-1)^p c^p \cup \partial^q(c^q)$$

*holds.*

*Proof.* We prove this result by computing $\partial^{p+q}(c^p \cup c^q)$ on any generator $\{x_0,\dots,x_{p+q+1}\}$ of $C_{p+q+1}(K)$.

$$\partial^{p+q}(c^p \cup c^q)(\{x_0,\dots,x_{p+q+1}\}) = (c^p \cup c^q)(\partial_{p+q+1}(\{x_0,\dots,x_{p+q+1}\}))$$

$$= (c^p \cup c^q)\left( \sum_{i=0}^{p+q+1} (-1)^i \{x_0,\dots,\widehat{x_i},\dots,x_{p+q+1}\} \right) ;$$

On the other hand,

$$\partial^p(c^p) \cup c^q(\{x_0,\dots,x_{p+q+1}\}) =$$
$$= \partial^p(c^p)(\{x_0,\dots,x_{p+1}\}) \times c^q(\{x_{p+1},\dots,x_{p+q+1}\})$$
$$= c^p\left( \sum_{i=0}^{p+1} (-1)^i \{x_0,\dots,\widehat{x_i},\dots,x_{p+1}\} \right) \times c^q(\{x_{p+1},\dots,x_{p+q+1}\})$$
$$= c^p \cup c^q\left( \sum_{i=0}^{p} (-1)^i \{x_0,\dots,\widehat{x_i},\dots,x_{p+q+1}\} \right) +$$
$$+ (-1)^{p+1} c^p(\{x_0,\dots,x_p\}) \times c^q(\{x_{p+1},\dots,x_{p+q+1}\})$$

and

$$c^p \cup \partial^q(c^q)(\{x_0,\dots,x_{p+q+1}\}) =$$
$$= c^p(\{x_0,\dots,x_p\}) \times c^q\left( \sum_{i=p}^{p+q+1} (-1)^i \{x_p,\dots,\widehat{x_i},\dots,x_{p+q+1}\} \right)$$
$$= c^p(\{x_0,\dots,x_p\}) \times c^q(\{x_{p+1},\dots,x_{p+q+1}\}) +$$
$$+ (-1)^p c^p \cup c^q\left( \sum_{i=p+1}^{p+q+1} (-1)^i \{x_0,\dots,\widehat{x_i},\dots,x_{p+q+1}\} \right) .$$

Therefore,

$$\partial^{p+q}(c^p \cup c^q)(\{x_0,\dots,x_{p+q+1}\}) =$$
$$= (\partial^p(c^p) \cup c^q + (-1)^p c^p \cup \partial^q(c^q))(\{x_0,\dots,x_{p+q+1}\}) . \quad \blacksquare$$

**(IV.2.2) Corollary.** *The graded Abelian group $Z^*(K;\mathbb{Z})$ is a graded ring with identity; the graded Abelian group $B^*(K;\mathbb{Z})$ is one of its (bilateral) ideals.*

*Proof.* It is enough to prove that the cup product has the following properties:

$$\cup: Z^p(K;\mathbb{Z}) \times Z^q(K;\mathbb{Z}) \longrightarrow Z^{p+q}(K;\mathbb{Z}) ,$$

$$\cup: Z^p(K;\mathbb{Z}) \times B^q(K;\mathbb{Z}) \longrightarrow B^{p+q}(K;\mathbb{Z}) ,$$

$$\cup: B^p(K;\mathbb{Z}) \times Z^q(K;\mathbb{Z}) \longrightarrow B^{p+q}(K;\mathbb{Z}) .$$

We prove only the last one. Let $c^p = \partial^{p-1}(c^{p-1}) \in B^p(K;\mathbb{Z})$ and $c^q \in Z^q(K;\mathbb{Z})$ be given arbitrarily. Then,

$$c^p \cup c^q = \partial^{p+q-1}(c^{p-1} \cup c^q) . \qquad \blacksquare$$

The quotient ring

$$H^*(K;\mathbb{Z}) = Z^*(K;\mathbb{Z})/B^*(K;\mathbb{Z})$$

is called *cohomology ring* of the (finite and oriented) simplicial complex $K$ with coefficients in $\mathbb{Z}$. This ring is skew-commutative; in fact, we have the following result:

**(IV.2.3) Theorem.** *For every $x \in H^p(K;\mathbb{Z})$ and $y \in H^q(K;\mathbb{Z})$,*

$$x \cup y = (-1)^{pq} y \cup x .$$

*Proof.* Let us suppose that

$$x = c^p + B^p(K;\mathbb{Z}) \text{ and } y = c^q + B^q(K;\mathbb{Z}) ;$$

we wish to prove that, for every generator $\{x_0, \ldots, x_{p+q}\}$ of $C_{p+q}(K)$,

$$c^p \cup c^q(\{x_0, \ldots, x_{p+q}\}) = (-1)^{pq} c^q \cup c^p(\{x_0, \ldots, x_{p+q}\}) .$$

Since we are working with oriented simplexes, according to our rules we have, for every $\ell$-simplex $\{x_0, x_1, \ldots, x_\ell\}$,

$$\{x_0, x_1, \ldots, x_\ell\} = (-1)^{\frac{1}{2}\ell(\ell-1)}\{x_\ell, x_{\ell-1}, \ldots, x_0\}$$

and so

$$c^p \cup c^q(\{x_0, x_1, \ldots, x_{p+q}\}) =$$
$$= c^p \cup c^q((-1)^{\frac{1}{2}(p+q)(p+q-1)}(\{x_{p+q}, \ldots, x_0\}))$$
$$= (-1)^{\frac{1}{2}(p+q)(p+q-1)} c^p(\{x_{p+q}, \ldots, x_p\}) \times c^q(\{x_p, \ldots, x_0\})$$
$$= (-1)^{\frac{1}{2}(p+q)(p+q-1)} c^q(\{x_p, \ldots, x_0\}) \times c^p(\{x_{p+q}, \ldots, x_p\})$$

(for $\mathbb{Z}$ is commutative)

$$= (-1)^{\frac{1}{2}(p+q)(p+q-1)}(-1)^{\frac{1}{2}q(q-1)}(-1)^{\frac{1}{2}p(p-1)}$$
$$= c^q(\{x_0,\ldots,x_q\}) \times c^p(\{x_q,\ldots,x_{p+q}\})$$
$$= (-1)^{pq}c^q(\{x_0,\ldots,x_q\}) \times c^p(\{x_q,\ldots,x_{p+q}\}) . \blacksquare$$

Hence, $H^*(K;\mathbb{Z})$ is a *graded skew-commutative* ring with identity. We now prove that $H^*(-;\mathbb{Z})$ is a contravariant functor from the category of simplicial complexes **Csim** to the category of graded commutative rings with identity.

**(IV.2.4) Lemma.** *For any* $c^p \in C^p(L;\mathbb{Z})$, $c^q \in C^q(L;\mathbb{Z})$, *and any simplicial function* $f\colon K \to L$, *the equality*

$$C^{p+q}(f)(c^p \cup c^q) = C^p(f)(c^p) \cup C^q(f)(c^q)$$

*holds.*

*Proof.* For each $\{x_0,\ldots,x_p,\ldots,x_{p+q}\} \in C_{p+q}(K)$,

$$C^{p+q}(f)(c^p \cup c^q)(\{x_0,\ldots,x_p,\ldots,x_{p+q}\})$$

$$= (c^p \cup c^q)C_{p+q}(f)(\{x_0,\ldots,x_p,\ldots,x_{p+q}\})$$

$$= \begin{cases} c^p(\{f(x_0),\ldots,f(x_p)\}) \times c^q(\{f(x_p),\ldots,f(x_{p+q})\}) & (\forall i \neq j)\, f(x_i) \neq f(x_j) \\ 0 & \text{otherwise} . \end{cases}$$

On the other hand,

$$(C^p(f)(c^p) \cup C^q(f)(c^q))(\{x_0,\ldots,x_p,\ldots,x_{p+q}\})$$

$$= (c^p C_p(f) \cup c^q C_q(f))(\{x_0,\ldots,x_p,\ldots,x_{p+q}\})$$

$$= \begin{cases} c^p(\{f(x_0),\ldots,f(x_p)\}) \times c^q(\{f(x_p),\ldots,f(x_{p+q})\}) & (\forall i \neq j)\, f(x_i) \neq f(x_j) \\ 0 & \text{otherwise} . \end{cases}$$

$\blacksquare$

Consequently, $C^*(f)\colon C^*(L;\mathbb{Z}) \to C^*(K;\mathbb{Z})$ preserves cup products; since the homomorphisms $C^p(f)$ commute with the appropriate coboundary operators, the simplicial function induces a ring homomorphism $H^*(f)\colon H^*(L;\mathbb{Z}) \to H^*(K;\mathbb{Z})$.

Regarding the cohomology of polyhedra, given any polyhedron $|K|$, we may define in $H^*(|K|;\mathbb{Z})$ a structure of ring with identity simply because $H^*(|K|;\mathbb{Z}) = H^*(K;\mathbb{Z})$. We must verify that $H^*(-;\mathbb{Z})$ is a contravariant functor from the category of polyhedra **P** to the category of graded commutative rings with identity. In fact, let $f\colon |K| \to |L|$ be a continuous function and $g\colon K^{(r)} \to L$ a simplicial approximation of $f$. We know that the homomorphism $H^*(\pi^r;\mathbb{Z})$ induced by the projection $\pi^r\colon K^{(r)} \to K$ is an isomorphism; in addition,

$$H^*(f;\mathbb{Z}) = H^*(\pi^r;\mathbb{Z})^{-1}H^*(g;\mathbb{Z})\colon H^*(|L|;\mathbb{Z}) \to H^*(|K|;\mathbb{Z}).$$

Since $g$ and $\pi^r$ are simplicial functions, $H^*(g;\mathbb{Z})$ and $H^*(\pi^r;\mathbb{Z})$ are ring homomorphisms; however, since $H^*(\pi^r;\mathbb{Z})$ is an isomorphism, also $H^*(\pi^r;\mathbb{Z})^{-1}$ preserves cup products. We conclude that if two polyhedra $|K|$ and $|L|$ are of the same homotopy type, then their cohomology rings are isomorphic (as rings).

We now wish to prove that the equality of the (co)homology groups is a necessary but not sufficient condition for the cohomology rings to be isomorphic. To this end, let us return to the cohomology of the bi-dimensional torus $T^2$ and of the space $X = S^2 \vee (S^1 \vee S^1)$. We have already proved (see Sect. III.4) that these two triangulable spaces have the same homology and, by the Universal Coefficient Theorem in Cohomology, also the same cohomology; moreover, we have already seen (see Sect. III.4 again) that $T^2$ and $X$ are not homeomorphic. We now compute the cohomology of $T^2$ and $X$ once more, by explicitly considering the ring structure.

We begin with $T^2$. Let us consider the oriented triangulation of $T^2$ used to compute $H_*(T^2;\mathbb{Z})$ (see Fig. II.10 in Sect. II.2, p. 63). We recall that if $z_1^1$ and $z_1^2$ are the two 1-cycles

$$z_1^1 = \{0,3\} + \{3,4\} + \{4,0\} \text{ and } z_1^2 = \{0,1\} + \{1,2\} + \{2,0\},$$

then

$$w_1^1 = \{1,6\} + \{6,5\} + \{5,8\} + \{8,7\} + \{7,2\} + \{2,1\}$$

is a 1-cycle homologous to $z_1^1$ (see how $H_*(T^2,\mathbb{Z})$ was computed) and

$$w_1^2 = \{3,7\} + \{7,5\} + \{5,8\} + \{8,6\} + \{6,4\} + \{4,3\}$$

is a 1-cycle homologous to $z_1^2$. For each 1-simplex $\{i,j\}$ of $T^2$, let

$$\{i,j\}^* \in \operatorname{Hom}(C_1(T^2),\mathbb{Z})$$

be the function such that

$$(\forall\{k,\ell\} \in T^2) \ \{i,j\}^*(\{k,\ell\}) = \begin{cases} 1 & \text{if } \{k,\ell\} = \{i,j\} \\ 0 & \text{if } \{k,\ell\} \neq \{i,j\}. \end{cases}$$

We now perform some calculations:

1. $\partial^1\{1,6\}^*(\{1,6,5\}) = (\{1,6\}^*\partial_2)(\{1,6,5\}) = \{1,6\}^*(\partial_2\{1,6,5\}) = 1$
2. $\partial^1\{1,6\}^*(\{1,2,6\}) = -1$

we conclude that $\partial^1\{1,6\}^* = \{1,6,5\}^* - \{1,2,6\}^*$. Similarly, we find that

$$\partial^1\{6,5\}^* = \{1,6,5\}^* - \{5,6,8\}^*, \ \partial^1\{5,8\}^* = \{5,8,7\}^* - \{5,6,8\}^*,$$
$$\partial^1\{8,7\}^* = \{5,8,7\}^* - \{7,8,2\}^*, \ \partial^1\{7,2\}^* = \{7,2,1\}^* - \{7,8,2\}^*,$$
$$\partial^1\{2,1\}^* = \{7,2,1\}^* - \{1,2,6\}^*,$$

and therefore

$$\partial^1(\{1,6\}^* - \{6,5\}^* + \{5,8\}^* - \{8,7\}^* + \{7,2\}^* - \{2,1\}^*) = 0,$$

that is to say, the cochain

$$\zeta_1^1 = \{1,6\}^* - \{6,5\}^* + \{5,8\}^* - \{8,7\}^* + \{7,2\}^* - \{2,1\}^*$$

is a cocycle. Likewise, we prove that

$$\zeta_2^1 = \{3,7\}^* - \{7,5\}^* + \{5,8\}^* - \{8,6\}^* + \{6,4\}^* - \{4,3\}^*$$

is a cocycle. These cocycles are not cohomologous to each other; therefore, their classes generate the cohomology group $H^1(T^2; \mathbb{Z})$. In order to compute the cup product $\zeta_1^1 \cup \zeta_2^1$, we need to apply it to the 2-simplexes of $T^2$; we are able to draw nontrivial conclusions only from $\{5,6,8\}$ and $\{8,7,5\}$; besides, by the distributivity of the cup product with respect to the sum, $\{5,6\}^* \cup \{6,8\}^*$ and $\{8,7\}^* \cup \{7,5\}^*$ are the parts of $\zeta_1^1 \cup \zeta_2^1$, which may yeld nontrivial results when applied to the two 2-simplexes that we have just singled out. In fact,

$$\{5,6\}^* \cup \{6,8\}^*(\{5,6,8\}) = 1 \text{ and } \{8,7\}^* \cup \{7,5\}^*(\{8,7,5\}) = 1$$

and so, $\zeta_1^1 \cup \zeta_2^1$ is not null.

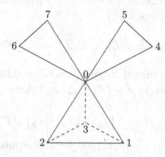

**Fig. IV.1**

We now look into the cohomology of $S^2 \vee (S^1 \vee S^1)$. We represent $S^2 \vee (S^1 \vee S^1)$ by the geometric realization of the simplicial complex with eight vertices, namely, 0, 1, 2, 3, 4, 5, 6, 7, depicted in Fig. IV.1, having the following simplexes (besides the vertices):

1-simplexes:

$$\{0,1\}, \{0,2\}, \{0,3\}, \{0,4\}, \{0,5\}, \{0,6\}, \{0,7\}, \{4,5\},$$
$$\{6,7\}, \{1,2\}, \{1,3\}, \{2,3\};$$

2-simplexes:

$$\{0,1,2\}, \{0,1,3\}, \{0,2,3\}, \{1,2,3\}.$$

It is easily proved that the only 1-cocycles are $\xi_1^1 = \{0,4\}^* + \{4,5\}^* + \{5,0\}^*$ and $\xi_2^1 = \{0,6\}^* + \{6,7\}^* + \{7,0\}^*$. These cocycles are indipendent and their cohomology classes generate $H^1(S^2 \vee (S^1 \vee S^1); \mathbb{Z})$; also easily proved is the fact that $\xi_1^1 \cup \xi_2^1 = 0$. The cohomology rings of $T^2$ and $S^2 \vee (S^1 \vee S^1)$ are therefore different; consequently, the polyhedra $T^2$ and $S^2 \vee (S^1 \vee S^1)$ cannot be of the same homotopy type.

## IV.3 The Cap Product

In this section, we study a product between homology and cohomology classes. More precisely, let $K$ be a simplicial complex that we once again assume to be finite and oriented. Consider the chain and the cochain groups associated with $K$

$$C_n(K; \mathbb{Z}) \quad \text{and} \quad C^n(K; \mathbb{Z}) = \text{Hom}(C_n(K), \mathbb{Z}) ,$$

where $n$ is an integer such that $0 \leq n \leq \dim K$. For every $d \in C^p(K; \mathbb{Z})$ and for every $(p+q)$-simplex $\{x_0, \ldots, x_p, \ldots, x_{p+q}\}$ of $K$ (that is to say, a generator of $C_{p+q}(K)$), we define

$$d \cap \{x_0, \ldots, x_p, \ldots, x_{p+q}\} := d(\{x_0, \ldots, x_p\})\{x_p, \ldots, x_{p+q}\} \in C_q(K);$$

we linearly extend the definition $d \cap c$ to every $(p+q)$-chain $c \in C_{p+q}(K; \mathbb{Z})$; in this way we obtain a bilinear relation

$$\cap \colon C^p(K; \mathbb{Z}) \times C_{p+q}(K; \mathbb{Z}) \longrightarrow C_q(K; \mathbb{Z})$$

called *cap product*. Note that the cap product is not defined if $q < 0$ or $q > \dim K - p$. In particular, if $p = q$, for every $d \in C^p(K; \mathbb{Z})$ and every $p$-simplex $\{x_0, \ldots, x_p, \}$, we have

$$d \cap \{x_0, \ldots, x_p\} = d(\{x_0, \ldots, x_p\})\{x_p\} \in C_0(K; \mathbb{Z}).$$

Let $\varepsilon \colon C_0(K; \mathbb{Z}) \to \mathbb{Z}$ be the augmentation homomorphism of the positive chain complex $C(K, \mathbb{Z})$; by definition,

$$\varepsilon(d \cap \{x_0, \ldots, x_p\}) = d(\{x_0, \ldots, x_p\})$$

and by linearity

$$\varepsilon(c \cap d) = d(c)$$

for every $c \in C_p(K; \mathbb{Z})$ and any $d \in C^p(K; \mathbb{Z})$. The next two results establish a relation between cap and cup products.

**(IV.3.1) Theorem.** *For every $c \in C_{p+q+r}(K; \mathbb{Z})$, $d \in C^p(K; \mathbb{Z})$, and $e \in C^q(K; \mathbb{Z})$, the equality*

$$d \cap (e \cap c) = (d \cup e) \cap c$$

*holds.*

*Proof.* Let us suppose that $c = \{x_0, \ldots, x_p, \ldots, x_{p+q}, \ldots, x_{p+q+r}\}$; then

$$
\begin{aligned}
d \cap (e \cap c) &= d \cap (e \cap \{x_0, \ldots, x_p, \ldots, x_{p+q}, \ldots, x_{p+q+r}\}) \\
&= d \cap e(\{x_0, \ldots, x_q\})\{x_q, \ldots, x_{p+q+r}\} \\
&= e(\{x_0, \ldots, x_q\})d(\{x_q, \ldots, x_{q+p}\})\{x_{q+p}, \ldots, x_{q+p+r}\} \, .
\end{aligned}
$$

On the other hand,

$$
\begin{aligned}
(d \cup e) \cap c &= (d \cup e) \cap \{x_0, \ldots, x_p, \ldots, x_{p+q}, \ldots, x_{p+q+r}\} \\
&= e(\{x_0, \ldots, x_q\})d(\{x_q, \ldots, x_{q+p}\})\{x_{q+p}, \ldots, x_{q+p+r}\} \, .
\end{aligned}
$$

The general result follows by linearity.                                              ■

**(IV.3.2) Theorem.** *For every $c \in C_{p+q}(K; \mathbb{Z})$, $d \in C^p(K; \mathbb{Z})$, and $e \in C^q(K; \mathbb{Z})$, the equality*

$$
(d \cup e)(c) = e(d \cap c)
$$

*holds.*

*Proof.* This proof is similar to the previous one: we first show that the equation is true for $c = \{x_0, \ldots, x_{p+q}\}$ and then use linearity for completing the proof.      ■

**(IV.3.3) Theorem.** *For every $c \in C_{p+q}(K; \mathbb{Z})$ and $d \in C^p(K; \mathbb{Z})$,*

$$
\partial_q(d \cap c) = (-1)^p(d \cap \partial_{p+q}(c) - \partial^p(d) \cap c) \, .
$$

*Proof.* We begin by noting that an element $x \in C_{p+q}(K; \mathbb{Z})$ is null if and only if, for every $e \in C^{p+q}(K; \mathbb{Z})$, we have $e(x) = 0$. Hence, it is enough to prove that

$$
e(\partial_q(d \cap c)) = e((-1)^p(d \cap \partial_{p+q}(c) - \partial^p(d) \cap c))
$$

for any $e$. In fact, by Remark (IV.1.2) and by Theorem (IV.3.2),

$$
e(\partial_q(d \cap c)) = (\partial^{q-1}e)(d \cap c) = (d \cup \partial^{q-1}(e))(c) \, .
$$

On the other hand,

$$
\begin{aligned}
e((d \cap \partial_{p+q}(c) - \partial^p(d) \cap c)) &= ((d \cup e)(\partial_{p+q}(c)) - (\partial^p(d) \cup e)(c)) \\
&= \partial^{p+q-1}(d \cup e)(c) - (\partial^p(d) \cup e)(c) \, ;
\end{aligned}
$$

however, by Theorem (IV.2.1), this last expression equals

$$
(\partial^p(d) \cup e)(c) + (-1)^p(d \cup \partial^{q-1}(e))(c) - (\partial^p(d) \cup e)(c) = (-1)^p(d \cup \partial^{q-1}(e))(c) \, .
$$
                                                                                      ■

**(IV.3.4) Corollary.** *The cap product induces the following homomorphisms:*

$$
\cap \colon Z^p(K; \mathbb{Z}) \times Z_{p+q}(B; \mathbb{Z}) \longrightarrow Z_q(K; \mathbb{Z}) \, ,
$$

$$\cap:\ B^p(K;\mathbb{Z})\times Z_{p+q}(B;\mathbb{Z})\longrightarrow B_q(K;\mathbb{Z})\ ,$$

$$\cap:\ Z^p(K;\mathbb{Z})\times B_{p+q}(B;\mathbb{Z})\longrightarrow B_q(K;\mathbb{Z})\ .$$

*Proof.* The first assertion follows directly from the theorem.

Given $d = \partial^{p-1}(d') \in B^p(K;\mathbb{Z})$ and $c \in Z_{p+q}(K;\mathbb{Z})$,

$$\partial_{q+1}(d'\cap c) = (-1)^p(d'\cap\partial_{p+q}(c) - \partial^{p-1}(d')\cap c) = (-1)^p(d\cap c)$$

and, therefore,

$$d\cap c = (-1)^p\partial_{q+1}(d'\cap c) \in B_q(K;\mathbb{Z})\ .$$

The last case is proved in a similar way.                                       ∎

By what we have showed, the cap product of $p$-cochains and $(p+q)$-chains becomes a bilinear relation in (co)homology

$$\cap:\ H^p(K;\mathbb{Z})\times H_{p+q}(K;\mathbb{Z})\longrightarrow H_q(K;\mathbb{Z})$$

which is also called *cap product*. The remarks on cohomology of polyhedra (see Sect. IV.1.1) allow us to define

$$\cap:\ H^p(|K|;\mathbb{Z})\times H_{p+q}(|K|;\mathbb{Z})\longrightarrow H_q(|K|;\mathbb{Z})$$

for any polyhedron $|K| \in \mathbf{P}$.

We now consider the morphisms. Let a simplicial function $f\colon K \to L$, an element $d$ of $C^p(L)$, and $c \in C_{p+q}(K)$ be given. Then,

$$C_{p+q}(f)(c) \in C_{p+q}(L) \text{ and } C^p(f)(d) \in C^p(K)$$

(recall that

$$C^p(f) = \mathrm{Hom}(f;\mathbb{Z})\colon\ \mathrm{Hom}(C_p(L);\mathbb{Z}) = C^p(L) \to \mathrm{Hom}(C_p(K);\mathbb{Z})\ ).$$

Therefore, $d\cap C_{p+q}(f)(c) \in C_q(L)$ and $C^p(f)(d)\cap c \in C_q(K)$.

**(IV.3.5) Theorem.** *Given a simplicial function* $f\colon K \to L$, $d \in C^p(L)$, *and* $c \in C_{p+q}(K)$,

$$d\cap C_{p+q}(f)(c) = C_q(f)(C^p(f)(d)\cap c)\ ,$$

$$C^q(f))(d\cap C_{p+q}(f)(c)) = C^p(f)(d)\cap c\ .$$

*Proof.* We only prove the first equality. Let us suppose that

$$c = \{x_0,\dots,x_p,\dots,x_{p+q}\}$$

and that for any $i \neq j$ between $0$ and $p+q$, $f(x_i) \neq f(x_j)$; it follows that

$$C_{p+q}(f)(c) = \{f(x_0),\dots,f(x_p),\dots,f(x_{p+q})\}\ .$$

Therefore

$$d \cap C_{p+q}(f)(c) = d(\{f(x_0), \ldots, f(x_p)\})\{f(x_p), \ldots, f(x_{p+q})\} .$$

On the other hand, $C^p(f) = dC_p(f)$ and so,

$$C_q(f)(C^p(f)(d) \cap c) = d(\{f(x_0), \ldots, f(x_p)\})\{f(x_p), \ldots, f(x_{p+q})\} .$$

If there are $i$, $j$ for which $f(x_i) = f(x_j)$, then $C_{p+q}(f)(c) = 0$ and at least one of the two assertions

$$C^p(f)(d) = 0 , \; C_q(f)(\{x_p, \ldots, x_{p+q}\}) = 0$$

is true; therefore, the theorem holds also in this case. ∎

Theorem (IV.3.5) and what we have previously done enable us to state the following result:

**(IV.3.6) Theorem.** *Let* $|L|, |K| \in \mathbf{P}$ *and a continuous function* $f \colon |K| \to |L|$ *be given. The following diagram:*

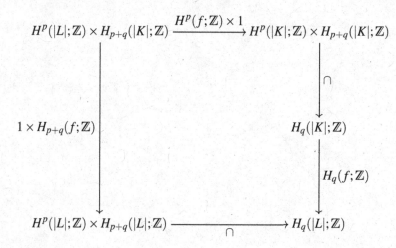

*is commutative.*

*Proof.* We leave it to the reader. ∎

# Exercises

**1.** Let $f \colon |K| \to |L|$ be a given map, $z \in H_n(|K|, \mathbb{Z})$, and $u \in \check{H}^q(|L|, \mathbb{Z})$. Prove that

$$H_{n-q}(f)(u \cap H_n(f)(z)) = H^q(f)(u) \cap z .$$

**2.** Prove that

$$(\forall z \in H_n(|K|, \mathbb{Z})) \; 1 \cap z = z .$$

# Chapter V
# Triangulable Manifolds

## V.1 Topological Manifolds

A Hausdorff topological space $X$ is called *an n-dimensional manifold*[1] or simply *an n-manifold*, if for every point $x \in X$ there exists an open set $U$ of $X$ that contains $x$ and is homeomorphic to an open set of $\mathbb{R}^n$. Hence, an $n$-manifold $X$ is characterized by a set $\mathfrak{A} = \{(U_i, \phi_i) \mid i \in J\}$, where $U_i$ are open sets covering $X$, and $\phi_i$ is a homeomorphism from $U_i$ onto an open set of $\mathbb{R}^n$. The set $\mathfrak{A}$ is the *atlas* of $X$ and each pair $(U_i, \phi_i)$ is a *chart* of $X$.

Even before we give some examples of manifolds, we note that the condition that $X$ be Hausdorff is an integral part of the definition and does not depend on the other conditions. In fact, consider

$$X = ]-1, 2] = \{x \in \mathbb{R} \mid -1 < x \leq 2\}$$

with the topology given by the set $\mathfrak{U}$ of open sets $U$, where $U \in \mathfrak{U}$ if and only if one of the following conditions holds true: (a) $U = X$; (b) $U = \emptyset$; (c) $U$ is any union of sets such as $]\alpha, \beta[$ with $-1 \leq \alpha < \beta \leq 2$ or $]\alpha, 0[\cup]\beta, 2]$, where $-1 \leq \alpha < 0$ and $-1 \leq \beta < 2$. This is not a Hausdorff space since any open set containing 0 intersects, any open set containing 2. On the other hand, any $x \in X \smallsetminus \{2\}$ is contained by an open set homeomorphic to an open set of $\mathbb{R}$; regarding the point $x = 2$, the reader may verify that the open set $U = ]-\frac{1}{2}, 0[\cup]\frac{3}{2}, 2]$ is homeomorphic to an open interval of $\mathbb{R}$; hence, $X$ has the properties of 1-manifold except for the Hausdorff separation property. Here is a simple and useful result:

**(V.1.1) Lemma.** *Let $V$ be an open set in an n-manifold $X$. Then $V$ is an n-manifold.*

*Proof.* For any $x \in V \subset X$, let $(U, \phi)$ be a chart of $X$ containing $x$. Then $U \cap V$ is open in $V$ containing $x$ and $\phi(U \cap V)$ is an open set of $\mathbb{R}^n$ homeomorphic to $U \cap V$. ■

---

[1] In the literature, it is sometimes required that the topological space $X$ fulfill other conditions to be defined as a manifold (e.g., second countable, paracompact).

D.L. Ferrario and R.A. Piccinini, *Simplicial Structures in Topology*,
CMS Books in Mathematics, DOI 10.1007/978-1-4419-7236-1_V,
© Springer Science+Business Media, LLC 2011

**(V.1.2) Lemma.** *The Cartesian product of an n-manifold X by an m-manifold Y is an $(n+m)$-manifold.*

*Proof.* For any $(x,y) \in X \times Y$, we choose two charts $(U, \phi)$ and $(V, \psi)$ of $x \in X$ and $y \in Y$, respectively; we note that $(U \times V, \phi \times \psi)$ is a chart of the Cartesian product of the manifolds: in fact, $\phi(U) \times \psi(V)$ is an (elementary) open set of $\mathbb{R}^n \times \mathbb{R}^m \cong \mathbb{R}^{n+m}$.                                                                                     ∎

We now give some examples. It is easily proved that, for every $n > 0$, the Euclidean space $\mathbb{R}^n$ is an $n$-manifold. The circle $S^1 \subset \mathbb{R}$, with the topology induced by the Euclidean topology of $\mathbb{R}$, is a 1-manifold; this is readily proved. The sphere $S^2$ with the topology induced by $\mathbb{R}^3$ is a 2-manifold (or *surface*): in fact, $S^2$ is Hausdorff and its atlas is the set

$$\mathfrak{A} = \{(S^2 \smallsetminus \{-x\}, \phi_x) \mid x \in S^2\}$$

where $\phi_x$ is the stereographic projection of $S^2 \smallsetminus \{-x\}$ from the point $-x$ on the plane $T_x$ tangent to $S^2$ at $x$. A similar result holds also for hyperspheres $S^n$, $n > 2$.

An immediate consequence from Lemma (V.1.2) is that the *torus* $T^2 = S^1 \times S^1$ is a 2-manifold.

To prove that the real projective space $\mathbb{R}P^n$ is an $n$-manifold, we use the following theorem.

**(V.1.3) Theorem.** *Let X be a compact n-manifold and let G be a finite topological group freely acting on X. Then, the quotient space $X/G$ is an n-manifold.*

*Proof.* Let $G$ be the finite group whose elements are $g_1 = 1_G, g_2, \ldots, g_p$; then, the orbit of any element $x \in X$ consists in $p$ (distinct) elements $x = xg_1, xg_2, \ldots, xg_p$. For each pair $(x, xg_i)$, $i = 2, \ldots, p$, we take a pair $(U_i, V_i)$ of open sets of $X$ such that

$$x \in U_i, \ xg_i \in V_i \text{ and } U_i \cap V_i = \emptyset.$$

Since $V_i$ contains $xg_i$, $V_i g_i^{-1}$ surely contains $x$; therefore, the set

$$U = \bigcap_{i=2}^{p} \left( U_i \cap V_i g_i^{-1} \right)$$

is an open set of $X$ containing $x$, is disjoint from all open sets $V_i$, $i = 2, \ldots, p$ and, consequently, from all $U g_i$, $i = 2, \ldots, p$ (because

$$U g_i \subset \left( U_i \cap V_i g_i^{-1} \right) g_i \subset V_i g_i^{-1} g_i = V_i$$

and $V_i \cap U = \emptyset$ for each $i = 2, \ldots, p$). The restriction to $U$ of the canonical epimorphism $q \colon X \to X/G$, that is to say,

$$q|U \colon U \to q(U) \subset X/G$$

is a bijection. By Lemma (I.3.2), the projection $q$ (and so also its restriction $q|U$) is a map both open and closed; therefore, $q|U$ is a homeomorphism. Since $U$, as an open set of $X$, is an $n$-manifold, (see Lemma (V.1.1)), it follows that the point $[x] \in X/G$ belongs to an open set of $X/G$ homeomorphic to an open set of $\mathbb{R}^n$. To reach the conclusion that $X/G$ is an $n$-manifold, we need to demonstrate that $X/G$ is a Hausdorff space. But this is a direct consequence of Theorem (I.1.29), because of the hypotheses. ∎

Since the real projective space may be written as the quotient $\mathbb{R}P^n = S^n/\mathbb{Z}_2$, it is a consequence of this theorem that $\mathbb{R}P^n$ is an $n$-manifold.

## V.1.1 Triangulable Manifolds

An $n$-manifold $X$ is *triangulable* if there exists a polyhedron $|K|$ homeomorphic to $X$. The real projective space $\mathbb{R}P^n$ is an example of a triangulable manifold (see Sect. III.5). An open disk $\mathring{D}^n_r(x) \subset \mathbb{R}^n$ with radius $r$ and center $x \in \mathbb{R}^n$ is triangulable.

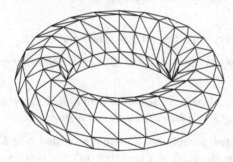

**Fig. V.1** Triangulated torus

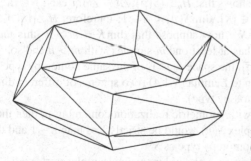

**Fig. V.2** Triangulated torus (with fewer triangles)

We now prove that the dimension of the simplicial complex $K = (X, \Phi)$ equals the dimension of the manifold $X$; moreover, if $X$ is connected, the spaces $|\bar{\sigma}| \subset |K|$ appear in a particular way, in other words, each two of them either intersect each other at an $n-1$ face or they are linked by a chain of $n$-simplexes. The proof of these results is based on an argument found in the next lemma; we ask the reader to remember that we may associate the closed subspace $S(p)$, that is to say, the boundary of the space

$$D(p) = \bigcup_{\sigma \in B(p)} |\overline{\sigma}|,$$

with each $p \in |K|$, where $B(p)$ is the set of all $\sigma \in \Phi$ such that $p \in |\overline{\sigma}|$ (see Sect. II.2).

**(V.1.4) Lemma.** *Let $X \cong |K|$ be a triangulable $n$-manifold; then for whatever $p \in |K|$, $S(p)$ is of the same homotopy type as the sphere $S^{n-1}$.*

*Proof.* Let $(U, \phi)$ be a chart of $|K|$ containing the point $p$. Since $\phi(U)$ is homeomorphic to an open set $W \subset \mathbb{R}^n$, there is a closed disk $D_\varepsilon^n(\phi(p))$, with center $\phi(p)$ and radius $\varepsilon$, contained in $W$. The restriction of $\phi^{-1}$ to $D_\varepsilon^n(\phi(p))$ is a homeomorphism from $D_\varepsilon^n(\phi(p))$ into a subspace of $|K|$ containing $p$. We recall that $D_\varepsilon^n(\phi(p))$ is a triangulable space; let $L$ be a simplicial complex whose geometric realization is identified to $D_\varepsilon^n(\phi(p))$. We now consider the homeomorphism $\phi^{-1}: |L| \to |K|$ and apply Theorem (II.2.11) to conclude that $S(\phi(p)) \sim S(p)$. We complete the proof by noting that $S(\phi(p)) \cong S^{n-1}$. ∎

The following result is very important (cf. [24, Theorem 5.3.3]).

**(V.1.5) Theorem.** *Any triangulable $n$-manifold $|K|$ has the following properties:*

1. *$\dim K = n$*
2. *For every vertex $p \in |K|$, there is an $n$-simplex $\sigma$ of $K$ such that $p \in |\overline{\sigma}|$*
3. *Every $(n-1)$-simplex of $K$ is a face of exactly two $n$-simplexes*

*Moreover, if $|K|$ is connected, for any two $n$-simplexes $\sigma$ and $\tau$ of $K$, there is a sequence of $n$-simplexes $\sigma = \sigma_1, \ldots, \sigma_r = \tau$ such that $\sigma_i \cap \sigma_{i+1}$ is an $(n-1)$-simplex for every $i = 1, \ldots, r-1$.*

*Proof.* 1. Lemma (V.1.4) shows that $H_{n-1}(S(p); \mathbb{Z}) \cong \mathbb{Z}$ for each $p \in |K|$. Hence, if $\dim K < n$, for every $p \in |K|$, $\dim S(p) < n-1$; therefore, $H_{n-1}(S(p); \mathbb{Z}) = 0$, contradicting the lemma. We now suppose that $\dim K = m > n$; this means that the simplicial complex $K$ has at least one $m$-simplex with $m > n$ and so, for every point $p$ in the interior of $|\overline{\sigma}|$, the space $S(p)$ is of the same homotopy type as $S^{m-1}$, once again in contradiction to Lemma (V.1.4) (two spheres of different dimensions cannot be of the same homotopy type).

2. If $K$ had no $n$-simplex with geometric realization containing $p$, then the dimension of the simplicial complex $S(p)$ would be strictly less than $n-1$ and thus $S(p)$ could not be of the same homotopy type as $S^{n-1}$.

3. Let us suppose $\sigma_{n-1}$ to be a face of $r$ $n$-dimensional simplexes $\tau_n^i, i = 1, \ldots, r$. Let

$$\text{Int}|\overline{\sigma_{n-1}}| = |\overline{\sigma}_{n-1}| \smallsetminus |\dot{\overline{\sigma}}_{n-1}|$$

be the interior of $|\overline{\sigma_{n-1}}|$; for every $i = 1, \ldots, r$, we define

$$|K_i| = |\dot{\tau}_n^i| \smallsetminus \text{Int}\,|\overline{\sigma}_{n-1}|$$

(remember that $\dot{\tau}_n^i$ is the boundary of $\tau_n^i$ – see definition in Sect. II.2).

We note that for each $p \in \text{Int}\,|\overline{\sigma}_{n-1}|$, the space $S(p)$ is written as the union $S(p) = \bigcup_{i=1}^r |K_i|$; besides, $|K_i| \cap |K_j| = |\dot{\sigma}_{n-1}|$ for each pair of distinct exponents $(i,j)$ with $i, j \in \{1, 2, \ldots, n\}$. By applying the Mayer–Vietoris sequence to the pair $(|K_1|, |K_2|)$, we obtain

$$H_{n-1}(|K_1| \cup |K_2|; \mathbb{Z}) \cong H_{n-2}(|\dot{\sigma}_{n-1}|; \mathbb{Z}) \cong \mathbb{Z}$$

(note that $|\dot{\sigma}_{n-1}| \cong S^{n-2}$); we next apply the Mayer–Vietoris sequence to the pair $(|K_1| \cup |K_2|, |K_3|)$ and so on, to get the free group with $r - 1$ generators

$$H_{n-1}\left(\bigcup_{i=1}^r |K_i|; \mathbb{Z}\right);$$

however,

$$H_{n-1}\left(\bigcup_{i=1}^r |K_i|; \mathbb{Z}\right) \cong H_{n-1}(S(p); \mathbb{Z}) \cong \mathbb{Z}$$

and so, $r - 1 = 1$, that is to say, $r = 2$.

We now assume that $|K|$ is connected. Then by Lemma (II.4.4), $K$ is connected, in the meaning of the definition given in Sect. II.4 (from the topological point of view, $|K|$ is connected if and only if it is path-connected).

Let $\sigma_n \in \Phi$ be any $n$-simplex. We view the simplicial complex $K$ as the union of two subcomplexes: (a) the simplicial complex $L$ defined by all $n$-simplexes of $K$ (and their subsimplexes) that may be linked to $\sigma_n$ by a sequence of $n$-simplexes as mentioned in 3; (b) the simplicial complex $M$ of all $n$-simplexes of $K$ for which this condition does not hold. Note that $K = L \cup M$. If $M = \emptyset$, the assertion is proved; this occurs when $\dim |K| = 1$ because $K$ is connected; incidentally note that since $K$ is connected $L \cap M \neq \emptyset$. We now consider $M$ nonempty and $\dim |K| \geq 2$. Then, $\dim(L \cap M) \leq n - 2$; in fact, if $\sigma_{n-1}$ were an $(n-1)$-simplex of both $L$ and $M$, $\sigma_{n-1}$ would be a face of both an $n$-simplex of $L$ and an $n$-simplex of $M$, which is impossible because of the definitions. For any vertex $x$ of $L \cap M$,

$$\dim(S(x) \cap L) = \dim(S(x) \cap M) = n \quad 1$$

and $\dim(S(x) \cap L \cap M) \leq n - 3$. However, because of 2., every $(n-2)$-simplex of $S(x)$ is a face shared by two $(n-1)$-simplexes; thus if $(S(x) \cap L)_{n-1}$ and $(S(x) \cap M)_{n-1}$ are the sets of all $(n-1)$-simplexes of $S(x) \cap L$ and $S(x) \cap M$, respectively, the chains

$$c_L = \sum_{\tau \in (S(x) \cap L)_{n-1}} \tau,$$

$$c_M = \sum_{\tau \in (S(x) \cap M)_{n-1}} \tau$$

are linearly independent cycles of $C_{n-1}(S(x)) \otimes \mathbb{Z}_2$, because the $(n-2)$-simplexes of $(d_{n-1} \otimes 1_{\mathbb{Z}_2})(c_L)$ and $(d_{n-1} \otimes 1_{\mathbb{Z}_2})(c_M)$ appear always twice; consequently, the vector space $H_{n-1}(S(x); \mathbb{Z}_2)$ has dimension $\geq 2$, a fact that contradicts the well-known result $H_{n-1}(S(x); \mathbb{Z}_2) \cong \mathbb{Z}_2$ (see Theorem (II.5.5)). Therefore, $M = \emptyset$ and $K = L$.  ∎

## Exercises

**1.** Show by means of an example that a closed subspace of an $n$-manifold is not necessarily an $n$-manifold.

**2.** Let $p$ be a positive integer, $p \geq 2$, and let $C_p$ be the (multiplicative) cyclic group of order $p$ of the $p$-th roots of the identity in $\mathbb{C}$,

$$C_p = \left\{ \zeta_p^j \mid j = 1 \ldots p \right\},$$

where $\zeta_p = e^{2\pi i/p}$. If $q$ is an integer prime to $p$, then $C_p$ acts on $S^3 = \{(z, w) \in \mathbb{C}^2 \mid |z|^2 + |w|^2 = 1\} \subset \mathbb{C}^2$ by setting

$$\left( (z, w), \zeta_p^j \right) \mapsto \left( z\zeta_p^j, w\zeta_p^{qj} \right)$$

for every $j$. Prove that the quotient space (called *lens space* and denoted by $L(p, q)$) is a 3-manifold.

**3.** Let $G$ be a finite group that acts freely on a $G$-space $X$. Prove that if $X/G$ is an $n$-manifold, then also $X$ is an $n$-manifold.

**4.** Prove that each compact manifold is homeomorphic to a subspace of an Euclidean space $\mathbb{R}^N$, for some $N$.

## V.2  Closed Surfaces

In this section we study the compact connected topological 2-manifolds which we shall henceforth call *closed surfaces*. In the previous section we gave some examples of closed surfaces: $S^2$, $\mathbb{R}P^2$, $T^2 \cong S^1 \times S^1$. But there are more, like the complex projective line (which, nevertheless and as we shall presently see, is homeomorphic to the sphere $S^2$) and the Klein bottle. We begin by studying these examples in more detail.

In order to construct the complex projective line $\mathbb{C}P^1$, we may proceed as follows: we consider the space $\mathbb{C}^2 \smallsetminus \{(0,0)\}$ and the equivalence relation

$$(z_0, z_1) \equiv (z_0', z_1') \iff (\exists z \in \mathbb{C} \smallsetminus \{0\}) \, z_0' = zz_0, \, z_1' = zz_1;$$

we define

$$\mathbb{C}P^1 := (\mathbb{C}^2 \smallsetminus \{0\})/\equiv .$$

In what follows, the equivalence class of $(z_0, z_1) \in \mathbb{C}^2 \smallsetminus \{(0,0)\}$ is denoted by square brackets $[(z_0, z_1)]$. We consider $S^2$ contained in $\mathbb{C} \times \mathbb{R} \cong \mathbb{R}^3$. The function

$$\eta : \mathbb{C}P^1 \to S^2$$

defined by setting

$$\eta \left( [(z_0, z_1)] \right) = \left( \frac{2 z_1 \bar{z}_0}{|z_0|^2 + |z_1|^2}, \frac{|z_1|^2 - |z_0|^2}{|z_0|^2 + |z_1|^2} \right) \subset \mathbb{C} \times \mathbb{R} \cong \mathbb{R}^3$$

has image $S^2 = \{(w, x) \in \mathbb{C} \times \mathbb{R} : |w|^2 + x^2 = 1\}$. It is easily seen that $\eta$ is a homeomorphism from $\mathbb{C}P^1$ onto $S^2$ and so that

$$\mathbb{C}P^1 \cong S^2$$

is a closed surface. We recall that the *Klein bottle* $K$ is obtained from a rectangular band with vertices $(0,0)$, $(0,1)$, $(1,0)$, $(1,1)$ through the identifications

$$(t, 0) \equiv (t, 1), \ 0 \le t \le 1,$$

$$(0, s) \equiv (1, 1 - s), \ 0 \le s \le 1.$$

The reader can easily prove that $K$ is a closed 2-manifold by noting, for instance, that if the group of order 2 acts on the torus $T^2 = S^1 \times S^1$ by the transformation (defined on pairs $(t, s)$ of real numbers *modulo* 1)

$$(t, s) \mapsto (t + 1/2, 1 - s),$$

then the quotient is a surface homeomorphic to $K$ (by Theorem (V.1.3) on p. 172). Figure V.3 shows a possible embedded triangulation (with auto-intersection).

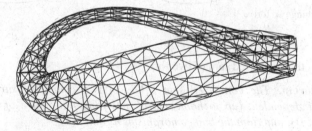

**Fig. V.3** Klein Bottle

We intend to define an operation for 2-manifolds, the so-called *connected sum*. For every point $x$ of a 2-manifold $S$, there exist a closed disk $D$ in $S$, such that $x \in D$, and a homeomorphism $h: D^2 \to D$. We now consider two 2-manifolds $S_1$ and $S_2$

and take the closed disks $D_1 \subset S_1$ and $D_2 \subset S_2$, together with the homeomorphisms $h_1 \colon D^2 \to D_1$ and $h_2 \colon D^2 \to D_2$; let $S^1 = \partial D^2$ be the boundering circle of the unit disk $D^2$. The restrictions of the homeomorphisms $h_1$ and $h_2$ to $S^1$ are homeomorphisms from $S^1$ onto the boundaries $\partial D_1$ and $\partial D_2$, respectively. We now define the maps

$$h_1' \colon S^1 \xrightarrow{h_1|S^1} \partial D_1 \hookrightarrow S_1 \smallsetminus \mathrm{Int} D_1$$

$$h_2' \colon S^1 \xrightarrow{h_2|S^1} \partial D_2 \hookrightarrow S_2 \smallsetminus \mathrm{Int} D_2$$

and construct the pushout of the pair of morphisms $(h_1', h_2')$ in **Top**. In this manner, we obtain the space (unique up to homeomorphism)

$$S_1 \# S_2 := S_1 \smallsetminus \mathrm{Int} D_1 \underset{h_1', h_2'}{\bigsqcup} S_2 \smallsetminus \mathrm{Int} D_2$$

named *connected sum* of $S_1$ and $S_2$. Figures V.4, V.5, and V.6 outline a procedure

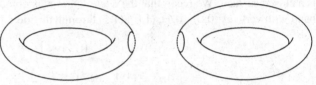

**Fig. V.4**  Two tori less a disk

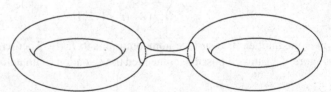

**Fig. V.5**  Attaching two tori

for obtaining the connected sum of two tori.

**(V.2.1) Theorem.** *The connected sum $S_1 \# S_2$ of two connected surfaces is a 2-manifold independent (up to homeomorphism) from the choice of the closed disks $D_1$ and $D_2$, and from the homeomorphisms $h_1$ and $h_2$.*

*Proof.* It is easily proved that $S_1 \# S_2$ is a 2-manifold; we only need to find local charts for the points in the gluing zone. It does not depend on the choice of the disks $D_1$ and $D_2$ and on the homeomorphisms $h_1 \colon D^2 \to D_1$ and $h_2 \colon D^2 \to D_2$ due to the following result: if $S$ is a surface, then for every pair of homeomorphisms $h_1 \colon D^2 \to S$ and $h_2 \colon D^2 \to S$ (on the images), there is a homeomorphism $f$ between

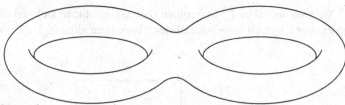

**Fig. V.6** Connected sum of
two tori: the torus of genus 2

their images such that $h_2 = fh_1$. The independence from the choice of $h_1$ and $h_2$
derives from the universal property of pushouts. The details of this proof are left to
the reader.                                                                          ■

If $\mathfrak{S}$ represents the set of all classes of homeomorphisms of closed connected
surfaces, we have the following result whose proof we shall omit:

**(V.2.2) Proposition.** *The connected sum determines an operation*

$$\#: \mathfrak{S} \times \mathfrak{S} \to \mathfrak{S}$$

*which is associative, commutative, and has a neutral element (the sphere $S^2$).*

The sphere $S^2$, the torus $T^2$, the real projective plane $\mathbb{R}P^2$, and the connected
sums of these spaces have particularly interesting representations.  We begin
with $S^2$.  In Sect. I.1 we have proved that the sphere $S^2$ is homeomorphic to the
quotient space obtained from $D^2$ by identifying each point $(x,y) \in S^1$ to $(x,-y)$
(see Example (I.1.7)). So if, as in Fig. V.7, $a$ denotes the semicircle from $(-1,0)$

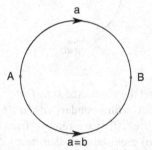

**Fig. V.7** Identifying polygon
for $S^2$

to $(1,0)$ through $(0,1)$ and $b$, the semicircle from $(-1,0)$ to $(1,0)$ through $(0,-1)$
(both from $(-1,0)$ to $(1,0)$ ), $S^2$ may be viewed as the disk $D^2$ with two vertices
$A = (-1,0)$, $B = (1,0)$ and two identified arcs (sides), that is to say, a disk with
two vertices and only one side $a = b$; if the boundary is travelled clockwise from
$A$, side $b$ is travelled against the direction we have given it and we may therefore
view $S^2$ as a disk with boundary $a^1a^{-1}$ (we assign the exponent 1 to side $a$ when
following the chosen direction, and the exponent $-1$ when following the opposite
direction).

We view the torus $T^2$ as a square (homeomorphic to a closed disk) whose horizontal sides, as well as its vertical sides, have been identified. See Fig. V.8. In other

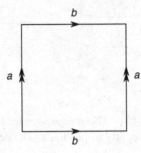

**Fig. V.8**  Identifying polygon for the torus $T^2$

words, we interpret $T^2$ as a square (closed disk) with a single vertex $A$ and two sides $a$, $b$. If we travel clockwise the boundary of the square that represents $T^2$, we read such boundary as the word $a^1b^1a^{-1}b^{-1}$.

The real projective plane $\mathbb{R}P^2$ is viewed as a closed disk with a single vertex $A$; its boundary is given by $a^1a^1$ (we identify the antipodal points of the boundary), as in Fig. V.9.

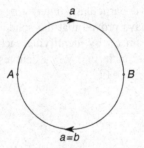

**Fig. V.9**  Projective plane

Let us now try to interpret the connected sum $T^2\#T^2$. We suppose the first torus to be represented by a square with boundary $a^1b^1a^{-1}b^{-1}$ and the second one, by a square with boundary $c^1d^1c^{-1}d^{-1}$; we remove from each square the interior of (a portion corresponding to) a closed disk that meets the boundary of the square exactly at one of its vertices, as shown in Fig. V.10. We obtain two closed polygons with boundaries $a^1b^1a^{-1}b^{-1}e$ and $c^1d^1c^{-1}d^{-1}f$, respectively. By identifying $e$ and $f$, we may interpret $T^2\#T^2$ as an octagon whose vertices are all identified into a single one, and with boundary $a^1b^1a^{-1}b^{-1}c^1d^1c^{-1}d^{-1}$, as in Fig. V.11.

How should we interpret the connected sum $\mathbb{R}P^2\#\mathbb{R}P^2$? We assume the first (respectively, the second) of these projective planes to be a closed disk with two identified antipodal vertices and two sides $a^1a^1$ (respectively, $b^1b^1$) whose antipodal points have been identified. We remove, from each of these disks, the interior of a small closed disk tangent to the boundary of the larger disk, at one of the two

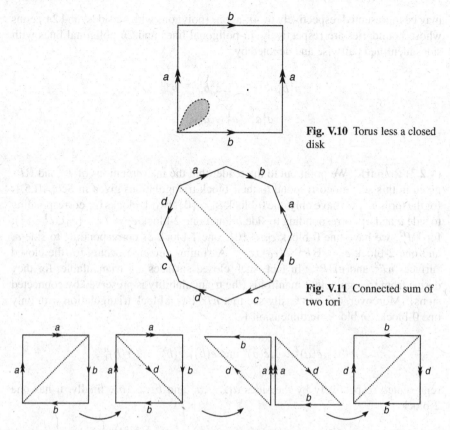

**Fig. V.10** Torus less a closed disk

**Fig. V.11** Connected sum of two tori

**Fig. V.12** Connected sum of two projective planes: the Klein bottle

chosen vertices; in this manner, we obtain two triangles with boundaries $a^1a^1c^1$ and $b^1b^1c^{-1}$; these triangles are put together by identifying $c$ (in other words, by taking their connected sum) and getting a square with boundary $a^1a^1b^1b^1$; this square, with the necessary identifications, is homeomorphic to $\mathbb{R}P^2\#\mathbb{R}P^2$. By cutting the square along the diagonal $d$, we are now left with two right triangles, both having a side $a$ and a side $b$; we have thus obtained the right triangles $a^1b^1d^{-1}$ and $b^1a^1d^1$. Next, we glue the triangles along the (oriented) side $a$ in order to obtain a square with boundary $d^1b^1d^1b^{-1}$ which, with the necessary identifications, corresponds to a Klein bottle. Figure V.12, illustrates the steps for the procedure that we have just described.

We close by noting that the surfaces

$$nT^2 := \underbrace{T^2\#\ldots\#T^2}_{n\ \text{times}}$$

$$n\mathbb{R}P^2 := \underbrace{\mathbb{R}P^2\#\ldots\#\mathbb{R}P^2}_{n\ \text{times}}$$

may be represented respectively by $4n$-agons (polygons with $4n$ sides) and $2n$-agons whose boundaries are respectively $4n$-poligonal lines and $2n$-poligonal lines with sides identified pairwise and denoted by

$$a_1^1 b_1^1 a_1^{-1} b_1^{-1} \ldots a_n^1 b_n^1 a_n^{-1} b_n^{-1} ,$$

$$a_1^1 a_1^1 a_2^1 a_2^1 \ldots a_n^1 a_n^1 .$$

**(V.2.3) Remark.** We point out to the reader that the interpretations of $T^2$ and $\mathbb{R}P^2$ given in this section correspond to their block triangulations given in Sect. III.5.1: for the torus $T^2$, we have only one 0-block $e_0 = \{0\}$, two 1-blocks ($e_1^1$ corresponding to side $a$ and $e_1^2$ corresponding to side $b$), and one 2-block $e_2 = T^2 \smallsetminus (e_0 \cup e_1^1 \cup e_1^2)$; for $\mathbb{R}P^2$, we have one 0-block $e_0 = \{0\}$, one 1-block $e_1$ corresponding to side $a$, and one 2-block $e_2 = \mathbb{R}P^2 \smallsetminus (e_0 \cup e_1)$. A similar situation occurs for the closed surfaces $nT^2$ and $n\mathbb{R}P^2$. In fact, such closed surfaces are triangulable, for they derive from triangulable 2-manifolds (the triangulability is preserved by connected sums). Moreover, one can easily see that $nT^2$ has a block triangulation with only one 0-block, $2n$ blocks in dimension 1,

$$e(a)_1^1, e(a)_1^2, \ldots, e(a)_1^n \text{ and } e(b)_1^1, e(b)_1^2, \ldots, e(b)_1^n ,$$

represented respectively by the sides $a_1, \ldots, a_n$ and $b_1, \ldots, b_n$; finally, it has one 2-block

$$e_2 = nT^2 \smallsetminus (e_0 \cup (\cup_{i=1}^n e(a)_1^i) \cup (\cup_{i=1}^n e(b)_1^i)) .$$

As for $n\mathbb{R}P^2$, we have a block triangulation with only one 0-block $e_0$, $n$ 1-blocks $e_1^1, \ldots, e_1^n$ corresponding to the $a_1, \ldots, a_n$, and only one 2-block $e_2 = n\mathbb{R}P^2 \smallsetminus (e_0 \cup (\cup_{i=1}^n e_1^i))$.

As a consequence, we have the following

**(V.2.4) Theorem.** *The closed surfaces $S^2$, $nT^2$, and $n\mathbb{R}P^2$ (with $n \geq 1$) are, pairwise, not homeomorphic.*

*Proof.* In order to prove this result, we compute the homology groups $H_1(S^2; \mathbb{Z})$, $H_1(nT^2; \mathbb{Z})$, $H_1(n\mathbb{R}P^2; \mathbb{Z})$ and show that no two of them are isomorphic.

We know that $H_1(S^2; \mathbb{Z}) \cong 0$. We now compute the first homology group with coefficients in $\mathbb{Z}$ of the other two manifolds through the block triangulation that we have just described. Beginning with $nT^2$, we give $\overline{e(a)_1^i}$ and $\overline{e(b)_1^i}$ an orientation, by choosing a generator $\beta(a)_1^i$ and $\beta(b)_1^i$, respectively, for each

$$H_1(\overline{e(a)_1^i}, e(a)_1^i; \mathbb{Z}) \text{ and } H_1(\overline{e(b)_1^i}, e(b)_1^i; \mathbb{Z})$$

where $i = 1, \ldots, n$ (we note that the generators $\beta(a)_1^i$ and $\beta(b)_1^i$ are cyclic – see the definition of block homology); similarly, the simplicial complex $\overline{e_2}$ is given an orientation by choosing a generator $\beta_2$. The chain group $C_1(e(nT^2))$ is the free Abelian group generated by $\beta(a)_1^i$ and $\beta(b)_1^i$, $i = 1, \ldots, n$; on the other hand, $d_2(\beta_2) = 0$, since the 1-blocks appear twice, and in opposite directions. Therefore,

$$H_1(nT^2; \mathbb{Z}) \cong \mathbb{Z}^{2n}$$

(see Theorem (III.5.9)). In the case of $n\mathbb{R}P^2$, we give the 1-blocks $e_1^i$ an orientation by choosing a generator $\beta_1^i$ for each $H_1(\overline{e_1^i}, \dot{e}_1^i; \mathbb{Z})$, and a generator

$$\beta_2 \in H_2(\overline{e_2}, \dot{e}_2; \mathbb{Z}) \cong \mathbb{Z} \,.$$

In this case, we have $d_2(\beta_2) = 2(\sum_{i=1}^{n} \beta_1^i)$; then, $C_1(e(n\mathbb{R}P^2))$ is the Abelian group generated by $\beta_1^1, \beta_1^2, \ldots, \beta_1^{n-1}, \sum_{i=1}^{n} \beta_1^i$ and since $2(\sum_{i=1}^{n} \beta_1^i)$ is a boundary,

$$H_1(n\mathbb{R}P^2; \mathbb{Z}) \cong \mathbb{Z}^{n-1} \times \mathbb{Z}_2 \,. \qquad \blacksquare$$

The connected sum enables us to join two surfaces, but there are other similar constructions. Let us consider the cylinder $I \times S^1$ and two immersions of the disk $h_1, h_2 \colon D^2 \to S$ in the surface $S$ in such a way that $D_1 = h_1(D^2)$ and $D_2 = h_2(D^2)$ are disjoint. The space

$$S' = S \setminus (\mathrm{Int}\, D_1 \cup \mathrm{Int}\, D_2) \sqcup_h (I \times S^1),$$

obtained by joining $S \setminus (\mathrm{Int}\, D_1 \cup \mathrm{Int}\, D_2)$ and the cylinder $I \times S^1$ through the map

$$h \colon \{0,1\} \times S^1 \to S \setminus (\mathrm{Int}\, D_1 \cup \mathrm{Int}\, D_2),$$

defined by $h(0,t) = h_1(t)$ and $h(1,t) = h_2(t)$ for every $t \in S^1 \subset D^2$, is still a surface.

We say that $S'$ is obtained by *attaching a handle* to $S$. Unlike the connected sum, the attachment of a handle depends on the homotopy class of the attaching function $h$. In Figs. V.13 and V.14, we see two different procedures for attaching the

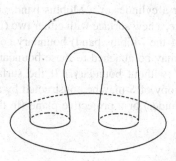

Fig. V.13 Attaching a handle (first method)

handle. It can be proved (Exercise 2) that the attaching in Fig. V.13 (first method) is equivalent to the connected sum with a torus, whereas the one in Fig. V.14

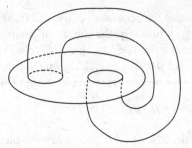

**Fig. V.14** Attaching a handle
(second method)

(second method), to the connected sum with a Klein bottle. The connected sum
with a projective plane, pictured in Fig. V.15, will be one of the fundamental steps

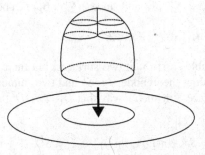

**Fig. V.15** Attaching a projective plane

in the proof of the next theorem, known as the *Fundamental Theorem of Closed
Surfaces*, which classifies all closed surfaces. It tells us that the only closed surfaces are exactly the sphere $S^2$, $nT^2$, and $n\mathbb{R}P^2$. Its proof is based on the fact
that any closed surface is triangulable. The existence of a triangulation for closed
surfaces was first proved by Tibor Radó in [29]; we assume this result to be well
known.[2] Here is another fundamental step in the proof: if $\gamma \subset S$ is a (simplicial)
simple closed curve in $S$, then, there exists a neighbourhood of $\gamma$ in $S$, given by
the union of triangles in a simplicial subdivision of a triangulation of $S$, which
is homeomorphic to either a cylinder or a Möbius band. By cutting the surface
along $\gamma$ we obtain, therefore, a new surface with either two (in the case of the cylinder) or one (in the case of the Möbius band) boundary component. Finally, either one or two disks $D^2$ may be attached to these boundaries, in order to obtain
a (triangulated) surface $S'$ without boundary. If the surface $S'$ is in turn connected, a homeomorphic copy of $S$ may be constructed by attaching either a handle (in the case of the cylinder) or a projective plane (in the case of the Möbius
band).

---

[2] See [8] for an elementary proof, based essentially on the Jordan–Schoenflies Theorem: a simple
closed curve $J$ on the Euclidean plain divides it into two regions and there exists a homeomorphism
from the plane in itself that sends $J$ into a circle.

Before proceeding with this Theorem, we note that the real projective plane may be obtained by the adjunction of a disk to a Möbius band. In fact, the Möbius band $M$ is obtained from a square by identifying the vertical sides with opposite orientations: for instance, $M$ comes from the square $I \times I$ through the identifications:

$$(0,t) \equiv (1, 1-t) , \ 0 \leq t \leq 1 .$$

Let $D^2$ be the unity disk of $\mathbb{R}^2$ and let $S^1 = \partial D^2$. We define the function

$$f \colon S^1 \to M, \ e^{2\pi i t} \mapsto \begin{cases} (2t, 0) & 0 \leq t \leq \frac{1}{2} \\ (2t-1, 1) & \frac{1}{2} \leq t \leq 1; \end{cases}$$

the reader may easily verify that $\mathbb{R}P^2$ is given by the pushout

$$
\begin{array}{ccc}
S^1 & \xrightarrow{\ f\ } & M \\
\downarrow & & \downarrow \\
D^2 & \xrightarrow{\ \tilde{f}\ } & \mathbb{R}P^2
\end{array}
$$

The reader could study the problem also backwards: the real projective plane has a closed disk $D^2$ with boundary $a^1 a^1$ as a model; let $O_a$ be the origin of $a$ and $D$ a small closed disk contained in $D^2$, with centre at $O_a$; it is easily seen that

$$D^2 \smallsetminus \mathring{D} \cong M$$

and so $\mathbb{R}P^2$ is homeomorphic to the adjunction space of $M$ and a closed disk. Note that $M$ is a *surface with boundary*.

We finally note that, as a consequence of the preceding results, the connected sum of a torus and a real projective plane is homeomorphic to the connected sum of three real projective planes; for its proof, it is enough to see that

$$T^2 \# \mathbb{R}P^2 \cong K \# \mathbb{R}P^2 .$$

Initially, we have noted that the connected sum of two surfaces does not depend on the position of the disks that we remove for gluing the two surfaces; therefore, we may glue $T^2$ and $K$ to the component of $\mathbb{R}P^2$ given by the Möbius band $M$. In order to attach $T^2$ to $M$, we have to remove an open disk of $T^2$ and an open disk of $M$, and glue the two spaces (by means of a pushout!) on the boundaries created by this removal; as we have seen, this is like attaching a *handle* to $M$. However, the space that we obtain is exactly the same as the one we would get by removing two distinct open disks from $M$ and gluing a cylinder to $M$, the two circles that bound the cylinder being identified with the same orientation to the boundaries formed by the removal of the two open disks (if we travel clockwise the two circles limiting the cylinder, both attachments will be made clockwise). For attaching a Klein bottle $K$ to $M$, we take the cylinder that gives rise to $K$ and glue it to $M$ in such a manner that the circles that bound the cylinder are glued in opposite directions (one clockwise, the other anticlockwise, as seen in Fig. V.14). Now, the spaces obtained by the addition of a cylinder to $M$ by these two procedures are homeomorphic to each other.

**(V.2.5) Theorem.** *Any closed surface is homeomorphic to one of the following surfaces:*

*(i)* $S^2$
*(ii)* $nT := T^2\#\ldots\#T^2$ *(n times)*
*(iii)* $n\mathbb{R}P^2 := \mathbb{R}P^2\#\ldots\#\mathbb{R}P^2$ *(n times)*

*Proof.* Let $S$ be a surface and $K$ one of its triangulations. The complex $K$ has a finite number of vertices, 1-simplexes, and 2-simplexes; for simplicity of notation, we denote by $T_i$ (respectively, $\ell_i$) the geometric realizations of the 2-simplexes (respectively, the 1-simplexes) of $K$; these are called *triangles* and *sides*. Let us go through some preliminary considerations.

We begin by observing that each side of $|K|$ is only a side of two triangles (see Theorem (V.1.5)). Now, let $v$ be a vertex of $K$; then the triangles with the vertex $v$ may be arranged in cyclic order $T_1, T_2, \ldots, T_r = T_0$ so that for each $i \in \{1, \ldots, r\}$, the intersection $T_{i-1} \cap T_i$ is the side $\ell_i$. In fact, we note that, if there were two such sets of triangles around $v$, they would have only the point $v$ in common; therefore, by removing $v$ from their union, we would have a non-connected space; on the other hand, since $S$ is a surface, there would be an open set $U \subset S$ that would contain $v$ and be homeomorphic to an open disk of $\mathbb{R}^2$; hence $U \smallsetminus v$ would be connected, a contradiction. It follows that the neighbourhood of every simple cycle of sides in $|K|$ is homeomorphic to the cylinder $I \times S^1$ or the Möbius band.

We now consider the one-dimensional simplicial complex $G$ (namely, the graph) defined as follows: the vertices of $G$ are all triangles of $K$, and for each pair of adjacent triangles (that is to say, with one side in common) $T_1$ and $T_2$ of $K$, we add the 1-simplex $\{T_1, T_2\}$ to $G$. The complex $G$ may be embedded in $S \cong |K|$ by sending the triangle $T_i$ (viewed as a vertex of $G$) to its barycentre $b(T_i)$ (viewed as a point of $|K|$). Through a barycentric subdivision of $K$, the 1-simplexes of $G$ will be represented by a broken line that joins the two barycentres and crosses the common side of $T_1$ and $T_2$.

We now recall the concept of *spanning tree* (or *maximal tree*). A *tree* of $|K|$ is a contractible unidimensional subpolyhedron $|L| \subset |K|$; the existence of trees in a polyhedron $|K|$ is assured by the fact that the geometric realization of any 1-simplex of $K$, being homeomorphic to the disk $D^1$, is contractible. The trees of a complex are partially ordered by inclusion; a tree is called *spanning* if it is not contained in a tree strictly larger; since $K$ is finite, the existence of a spanning tree is also certain.

**(V.2.6) Lemma.** *If $|K|$ is path-connected and $|L|$ is a spanning tree, then $L$ contains all vertices of $K$.*

*Proof.* We suppose $a$ to be a vertex of $K$ but not of $L$. Let us take any vertex $b$ of $L$; since $|K|$ is path-connected, there is a path $\gamma\colon I \to |K|$ joining $a$ and $b$. By the

Simplicial Approximation Theorem, there are a subdivision of $I$ and a simplicial approximation $\gamma' \colon I \to |K|$ of $\gamma$ that may be viewed as a path of 1-simplexes of $K$, say,

$$\alpha = \{b, x_0\}.\{x_0, x_1\}.\dots.\{x_n, a\}.$$

Let $x_r$ be the last vertex of $\alpha$ in $L$; $\{x_r, x_{r+1}\}$ is then a 1-simplex that is not in $L$ (we may assume $x_r \neq x_{r+1}$ because $a \notin L$). It follows that

$$|\bar{L}| = |L| \cup |\{x_r, x_{r+1}\}|$$

is a unidimensional subcomplex of $|K|$, strictly larger than $|L|$ and contractible, as the union of two contractible spaces with a point in common. Hence, $|L|$ is not spanning, against the hypotesis. ∎

We now consider a spanning tree $T \subset G$, that has the following properties:

(a) $T$ contains all vertices of $G$.
(b) $T$ is a tree, in other words, $|T|$ is contractible to a point.

We finally define the subcomplex $K_T \subset K$ whose vertices are precisely those of $K$ and whose edges are *all edges of $K$ not crossed by any side of $G$*. Since $T$ is a tree, it is easily shown that $K_T$ is connected. We see in Fig. V.16 the triangulation for the

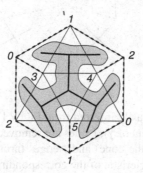

Fig. V.16 The tree $T$ and the graph $K_T$ for the projective plane

projective plane with the two graphs, $T$ (in thicker line) and $K_T$ (in broken line). If we now consider the second barycentric subdivision of $K$, we may have two open sets $U \supset |T|$ and $V \supset |K_T|$ such that

(a) $U \cap V = \emptyset$ and $\bar{U} \cup \bar{V} = |K|$
(b) $\mathring{U} = \mathring{V}$

as shown in Fig. V.16.[3] We note that $\bar{U}$ is homeomorphic to the disk $D^2$, since $T$ is contractible.

We now consider the Euler–Poincaré characteristic $\chi(K_T)$ (in this regard, see also Exercise 6 on p. 88). We have $\chi(K_T) \leq 1$ and $\chi(K_T) = 1$ if and only if $K_T$ is

---

[3] For instance, $U$ may be defined as $|T|$ together with the interior of all triangles and the sides of the second barycentric subdivision of $K$ that intersect $T$.

a tree. When $\chi(K_T) = 1$, $\overline{V}$ is then homeomorphic to the disk $D^2$ and $S$ is obtained by gluing two disks on the boundaries; it follows that $S \cong S^2$. Instead, if $\chi(K_T) < 1$, then the homology $H_1(K_T)$ is non-trivial and so there is at least one non-trivial simple cycle, denoted by $\gamma$, in the graph $K_T$. Let us consider the simplicial complex $K_\gamma$ obtained by *cutting K* along $\gamma$ (in other words, we duplicate all vertices and edges of $\gamma$). The resulting triangulated surface $S_\gamma = |K_\gamma|$ is still connected: indeed, the tree $T$ and the cycle $\gamma$ are disjoint, but $T$ has a vertex in the interior of each triangle of $K$, and therefore also of $K_\gamma$. It is a surface with boundary: if $\gamma$ has a neighbourhood homeomorphic to a Möbius band, then its boundary has one component; on the other hand, if the neighbourhood of $\gamma$ is a cylinder, there are two components. Let us consider the cones on the components of the boundary of $K_\gamma$ and attach them to the holes that we have created: we end up with a new triangulated surface $S' \cong |K'|$. The tree $T$ is easily extended to a tree $T' \subset K'$ by adding a vertex to the barycentre of every triangle of the attached cones and an edge crossing the edge of the boundary of the attachment, as in Fig. V.17. Let $l$ be the number of edges of the component of

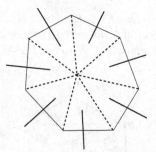

**Fig. V.17** Extension of $T$ to the cone on the boundary component

such a boundary. As it is easily inferred from Fig. V.17, by attaching a triangulated disk (cone on the component of $\gamma$), $l$ sides are removed from the graph $K_T$, whereas one vertex (the centre of the cone) and $l$ edges through such a vertex are added. The Euler–Poincaré characteristic of the corresponding graph $K'_{T'}$, obtained in this manner, is

$$\chi(K'_{T'}) = \begin{cases} \chi(K_T) + 1 & \text{if } K_\gamma \text{ has one boundary component} \\ \chi(K_T) + 2 & \text{if } K_\gamma \text{ has two boundary components.} \end{cases}$$

We note that the process of cutting along $\gamma$ may be reversed: $|K|$ is obtained by attaching a handle (in the case of two components) to $|K'|$ or a projective plane (in the case of a single component). By repeating this procedure, we must end up with nothing other than a sphere.[4]

We have therefore proved that every surface is the connected sum of a sphere, a certain number $n_T$ of tori, a certain number $n_K$ of Klein bottles, and a certain number $n_P$ of projective planes. However, since the Klein bottle is homeomorphic

---

[4] For a proof based on identifying polygons, see William Massey [25, Theorem 1.5.1].

to the connected sum of two projective planes, and the connected sum is associative, every surface is the connected sum of $S^2$, $n_T$ tori, and $2n_K + n_P$ projective planes. If $2n_K + n_P = 0$, then clearly $S = S^2$ when $n_T = 0$ and $S = T^2 \# T^2 \# \ldots \# T^2$ ($n_T$ times) if $n_T > 0$. Finally, if $2n_K + n_P > 0$, we may substitute $T^2 \# \mathbb{R}P^2$ with the connected sum $\mathbb{R}P^2 \# \mathbb{R}P^2 \# \mathbb{R}P^2$, iteratively, and conclude that $S$ is homeomorphic to the connected sum of $2n_K + n_P + 2n_T$ projective planes. ∎

## Exercises

**1.** Prove that, if $S_1$ and $S_2$ are two closed surfaces, the connected sum $S_1 \# S_2$ is a closed surface.

**2.** Prove that the attaching of a handle to a surface $S$ is equivalent to the connected sum of $S$ with either a torus or a Klein bottle.

**3.** Let the closed surfaces $S_1$ and $S_2$ be given; prove that

$$\chi(S_1 \# S_2) = \chi(S_1) + \chi(S_2) - 2 .$$

## V.3 Poincaré Duality

The reader must have noticed that the second integral homology group of the torus $T^2$ is isomorphic to $\mathbb{Z}$, whereas the corresponding homology group of $\mathbb{R}P^2$ is trivial. On the other hand, in the proof of the Fundamental Theorem of Closed Surfaces, we have noted that the real projective plane may be obtained by the adjunction of a 2-disk to the Möbius band, while the Klein bottle is the connected sum of two real projective planes. Well, everybody knows the story of the little man who walks on a Möbius band and, having gone around once, found himself upside down on the starting point (this is why the Möbius band was defined a "nonorientable" surface[5]). The reader could wonder whether the presence of the Möbius band in the real projective space – and consequently in the Klein bottle – is responsible for the "abnormal" behavior of the second homology group of these spaces. The answer is "yes" and has to do with the orientation of the simplexes of these triangulated spaces. Thus, we give the following definition.

**(V.3.1) Definition.** A triangulable manifold $V \cong |K|$ is *orientable* if its $n$-simplexes may be oriented *coherently*, that is to say, each $(n-1)$-simplex of $K$ inherits opposite orientations from its two adjacent $n$-simplexes (see Theorem (V.1.5)).

The reader is highly advised to draw triangulations for the spaces mentioned above and to verify that it is not possible to give the Klein bottle and the real projective plane coherent orientations, while this is possible in the case of the torus.

---

[5] Not in the sense of our definition of 2-manifold; it is really a 2-manifold with boundary.

The definition of orientability as given here is obviously incomplete and not very useful: in fact, a good definition must be independent from the triangulation and based on an invariant by homeomorphisms. The next result is meant to correct this flaw.

**(V.3.2) Theorem.** *Let $|K|$ be a triangulable n-manifold;[6] then*

$$H_n(|K|;\mathbb{Z}) \cong \mathbb{Z} \iff |K| \text{ is orientable.}$$

*Proof.* Let us suppose that $|K|$ is orientable and let $z = \sum \sigma$ be the formal sum of all $n$-simplexes of $K$. The definition of orientability implies that $z$ and all its integral multiples are $n$-cycles. On the other hand, if $z' \in Z_n(K)$ has a term which is a multiple $r\tau$ of an $n$-simplex $\tau$, then the terms of $z'$ are of the form $r\tau'$, where $\tau'$ runs over the set of all $n$-simplexes of $K$ intersecting $\tau$ in $(n-1)$-simplexes (otherwise, $z'$ would not be a cycle); but then, by part 3 of Theorem (V.1.5), $z'$ would equal the formal sum $r \sum \sigma = rz$, where $\sigma$ runs over the set of all $n$-simplexes of $K$. Therefore, $H_n(|K|;\mathbb{Z}) \cong \mathbb{Z}$.

With a similar argument we may prove that, conversely, given any orientation of the $n$-simplexes of $K$, the $n$-cycles of $C_n(K)$ are of the type $rz$ where $z = \sum \pm \sigma$ and $\sigma$ runs over the set of all $n$-simplexes. Since $\partial_n(z) = 0$, we conclude that it is possible to give an orientation to the $n$-simplexes, according to the Definition (V.3.1). ∎

Because of this last result, the definition of orientability of a triangulable $n$-manifold does not depend on the triangulation; furthermore, two triangulable connected $n$-manifolds of the same homotopy type are either both orientable or both nonorientable.

Let $V$ be a triangulable $n$-manifold and let us suppose that $V \cong |K|$, where $K = (X, \Phi)$. We consider the barycentric subdivision $K^{(1)} = (\Phi, \Phi^{(1)})$ and associate with every simplex $\sigma \in \Phi$ (that is to say, a vertex of $K^{(1)}$) the subset $B_\sigma$ of $\Phi^{(1)}$ defined by

$$e_\sigma = \{\{\sigma, \sigma^1, \dots, \sigma^r\} | \sigma \subset \sigma^1 \subset \dots \subset \sigma^r\}.$$

Note that, if $\dim \sigma = p$, the dimension of the simplexes of $e_\sigma$ is $\leq n - p$; in particular, if $\dim \sigma = n$, $e_\sigma$ coincides with the vertex $\sigma$ (namely, the barycenter $b(\sigma)$) of $K^{(1)}$.

We now consider the closure and the boundary of $e_\sigma$, that is to say, the simplicial subcomplexes $\overline{e_\sigma}$ and $\dot{e}_\sigma$ of $K^{(1)}$ (see Sect. III.5). The reader is asked to note that $\dot{e}_\sigma$ is the set of all simplexes of $\overline{e_\sigma}$ not having $b(\sigma)$ as a vertex; hence,

$$\overline{e_\sigma} = \dot{e}_\sigma * b(\sigma)$$

and, therefore, $|\overline{e_\sigma}|$ is contractible.

The reader is also asked to review Definitions (III.5.3) and (III.5.5) before turning to the next result.

---

[6] We remember that our definition of triangulable manifold requires $K$ to be connected by 1-simplexes.

**(V.3.3) Theorem.** *For every p-simplex $\sigma \in \Phi$, the block $\overline{e_\sigma}$ is a $(n-p)$-block of $K^{(1)}$; besides, the set defined by*

$$e(K^{(1)}) = \{\overline{e_\sigma} | \sigma \in \Phi\}$$

*is a block triangulation of $K^{(1)}$.*

*Proof.* We first have to prove that

$$H_r(\overline{e_\sigma}, \mathring{e}_\sigma; \mathbb{Z}) \cong \begin{cases} \mathbb{Z} & \text{if } r = n - p \\ 0 & \text{if } r \neq n - p. \end{cases}$$

As we have already seen, if $\sigma$ is an $n$-simplex, $e_\sigma$ is a simplicial subcomplex of $K^{(1)}$ having only the vertex $b(\sigma)$ and no other simplex; so, $\mathring{e}_\sigma = \emptyset$ and the statement above is true. We then assume that $\dim \sigma \leq n - 1$. Since $|K^{(1)}| \cong |K|$ and $\dim K^{(1)} = n$, the relative homology of the simplicial subcomplex $S(b(\sigma))$ of $K^{(1)}$ (see the definition given in Theorem (II.2.10)) is

$$\widetilde{H}_r(S(b(\sigma)); \mathbb{Z}) \cong \begin{cases} \mathbb{Z} & \text{if } r = n - 1 \\ 0 & \text{if } r \neq n - 1 \end{cases}$$

(see Lemma (V.1.4)).

We now prove that

$$S(b(\sigma)) = |(\sigma)^{(1)} * \mathring{e}_\sigma|$$

where $(\sigma)^{(1)}$ is the barycentric subdivision of the boundary of $\sigma$. In fact,

$$|\{\sigma^1, \ldots, \sigma^r\}| \in S(b(\sigma)) \iff \{\sigma^1, \ldots, \sigma^r\} \in \Phi^{(1)} \text{ with } \sigma \subset \sigma^1 \subset \ldots \subset \sigma^r$$

$$\iff (\exists t)\, \sigma^t \subset \sigma \subset \sigma^{t+1} \text{ with } \{\sigma^1, \ldots, \sigma^t\} \in (\sigma)^{(1)} \text{ and } \{\sigma^{t+1}, \ldots, \sigma^r\} \in \mathring{B}_\sigma.$$

On the other hand, $|(\sigma)^{(1)}| \cong S^{p-1}$ and consequently

$$S(b(\sigma)) \cong S^{p-1} * |\mathring{e}_\sigma|.$$

Let us now go back a step to note that if $L_1, \ldots, L_p$ are $p$ simplicial complexes, each of them having only two 0-simplexes (and no 1-simplex), then

$$|L_1 * \ldots * L_p| \cong S^{p-1}$$

and we conclude that $S(b(\sigma))$ is homeomorphic to the $p$th suspension of $|\mathring{e}_\sigma|$ (the suspension of an abstract simplicial complex $K$ is defined on p. 47 – it is easily seen that the suspension of the sphere $S^n$ is homeomorphic to the sphere $S^{n+1}$; the $p$th suspension of $K$ is defined by iteration). By Exercise 5 of Sect. II.4, we conclude that

$$\widetilde{H}_q(S(b(\sigma)); \mathbb{Z}) \cong \widetilde{H}_{q-p}(\mathring{e}_\sigma; \mathbb{Z})$$

and therefore

$$\tilde{H}_r(\dot{e}_\sigma;\mathbb{Z}) \cong \begin{cases} \mathbb{Z} & \text{if } r = n - p - 1 \\ 0 & \text{if } r \neq n - p - 1. \end{cases}$$

This result and the fact that the (reduced) homology of $\overline{e_\sigma}$ is trivial (for $\overline{e_\sigma}$ is contractible) applied to the exact sequence of the (reduced) homology of the pair $(\overline{e_\sigma}, \dot{e}_\sigma)$ enable us to conclude that

$$(\forall \sigma \in \Phi, \ \dim \sigma = p) \ e_\sigma \text{ is an } (n-p)\text{-block of } K^{(1)}.$$

We only need to prove that the set $\{\overline{e_\sigma} | \sigma \in \Phi\}$ is a block triangulation of $K^{(1)}$. In fact, let $\tilde{\sigma} = \{\sigma^0, \ldots, \sigma^r\}$ be any simplex of $K^{(1)}$; by the definition of $K^{(1)}$, we have $\sigma^0 \subset \ldots \subset \sigma^r$ and so

$$\tilde{\sigma} \subset \overline{e_{\sigma^0}} \setminus \dot{e}_{\sigma^0}.$$

Finally, for every $\sigma \in \Phi$,

$$\dot{e}_\sigma = \cup_{\tau \subset \sigma, \tau \neq \sigma} \overline{e_\tau}.$$    ∎

We now have all that is needed to prove the *Poincaré Duality Theorem*.

**(V.3.4) Theorem.** *Let $V$ be an oriented triangulable $n$-manifold, with triangulation given by the simplicial complex $K = (X, \Phi)$. Then, for every integer $p$ with $0 \leq p \leq n$,*

$$H^p(V;\mathbb{Z}) \cong H_{n-p}(V;\mathbb{Z}).$$

*Proof.* To prove this theorem, we must consider the first barycentric subdivision $K^{(1)}$ of $K$ and use both the projection $\pi \colon K^{(1)} \to K$ and the homomorphism of chain complexes $\aleph \colon C(K) \to C(K^{(1)})$ (review Sect. III.2). Let $z$ be the cycle of dimension $n$ defined as the formal sum of all $n$-simplexes of $K$; the homology class $[z]$ is a generator of $H_n(|K|;\mathbb{Z}) \cong \mathbb{Z}$ (see the proof of Theorem (V.3.2)).

For each generator $c_\sigma \in C^p(K;\mathbb{Z})$[7], we define the map

$$\beta_p^\sigma \colon C^p(K;\mathbb{Z}) \longrightarrow C_{n-p}(K^{(1)};\mathbb{Z})$$

$$\beta_p^\sigma(c_\sigma) := c_\sigma C_p(\pi) \cap \aleph_n(z).$$

Since $d_n^{K^{(1)}} \aleph_n = \aleph_{n-1} d_n^K$ and $z$ is a cycle, the $n$-chain $\aleph_n(z)$ is a cycle: in fact, it is the sum of all oriented $n$-simplexes of $K^{(1)}$ in agreement with the fact that $|K^{(1)}|$ is an oriented $n$-manifold. For any term of this sum, in other words, an $n$-simplex $\{\sigma_0, \ldots, \sigma^n\}$ of $K^{(1)}$, we have

$$(*) \qquad c_\sigma C_p(\pi) \cap \{\sigma_0, \ldots, \sigma^n\} = \pm\{\sigma_0, \ldots, \sigma^n\}$$

and so $\beta_p^\sigma(z) \in C_{n-p}(\overline{B_\sigma})$.

By Theorem (IV.3.3),

$$d_{n-p}^{K^{(1)}}(c_\sigma C_p(\pi) \cap \aleph_n(z)) = (-1)^p(c_\sigma C_p(\pi) \cap d_n^{K^{(1)}}(\aleph_n(z)) - d_p^K(c_\sigma C_p(\pi)) \cap \aleph_n(z))$$

$$= (-1)^p(c_\sigma C_p(\pi) \cap d_n^{K^{(1)}}(\aleph_n(z)));$$

---

[7] We remind the reader that $c_\sigma$ is the map that takes the $p$-simplex $\sigma$ to $1 \in \mathbb{Z}$ and all other $p$-simplexes to $0 \in \mathbb{Z}$.

therefore, $\beta_p^\sigma(z)$ is an element of $C_{n-p-1}(\mathring{e}_\sigma)$ and thus

$$\beta_p^\sigma(z) \in Z_{n-p}(\overline{e_\sigma}, \mathring{e}_\sigma; \mathbb{Z}).$$

The result (*) implies that $\beta_p^\sigma(z)$ is the sum of all $(n-p)$-simplexes of $\overline{B_\sigma}$ with coefficients $\pm 1$ and so, it is a cycle that generates $Z_{n-p}(\overline{e_\sigma}, \mathring{e}_\sigma; \mathbb{Z})$.

We now recall that the set

$$e(K^{(1)}) = \{e_\sigma | \sigma \in \Phi\}$$

is a block triangulation of $K^{(1)}$ (see Theorem (V.3.3)); moreover, the block homology of $K^{(1)}$ derives from the chain complex

$$C(e(K^{(1)})) = \{C_{n-p}(e(K^{(1)})), d_{n-p}^{e(K^{(1)})} | n-p \geq 0\}$$

where $C_{n-p}(B(K^{(1)}))$ is the free Abelian group defined by the generators

$$\beta_p^\sigma(z) \in Z_{n-p}(\overline{e_\sigma}, \mathring{e}_\sigma; \mathbb{Z})$$

(see Sect. III.5). Thus, the map

$$\beta_p \colon C_p(K) \to C_{n-p}(e(K^{(1)})) \, , \, \beta_p(\sigma) = \beta_p^\sigma(z)$$

defined on all $p$-simplexes $\sigma \in \Phi$ is an isomorphism. It follows that there exists an isomorphism

$$H(\beta_p) \colon H^p(K; \mathbb{Z}) \longrightarrow H_{n-p}(e(K^{(1)}); \mathbb{Z}) \cong H_{n-p}(K; \mathbb{Z}). \qquad \blacksquare$$

## Exercises

**1.** Compute the Euler–Poincaré characteristic of the closed surfaces.

**2.** An $n$-dimensional connected compact manifold without boundary is called a *closed $n$-manifold*. It is known that the closed 3-manifolds are triangulable (E. Moise, 1952). Prove that if $M$ and $N$ are two orientable closed 3-manifolds such that $\pi_1(M) \cong \pi_1(N)$, then $H_i(M; \mathbb{Z}) \cong H_i(N; \mathbb{Z})$ for $i = 0, 1, 2, 3$.[8]

---

[8] In particular, if $\pi_1(M) = 0$, then $M$ has the same homology groups as the 3-sphere $S^3$; in 1904, Poincaré asked the question: in this case, is it true that $M$ is homeomorphic to $S^3$? It was only recently that this famous "Poincaré conjecture" was proved affirmatively by Grigory Perelman, who used methods of differential geometry, specially, the *Ricci flow*.

# Chapter VI
# Homotopy Groups

## VI.1 Fundamental Group

The Fundamental Theorem of Surfaces assures us that any connected compact surface is homeomorphic to one of the following closed surfaces: the two-dimensional sphere, a connected sum of tori, or a connected sum of real projective planes. We have seen that the homology groups of such closed surfaces are not isomorphic, and therefore, the surfaces under discussion cannot be homeomorphic. It is possible to arrive at this same result by computing another algebraic invariant of the polyhedra, the so-called *fundamental group*, which is clearly related to the first homology group. In what follows, we shall study such concepts in detail.

In Sect. I.2, we have defined the concept of homotopy between two based maps $f,g \in \mathbf{Top}_*((X,x_0),(Y,y_0))$ and we have considered the set

$$[X,Y]_* = \mathbf{Top}_*(X,Y)/\sim;$$

we now turn our attention to the set $[X,Y]_*$ when $(X,x_0)$ is the unit circle $S^1$ with base point $\mathbf{e}_0 = (1,0)$.

We intend to define a group structure on the set $[S^1,Y]_*$; we must, therefore, define a multiplication in $[S^1,Y]_*$. We start by saying that a map $f\colon (S^1,\mathbf{e}_0) \to (Y,y_0)$ is a *loop* of $Y$ (based at $y_0$); we can now compose two loops $f$ and $g$ to form a loop $(f*g)$, obtained by traveling with double speed first the loop $f$ and then $g$; formally, this loop is given by the function

$$(f*g)(e^{2\pi t i}) = \begin{cases} f(e^{2\pi 2t i}) & 0 \le t \le \frac{1}{2} \\ g(e^{2\pi(2t-1)i}) & \frac{1}{2} \le t \le 1. \end{cases}$$

This is an intuitive approach to the multiplication in $[S^1,Y]_*$. Let us now go over this definition in such a way that we may prove the following results more systematically. Let us take the wedge product

$$S^1 \vee S^1 = \{\mathbf{e}_0\} \times S^1 \cup S^1 \times \{\mathbf{e}_0\} \subset S^1 \times S^1$$

D.L. Ferrario and R.A. Piccinini, *Simplicial Structures in Topology*,
CMS Books in Mathematics, DOI 10.1007/978-1-4419-7236-1_VI,
© Springer Science+Business Media, LLC 2011

and define the map (called *comultiplication* in $S^1$)

$$v \colon S^1 \to S^1 \vee S^1 \, , \ e^{2\pi t i} \longmapsto \begin{cases} (\mathbf{e}_0, e^{2\pi 2 t i}) & 0 \le t \le \frac{1}{2} \\ (e^{2\pi(2t-1)i}, \mathbf{e}_0) & \frac{1}{2} \le t \le 1; \end{cases}$$

besides, let us define the *folding map* $\sigma \colon S^1 \vee S^1 \to S^1$:

$$(\forall 0 \le t \le 1) \ (\mathbf{e}_0, e^{2ti\pi}) \longmapsto e^{2ti\pi} \, , \ (e^{2ti\pi}, \mathbf{e}_0) \longmapsto e^{2ti\pi}.$$

By these definitions, one can easily see that for every point $e^{2\pi t i} \in S^1$

$$(f * g)(e^{2\pi t i}) = \sigma(f \vee g)v(e^{2\pi t i});$$

moreover, we note that if $f' \sim f$ and $g' \sim g$, then $\sigma(f' \vee g')v$ is homotopic to $\sigma(f \vee g)v$. Finally, for any based space $(Y, y_0)$, we define the function (called *multiplication*)

$$[S^1, Y]_* \times [S^1, Y]_* \longrightarrow [S^1, Y]_*$$

$$(\forall [f], [g] \in [S^1, Y]_*) \ [f] \overset{v}{\times} [g] := [\sigma(f \vee g)v].$$

**(VI.1.1) Theorem.** *The set $[S^1, Y]_*$ with the multiplication $\overset{v}{\times}$ is a group whose unit element is the homotopy class of the constant function $c \colon S^1 \to Y$ (it takes each point of $S^1$ to $y_0$); besides, the inverse of $[f]$ is the homotopy class of the based map $h \colon S^1 \to Y$ defined, for each $e^{2\pi t i} \in S^1$, by the formula $h(e^{2\pi t i}) = f(e^{2\pi(1-t)i})$.*[1]

*Proof.* We first prove that the function $v \colon S^1 \to S^1 \vee S^1$ is *associative (up to homotopy)*, in other words, that the diagram

$$
\begin{array}{ccc}
S^1 & \overset{v}{\longrightarrow} & S^1 \vee S^1 \\
{\scriptstyle v}\big\downarrow & & \big\downarrow{\scriptstyle 1 \vee v} \\
S^1 \times S^1 & \underset{v \vee 1}{\longrightarrow} & S^1 \vee S^1 \vee S^1
\end{array}
$$

commutes up to homotopy (that is to say $(1 \vee v)v \sim (v \vee 1)v$). In fact, for every $e^{2\pi t i} \in S^1$,

$$(1 \vee v)v(e^{2\pi t i}) = \begin{cases} (\mathbf{e}_0, \mathbf{e}_0, e^{2\pi 2 t i}) & 0 \le t \le \frac{1}{2} \\ (\mathbf{e}_0, e^{2\pi 2(2t-1)i}, \mathbf{e}_0) & \frac{1}{2} \le t \le \frac{3}{4} \\ (\mathbf{e}_0, \mathbf{e}_0, e^{2\pi(4t-3)ti}) & \frac{3}{4} \le t \le 1 \end{cases}$$

$$(v \vee 1)v(e^{2\pi t i}) = \begin{cases} (\mathbf{e}_0, \mathbf{e}_0, e^{2\pi 4 t i}) & 0 \le t \le \frac{1}{4} \\ (\mathbf{e}_0, e^{2\pi(4t-1)i}, \mathbf{e}_0) & \frac{1}{4} \le t \le \frac{1}{2} \\ (e^{2\pi(2t-1)i}, \mathbf{e}_0, \mathbf{e}_0) & \frac{1}{2} \le t \le 1 \end{cases}$$

---

[1] Intuitively, the loop $h$ is the loop $f$, but traveled in the opposite direction.

The function

$$H \colon S^1 \times I \longrightarrow S^1 \vee S^1 \vee S^1,$$

defined for every $(e^{2\pi t i}, s) \in S^1 \times I$ by the formula

$$H(e^{2\pi t i}, s) = \begin{cases} (\mathbf{e}_0, \mathbf{e}_0, e^{2\pi(2s+2)ti}) , & 0 \leq t \leq \frac{2-s}{4} \\ (\mathbf{e}_0, e^{2\pi[4t-2(1-\frac{s}{2})]i}, \mathbf{e}_0) & \frac{2-s}{4} \leq t \leq \frac{3-s}{4} \\ (e^{2\pi[4(1-\frac{s}{2})(t-1)+1]i}, \mathbf{e}_0, \mathbf{e}_0) & \frac{3-s}{4} \leq t \leq 1, \end{cases}$$

is the desired homotopy.

We now prove that the multiplication $\overset{v}{\times}$ is associative. We have to prove that

$$(\forall f, g, h \in \mathbf{Top}_*(S^1, Y)) \; \sigma(\sigma(f \vee g) v \vee h) v \sim \sigma(f \vee \sigma(g \vee h) v) v.$$

In fact, the associativity of $v$ implies

$$\sigma(1_Y \vee \sigma)(f \vee g \vee h)(1 \vee v) v \sim \sigma(\sigma \vee 1_Y)(f \vee g \vee h)(v \vee 1) v;$$

on the other hand,

$$\sigma(1_Y \vee \sigma)(f \vee g \vee h)(1 \vee v) v = \sigma(f \vee \sigma(g \vee h) v) v$$
$$\sigma(\sigma \vee 1_Y)(f \vee g \vee h)(v \vee 1) v = \sigma(\sigma(f \vee g) v \vee h) v.$$

The homotopy class $[c]$ of the constant function at the base point $y_0 \in Y$ is the identity element of $[S^1, Y]_*$. We first prove that, if $i \colon S^1 \vee S^1 \to S^1 \times S^1$ is the inclusion map and $\Delta \colon S^1 \to S^1 \times S^1$ is the diagonal function $e^{2\pi t i} \mapsto (e^{2\pi t i}, e^{2\pi t i})$, then $iv \sim \Delta$; in fact, this assertion is ensured by the homotopy $H \colon S^1 \times I \to S^1 \times S^1$

$$H(e^{2\pi t i}, s) = \begin{cases} (e^{2\pi t s i}, e^{2\pi(t(2-s)i}) & 0 \leq t \leq \frac{1}{2} \\ (e^{2\pi[2t-1+s(1-t)]i}, e^{2\pi(st+1-s)i}) & \frac{1}{2} \leq t \leq 1. \end{cases}$$

We then conclude that, for every based map $f \colon S^1 \to Y$,

$$\sigma(f \vee c) v = \sigma(f \times c) iv \sim \sigma(f \times c) \Delta = f$$
$$\sigma(c \vee f) v = \sigma(c \times f) iv \sim \sigma(c \times f) \Delta = f.$$

Finally, we prove that every $[f] \in [S^1, Y]_*$ has an inverse. Indeed, for each $e^{2\pi t i}$, we define $h \colon S^1 \to Y$ by $h(e^{2\pi t i}) = f(e^{2\pi(1-t)i})$ and note that

$$\sigma(f \vee h) v(e^{2\pi t i}) = \begin{cases} f(e^{2\pi 2 t i}) & 0 \leq t \leq \frac{1}{2} \\ h(e^{2\pi(2t-1)i}) & \frac{1}{2} \leq t \leq 1; \end{cases}$$

the function

$$H(e^{2\pi t i}, s) = \begin{cases} y_0 & 0 \leq t \leq \frac{s}{2} \\ f(e^{2\pi(2t-s)i}) & \frac{s}{2} \leq t \leq \frac{1}{2} \\ f(e^{2\pi(2-2t-s)i}) & \frac{1}{2} \leq t \leq \frac{2-s}{2} \\ y_0 & \frac{2-s}{2} \leq t \leq 1 \end{cases}$$

is a homotopy $\sigma(f \vee h)v \sim c$; therefore, $[h]$ is a right inverse of $[f]$. Similarly, we prove that $[h]$ is also a left inverse of $[f]$.                                                                          ∎

The group $[S^1, Y]_* := \pi(Y, y_0)$ is the so-called *fundamental group of Y* (relative to the base point $y_0$). We leave to the reader the proof of the following result:

**(VI.1.2) Theorem.** *The construction of the fundamental group defines a covariant functor*

$$\pi : \mathbf{Top}_* \to \mathbf{Gr}.$$

By its definition, it is intuitive that the fundamental group of a space depends generally on the base point (see Exercise 1 on p. 210). We recall that two points $y_0, y_1 \in Y$ are joined by a *path* in $Y$ if there is a map $\lambda : I \to Y$ such that $\lambda(0) = y_0$ and $\lambda(1) = y_1$. The product $*$ may be easily defined also by the composition of two paths $\lambda$ and $\mu$, whenever $\lambda(1) = \mu(0)$, by setting

$$(\lambda * \mu)(t) = \begin{cases} \lambda(2t) & \text{if } 0 \leq t \leq \frac{1}{2} \\ \mu(2t - 1) & \text{if } \frac{1}{2} \leq t \leq 1. \end{cases}$$

For simplicity, we often write $\lambda\mu = \lambda * \mu$.

The following theorem holds true:

**(VI.1.3) Theorem.** *Let $y_0$ and $y_1$ be two points of Y joined by a path $\lambda : I \to Y$. Then $\lambda$ gives rise to a group isomorphism*

$$\phi_\lambda : \pi(Y, y_0) \to \pi(Y, y_1).$$

*Proof.* The homomorphism $\phi_\lambda$ is given by

$$(\forall [f] \in \pi(Y, y_0)) \; \phi_\lambda([f]) := [\lambda^{-1} * (f * \lambda)]$$

and its inverse $\phi_\lambda^{-1}$ is defined by

$$(\forall [g] \in \pi(Y, y_1)) \; \phi_\lambda^{-1}([g] = [\lambda * (g * \lambda^{-1})]$$

with $\lambda^{-1}(t) = \lambda(1 - t)$, for every $t \in I$. The result comes from

$$\lambda^{-1}\lambda \sim c_{y_0} \text{ and } \lambda\lambda^{-1} \sim c_{y_1}.$$                    ∎

**(VI.1.4) Corollary.** *The fundamental groups of a path-connected space do not depend on the base point (that is to say, they are isomorphic to each other).*

In general, two different paths of $Y$ joining two points $y_0, y_1 \in Y$ produce two distinct isomorphisms between $\pi(Y, y_0)$ and $\pi(Y, y_1)$; the next result tells us when two paths will produce the same isomorphism.

**(VI.1.5) Theorem.** *Let $\lambda, \mu : I \to Y$ be two paths such that $\lambda(0) = \mu(0) = y_0$ and $\lambda(1) = \mu(1) = y_1$. Then,*

$$\phi_\lambda = \phi_\mu \iff [\lambda * \mu^{-1}] \in Z\pi(Y, y_0)$$

*where $Z\pi(Y, y_0)$ is the center of the group $\pi(Y, y_0)$.*

*Proof.* Let us suppose that $\phi_\lambda = \phi_\mu$. Then, for every $[f] \in \pi(Y, y_0)$,

$$\lambda^{-1} * f * \lambda \sim \mu^{-1} * f * \mu$$
$$\lambda^{-1} * f * \lambda * \mu^{-1} \sim \mu^{-1} * f$$
$$(\mu * \lambda^{-1}) * f * (\lambda * \mu^{-1}) \sim f;$$

since $(\mu * \lambda^{-1})^{-1} = \lambda * \mu^{-1}$, we have that

$$f * (\lambda * \mu^{-1}) \sim (\lambda * \mu^{-1}) * f$$

and therefore

$$[\lambda * \mu^{-1}] \in Z\pi(Y, y_0).$$

Proving the converse is equally easy.                                                    ∎

**(VI.1.6) Corollary.** *The isomorphism* $\phi_\lambda : \pi(Y, y_0) \to \pi(Y, y_1)$ *associated with a path* $\lambda : I \to Y$ *from* $y_0$ *to* $y_1$ *does not depend on* $\lambda$ *when* $\pi(Y, y_0)$ *is Abelian.*

## VI.1.1 Fundamental Groups of Polyhedra

There are no fixed rules for computing the fundamental group of an arbitrary-based space; there is, however, a practical and efficient method for computing the fundamental group of a polyhedron based at a vertex. The reader is advised to review the material on simplicial complexes and their geometric realizations, also because of the notation. We recall that, by Lemma (V.2.6) on p. 186, there exists a one-dimensional subcomplex of each connected finite simplicial complex $K$ that contains all vertices and whose geometric realization is contractible; such a subcomplex is a *spanning tree* and is such that:

1. $\dim(L) = 1$.
2. $L$ contains all vertices of $K$ (that is to say, $X = Y$).
3. The polyhedron $|L|$ is contractible to a point.

**(VI.1.7) Theorem.** *Let* $K = (X, \Phi)$ *be a connected simplicial complex with a fixed vertex* $a_0$ *and let* $L = (Y, \Theta)$ *be a spanning tree in K. Let a symbol* $g_{ij}$ *be associated with every 1-simplex* $\{a_i, a_j\} \in \Phi$ *and let S be the set of all symbols* $g_{ij}$ *obtained in this manner; let*

$$R = \{g_{ij} \mid \{a_i, a_j\} \in \Theta\} \cup \{g_{ij}g_{jk}g_{ki} \mid \{a_i, a_j, a_k\} \in \Phi\}.$$

*Then,* $\pi(|K|, a_0)$ *is isomorphic to the group* $Gp(S; R)$ *generated by the set S with the relations R.*

*Proof.* First of all, let us order the finite set of the vertices of $K$:

$$a_0 < a_1 < \ldots < a_n.$$

Then, each $p$-simplex $\sigma$ of $|K|$ may be written as

$$\sigma = \{a_{i_0}, \ldots, a_{i_p}\}.$$

We now consider an arbitrary element $[f] \in \pi(|K|, a_0)$ represented by the loop $f: S^1 \to |K|$; by the Simplicial Approximation Theorem (III.2.4), there is a simplicial function

$$g: (S^1)^{(r)} \to K$$

such that, after identifying $(S^1)^{(r)}$ with $S^1$, $|g| \sim f$. Thus, the homotopy class $[f]$ can be represented by a loop based at $a_0$ and consisting exclusively of 1-simplexes of $K$. We now define a homomorphism

$$\psi: \pi(|K|, a_0) \longrightarrow Gp(S; R).$$

For each 1-simplex $\{a_i, a_j\}$ of $K$, we define the symbol

$$h_{ij} := \begin{cases} g_{ij} & \text{if } \{a_i, a_j\} \in K \setminus L \\ 1 & \text{if } \{a_i, a_j\} \in L; \end{cases}$$

we use these symbols for defining $\psi$ on any element $[f] \in \pi(|K|, a_0)$; in fact, $f$ may be represented, up to homotopy, by a closed path of 1-simplexes

$$\alpha = \{a_0, a_i\}\{a_i, a_j\} \ldots \{a_k, a_0\},$$

and so, we define

$$\psi([f]) := h_{0i}h_{ij} \ldots h_{k0} \in Gp(S; R).$$

It is necessary to prove that $\psi$ is well defined, in other words, that two simplicial maps $f$ and $f': S^1 \to |K|$, represented by two closed paths of 1-simplexes $\alpha = \{a_0, a_i\}\{a_i, a_j\} \ldots \{a_k, a_0\}$ and $\alpha' = \{a'_0, a'_i\}\{a'_i, a'_j\} \ldots \{a'_k, a'_0\}$ with $a'_0 = a_0$, induce the same element $\psi([f]) = \psi([f'])$. By the Simplicial Approximation Theorem (III.2.4) (on p. 113), it is possible to suppose that the homotopy between $f$ and $f'$ is given by a simplicial map. The equality between $h_{0i}h_{ij} \ldots h_{k0}$ and $h'_{0i}h'_{ij} \ldots h'_{k0}$ will follow after applying the relations in $Gp(S; R)$ a finite number of times.

Reciprocally, we define a group homomorphism

$$\phi: Gp(S; R) \longrightarrow \pi(|K|, a_0)$$

as follows: let $\{a_i, a_j\}$ be the simplex associated with the generator $g_{ij} \in S$; since $L$ contains all vertices of $K$, we choose two paths in $L$ originating at $a_0$:

1. $\alpha_i$, path in $L$ from $a_0$ to $a_i$
2. $\alpha_j$, path in $L$ from $a_0$ to $a_j$

We now define

$$\phi(g_{ij}) := [\alpha_i\{a_i,a_j\}\alpha_j^{-1}].$$

Notice that $\phi$ is independent from the choice of $\alpha_i$ and $\alpha_j$: a path of 1-simplexes contained in $L$ is contractible because $L$ is contractible; hence, if $\alpha_i'$ and $\alpha_j'$ were new paths of 1-simplexes from $a_0$ to $a_i$ and $a_j$, respectively, we would have $\alpha_i \sim \alpha_i'$ and $\alpha_j \sim \alpha_j'$ for they are contractible.

Suppose $\{a_i,a_j,a_k\}$ to be a 2-simplex of $K$; then

$$\phi(g_{ij})\phi(g_{jk})\phi(g_{ki}) = [\alpha_i\{a_i,a_j\}\alpha_j^{-1}][\alpha_j\{a_j,a_k\}\alpha_k^{-1}][\alpha_k\{a_k,a_i\}\alpha_i^{-1}]$$
$$= [\alpha_i\alpha_i^{-1}] = 1$$

and so $\phi$ may be extended to a group homomorphism.

Now, for every generator $g_{ij}$ of $Gp(S;R)$,

$$\psi\phi(g_{ij}) = \psi([\alpha_i\{a_i,a_j\}\alpha_j^{-1}]) = g_{ij}$$

as the paths $\alpha_i$ and $\alpha_j$ consist of 1-simplexes of $L$ (see the definition of $\phi$). Therefore,

$$\psi\phi = 1_{Gp(S;R)}.$$

On the other hand, for every 1-simplex $\{a_i,a_j\}$ of $K$

$$\phi\psi([\alpha_i\{a_i,a_j\}\alpha_j^{-1}]) = [\alpha_i\{a_i,a_j\}\alpha_j^{-1}]$$

(here $\alpha_i$ and $\alpha_j$ are paths of $L$ taken accordingly to the definition of $\phi$). Therefore, for a closed path of 1-simplexes

$$\alpha = \{a_0,a_i\}\{a_i,a_j\}\ldots\{a_k,a_0\},$$

we may write

$$\phi\psi(\alpha) = \phi\psi([\alpha_0\{a_0,a_i\}\alpha_i^{-1}])\ldots\phi\psi([\alpha_k\{a_k,a_0\}\alpha_0^{-1}])$$
$$= [\alpha_0\{a_0,a_i\}\alpha_i^{-1}]\ldots[\alpha_k\{a_k,a_0\}\alpha_0^{-1}] = \alpha,$$

and so $\phi\psi = 1_{\pi(|K|,a_0)}$. ∎

We now compute the fundamental groups of some spaces by way of examples.

**(VI.1.8) Example** $(\pi(S^1) \cong \mathbb{Z})$. We triangulate the circle with the simplicial complex $K$ of vertices $a_0,a_1,a_2$ and 1-simplexes

$$\{a_0,a_1\}, \{a_1,a_2\} \text{ and } \{a_2,a_0\}.$$

Here, we have the spanning tree $|L|$ given by the two simplexes $\{a_0,a_1\}$ and $\{a_1,a_2\}$. Hence, by Theorem (VI.1.7), $\pi(S^1,a_0)$ has only one generator.

**(VI.1.9) Example** $(\pi(S^2) = 0)$. It follows directly from Theorem (VI.1.7), when we triangulate $S^2$ as the boundary of a tetrahedron.

**(VI.1.10) Example** ($\pi(S^1 \vee S^1)$ is the free group with two generators). This group is not Abelian. It is an immediate consequence from Theorem (VI.1.7), when we triangulate each circle as the boundary of a triangle.

**(VI.1.11) Example** ($\pi(T^2) \cong \mathbb{Z} \times \mathbb{Z}$). We consider the triangulation of a torus with 9 vertices, 27 edges, and 18 faces (see figure in Sect. III.5.1). The spanning tree $|L|$ is given by the geometric realizations of the following 1-simplexes (to simplify the notation, we omit curly brackets and commas between two vertices):

$$01, 12, 23, 34, 46, 65, 57, 78$$

Sixteen of the twenty seven generators $g_{ij}$ become the identity in $\pi(T^2, 0)$:

$$g_{01}, g_{05}, g_{12}, g_{15}, g_{16}, g_{23}, g_{26}, g_{34}, g_{36}, g_{46}, g_{48}, g_{56}, g_{57}, g_{58}, g_{68}, g_{78};$$

but there are also the relations

$$g_{02}g_{28}g_{80} \text{ and } g_{17}g_{74}g_{41}.$$

All this leads us to the conclusion that we only have two generators in the fundamental group that are not reduced to the identity: $g_{02}$ and $g_{28}$; besides,

$$g_{02}g_{28}(g_{02})^{(-1)}(g_{28})^{(-1)} = 1$$

in $\pi(T^2, 0)$ which is, therefore, isomorphic to $\mathbb{Z} \oplus \mathbb{Z}$.[2]

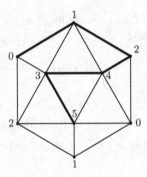

**Fig. VI.1** A triangulation of the real projective plane

**(VI.1.12) Example** ($\pi(\mathbb{R}P^2) \cong \mathbb{Z}_2$). Let us consider the triangulation of the real projective plane with 6 vertices, 15 edges, and 10 faces, as in Sect. III.5. The spanning tree $|L|$ is given by the geometric realizations of the following 1-simplexes: $01, 12, 24, 43, 35$, as shown in Fig. VI.1. In this example, the generators

$$g_{01}, g_{03}, g_{12}, g_{13}, g_{14}, g_{24}, g_{34}, g_{35}, g_{45}$$

---

[2] Another way to prove this result is as follows: if $X$ and $Y$ are topological spaces, then the fundamental group of the product $X \times Y$ coincides with the direct product of the fundamental groups $\pi(X)$ and $\pi(Y)$ (see Exercise 3 on p. 210).

are reduced to the identity of the fundamental group; besides,

$$g_{02} = g_{04} = g_{05} = g_{15} = g_{23} = g_{25}$$

and so we have only one generator, let us say, $g_{15}$, but with the property $(g_{15})^2 = 1$ in $\pi(\mathbb{R}P^2, 0)$.

Theorem (VI.1.7) enables us to compute the fundamental group of a path-connected polyhedron, but at a price: indeed, the number of generators and relations may be excessively large. So, we try to obtain results in a more economical way; with this in mind, let us consider a few things.

In Sect. I.2, we have proved that the category of groups is closed by pushouts. We now review this assertion. Let $G_1$ and $G_2$ be two groups given by generators and relations

$$G_1 = Gp(S_1, R_1) \text{ and } G_2 = Gp(S_2, R_2);$$

We now consider the homomorphisms $f\colon G \to G_1$ and $g\colon G \to G_2$, and form the pushout of the pair $(f, g)$ to obtain the group $G_1 *_{f,g} G_2$. This group is actually isomorphic to the group presented as $Gp(S_1 \cup S_2; R_1 \cup R_2 \cup R_{f,g})$, where $R_{f,g} = \{f(x)g(x)^{-1} | x \in G\}$. By definition of pushout, the group $G_1 *_{f,g} G_2$ has the following properties:

1. There exist two homomorphisms

$$\bar{g}\colon G_1 \longrightarrow G_1 *_{f,g} G_2$$
$$\bar{f}\colon G_2 \longrightarrow G_1 *_{f,g} G_2$$

such that $\bar{f}g = \bar{g}f$

2. For every group $H$ and homomorphisms

$$h\colon G_1 \to H \text{ and } k\colon G_2 \to H$$

such that $hf = kg$, there exists a *unique* homomorphism

$$\ell\colon G_1 *_{f,g} G_2 \longrightarrow H$$

such that

$$\ell\bar{f} = k \text{ and } \ell\bar{g} = h$$

Specifically, if $G = 0$ then $f = g = 0$, $R_{f,g} = 0$, and the group

$$G_1 *_{0,0} G_2 := Gp(S_1 \cup S_2; R_1 \cup R_2)$$

(also denoted by the symbol $G_1 * G_2$) is called *free product* of $G_1$ and $G_2$.

**(VI.1.13) Theorem** (Seifert–Van Kampen). *Let $|L|$ and $|M|$ be two path-connected polyhedra such that $|L \cap M|$ is path-connected and not empty. Let $a_0$ be a vertex shared by both polyhedra. The diagram*

$$
\begin{array}{ccc}
\pi(|L\cap M|,a_0) & \xrightarrow{\;\pi(i_L)\;} & \pi(|L|,a_0) \\
\Big\downarrow{\scriptstyle \pi(i_M)} & & \Big\downarrow{\scriptstyle \pi(i_M)} \\
\pi(|M|,a_0) & \xrightarrow[\;\pi(i_L)\;]{} & \pi(|L\cup M|,a_0)
\end{array}
$$

*(where $i_L$ and $i_M$ are the inclusion maps) is a pushout.*

*Proof.* We extend a spanning tree $|A(L\cap M)|$ of $|L\cap M|$ to the spanning trees $|A(L)|$ and $|A(M)|$ of $|L|$ and $|M|$, respectively; the union

$$|A(L)| \cup |A(M)| = |A(L\cup M)|$$

is a spanning tree of $|L\cup M|$ and so, by Lemma (V.2.6), it contains all vertices of $L \cup M$.

We now order the vertices of $L \cup M$; this automatically gives the vertices of $L$ and of $M$ an order. By Theorem (VI.1.1), we know that $\pi(|L\cup M|,a_0)$ is generated by the elements $g_{ij}$ corresponding to the ordered 1-simplexes $\{a_i,a_j\}$ of $L\cup M \smallsetminus A(L\cup M)$, with the relations $g_{ij}g_{jk}g_{ki}$ relative to the ordered 2-simplexes $\{a_i,a_j,a_k\}$ of $L\cup M$.

On the other hand, the pushout of groups $\pi(|L|,a_0)$ and $\pi(|M|,a_0)$ relative to the homomorphisms

$$\pi(i_L)\colon \pi(|L\cap M|,a_0) \to \pi(|L|,a_0),$$
$$\pi(i_M)\colon \pi(|L\cup M|,a_0) \to \pi(|M|,a_0),$$

is determined by the generators $g_{ij}$ and $h_{ij}$ corresponding, respectively, to the ordered 1-simplexes of $L\smallsetminus A(L)$ and $M\smallsetminus A(M)$, with the relations $g_{ij}g_{jk}g_{ki}$, $h_{ij}h_{jk}h_{ki}$ and

$$\pi(i_L)g_{ij}(\pi(i_M)h_{ji})^{-1}$$

whenever $g_{ij}=h_{ij}$ in $L\cap M$.

It is now easy to realize that the group $\pi(|L\cup M|,a_0)$ described as above in terms of generators and relations coincides with the pushout in question. ∎

**(VI.1.14) Corollary.** *If $\pi(|L\cap M|,a_0)=0$, then $\pi(|L\cup M|,a_0)$ is the free product*

$$\pi(|L|,a_0) * \pi(|M|,a_0).$$

Let us consider a few things before we turn to another result. Let $|K|$ be a connected polyhedron and let $\alpha$ be a closed path, based at a vertex $a_0 \in K$, that may be represented by the sequence of 1-simplexes:

$$\alpha = \{a_0,a_1\}.\{a_1,a_2\}.\;.\;.\;.\;.\{a_n,a_0\}.$$

We take a disk $D^2$ with center $c$ and divide its boundary $\partial D^2$ into $n+1$ parts by means of the (distinct) points $b_0,\ldots,b_n$. Let $D = (Y,\Psi)$ be the simplicial complex with vertices

$$Y = \{c,b_0,\ldots,b_n\}$$

and simplexes

$$\Psi = \{c,b_0,\ldots,b_n; b_0b_1, b_1b_2,\ldots,b_nb_0, cb_0, cb_1,\ldots,cb_n,$$

$$cb_0b_1, cb_1b_2,\ldots,cb_nb_0\}$$

(we omit the curly brackets for the simplexes). The one-dimensional simplicial complex $\partial D$ formed by the simplexes

$$\{b_0,\ldots,b_n; b_0b_1, b_1b_2,\ldots,b_nb_0\}$$

is a subcomplex of $D$; note that

$$|D| \cong D^2 \text{ and } |\partial D| \cong \partial D^2 \cong S^1.$$

The construction is nothing but a cone (with vertex $c$) on a simplicial subdivision of $S^1$. We now consider the simplicial function

$$f \colon \partial D \to K , \ (\forall i = 0,1,\ldots,n) \ f(b_i) = a_i$$

and an adjunction space $X$ defined by the pushout

$$
\begin{array}{ccc}
S^1 \cong |\partial D| & \xrightarrow{\ |f|\ } & |K| \\
\downarrow & & \downarrow \\
D^2 \cong |D| & \xrightarrow[\ |f|\ ]{} & X
\end{array}
$$

With these conditions we have the following:

**(VI.1.15) Theorem.** *The fundamental group $\pi(X,a_0)$ of the space $X$, obtained by adjoining a 2-cell to $|K|$ by means of the path $\alpha$, is obtained from the group $\pi(|K|,a_0)$ together with the relation defined by the homotopy class of the same path $\alpha$.*

*Proof.* We consider the barycentric subdivision of the triangulation of $D^2$ just described, in other words, before gluing $D^2$ to $K$, we refine the triangulation of $D^2$ by adding $2(n+1)$ new vertices

$$c_0,c_1,\ldots,c_n; d_0,d_1,\ldots,d_n$$

to the simplicial complex $D$, where each $c_i$ is the barycenter of the simplex $\{c,b_i\}$ and each $d_i$ is the barycenter of the triangle $\{c,b_i,b_{i+1}\}$ (here the index $i = 0\ldots n$ is understood as modulo $n$). See in Fig. VI.2 such a subdivision, for $n = 6$. Let then

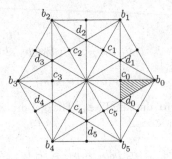

**Fig. VI.2** Barycentric subdivision of $D^2$

$M = (Z,\Theta)$ be the simplicial complex of the barycentric subdivision.

The space obtained by gluing $|M| \cong D^2$ to $|K|$ is still $X$; note that, naturally, $X$ may be viewed as the geometric realization of an abstract simplicial complex $W$. We now consider the 2-simplex $\sigma = \{b_0,c_0,d_0\}$ (indicated in Fig. VI.2) and the pushout

$$
\begin{array}{ccc}
|W \smallsetminus \sigma| \cap |\sigma| & \xrightarrow{\ i_{W \smallsetminus \sigma}\ } & |W \smallsetminus \sigma| \\[1.2em]
{\scriptstyle i_\sigma} \big\downarrow & & \big\downarrow {\scriptstyle \bar{i}_\sigma} \\[1.2em]
|\sigma| & \xrightarrow[\ \bar{i}_{W \smallsetminus \sigma}\ ]{} & X
\end{array}
$$

By Theorem (VI.1.13), we conclude that $\pi(X,a_0)$ is a pushout of the diagram

$$
\begin{array}{ccc}
\pi(|W \smallsetminus \sigma| \cap |\sigma|, a_0) & \xrightarrow{\ \pi(i_{W \smallsetminus \sigma})\ } & \pi(|W \smallsetminus \sigma|, a_0) \\[1.2em]
{\scriptstyle \pi(i_\sigma)} \big\downarrow & & \\[1.2em]
\pi(|\sigma|, a_0) & &
\end{array}
$$

and thus $\pi(X,a_0)$ is obtained from the free product

$$\pi(|\sigma|,a_0) * \pi(|W \smallsetminus \sigma|,a_0)$$

together with the relations

$$\pi(i_{W \smallsetminus \sigma})(c)(\pi(i_\sigma)(c))^{-1},$$

$c$ being the generator of $\pi(|\partial\sigma|, a_0)$ (notice that

$$|W \smallsetminus \sigma| \cap |\sigma| = |\partial\sigma|$$

and that $\partial\sigma$ is the one-dimensional simplicial complex defined by the 1-simplexes $\{b_0, c_0\}$, $\{c_0, d_0\}$, and $\{d_0, b_0\}$).

Since $\pi(|\sigma|, a_0) = 0$, the matter becomes much simpler: $\pi(X, a_0)$ derives from $\pi(|W \smallsetminus \sigma|, a_0)$ together with the relation

$$\pi(i_{W \smallsetminus \sigma})(c) = |\{b_0, c_0\}| \cdot |\{c_0, d_0\}| \cdot |\{d_0, b_0\}|.$$

The only thing left to do now is to compute the fundamental group of $|W \smallsetminus \sigma|$. To this end, we consider the barycenter $b(\sigma)$ of $\sigma$ and the radial projection of $|D \smallsetminus \sigma|$ on $|\partial D|$; we obtain a retraction that extends to a strong deformation retraction (see the definition given in Exercise 2 of Sect. I.2)

$$F \colon |W \smallsetminus \sigma| \longrightarrow |K|$$

where $F(|\{b_0, c_0\}| \cdot |\{c_0, d_0\}| \cdot |\{d_0, b_0\}|) = \alpha$. We conclude the proof by noting that $F$ induces an isomorphism among the homotopy groups concerned.  ∎

Let us recalculate the fundamental group of the real projective plane without using so many generators and relations as we did when we applied Theorem (VI.1.7). We know that $\mathbb{R}P^2$ is obtained from a disk $D^2$ by identifying the antipodal points of the boundary $\partial D^2$; in other words, $\mathbb{R}P^2$ is a pushout of the diagram

$$
\begin{array}{ccc}
S^1 & \xrightarrow{\ f\ } & S^1 \\
{\scriptstyle g}\downarrow & & \\
D^2 & &
\end{array}
$$

$$f \colon S^1 \to S^1 , \ e^{i\theta} \mapsto e^{2i\theta}.$$

We have only one generator (that of $\pi(S^1, a_0)$ given by the closed path $\alpha$) and only one relation $\alpha^2$; therefore, $\pi(\mathbb{R}P^2, a_0) \cong \mathbb{Z}_2$.

## VI.1.2  Polyhedra with a Given Fundamental Group

We intend to prove that, for every group $G$ given by a finite number of generators and relations, there exists a polyhedron whose fundamental group is $G$. More precisely:

**(VI.1.16) Theorem.** *Let $G = Gp(S; R)$ be a group with generators $S = \{g_1, \ldots, g_m\}$ and relations $R = \{r_1, \ldots, r_n\}$. Then there exists a two-dimensional polyhedron $|K|$ such that*

$$\pi(|K|, a_0) \cong G$$

*for some vertex $a_0 \in K$.*

*Proof.* For each generator $g_i$, we take a circle $S_i^1$ with a base point $e_i \in S_i^1$, $i = 1, \ldots, m$. We consider the set

$$Y = S_1^1 \vee S_2^1 \vee \ldots \vee S_m^1$$

of all $m$-tuples $(x_1, \ldots, x_m) \in S_1^1 \times \ldots S_m^1$ with no more than one coordinate $x_i$ different from its base point $e_i$; we give $Y$ the base point $a_0 = (e_1, \ldots, e_m)$ and the topology induced by the topology of the product space $S_1^1 \times \ldots \times S_m^1$. We pause here to interpret $Y$ in another way. We consider the pushout

$$
\begin{array}{ccc}
\{x\} & \xrightarrow{\ i_1\ } & S_1^1 \\
\downarrow{\scriptstyle i_2} & & \downarrow{\scriptstyle \overline{i_2}} \\
S_2^1 & \xrightarrow[\ \overline{i_1}\ ]{} & Z
\end{array}
$$

(with $i_j(x) = e_j$, $j = 1, 2$); note that $Z \cong S_1^1 \vee S_2^1$ and, by induction, also $S_1^1 \vee S_2^1 \vee \ldots \vee S_m^1$ may be viewed as the pushout space of an appropriate diagram. Besides, we observe that, due to Corollary (VI.1.14), $\pi(S_1^1 \vee S_2^1, a_0)$ is a free group with two generators and, in general, $\pi(S_1^1 \vee S_2^1 \vee \ldots \vee S_m^1, a_0)$ is a free group with $m$ generators.

We now return to our theorem. A relation, let us say, $r_j$ is a word

$$r_j = b_{i_1}^{\varepsilon_1} \ldots b_{i_p}^{\varepsilon_p}$$

with $\varepsilon_i = \pm 1$; we define a function

$$f_j \colon S^1 \longrightarrow Y,$$

corresponding to the word $r_j$, as follows: since the word $r_j$ has $p$ letters, we divide $S^1$ into $p$ equal parts; the $q$th arc is completely wrapped around the component $S_{i_q}^1$ of $Y$, clockwise if $\varepsilon_q = +1$ and counter-clockwise if $\varepsilon_q = -1$. Analytically, such a function is described as follows:

$$
f_j(e^{i\theta}) =
\begin{cases}
e^{i(p\theta - 2(q-1)\pi)} & \text{if } \varepsilon_q = 1 \\
e^{i(2q\pi - p\theta)} & \text{if } \varepsilon_q = -1,
\end{cases}
$$

where $2(q-1)\pi/p \le \theta \le 2q\pi/p$, $1 \le q \le p$ (look up the function $f \colon S^1 \to S^1$ used for computing the fundamental group of the real projective plane, just before this subsection).

We now construct the pushout

$$\partial W = \vee_{j=1}^{n} S_j^1 \xrightarrow{\ \vee f_j\ } Y$$

$$W = \vee_{j=1}^{n} D_j^2 \xrightarrow[\vee f_j]{} X$$

In other words, we construct a space $X$ by gluing $n$ two-dimensional disks $D_j^2$ to $Y$ by means of the functions $f_j$, $j = 1,\ldots,n$, one for each relation $r_j$.

We may consider the space $Y$ as the geometric realization of a one-dimensional simplicial complex with $2m + 1$ vertices

$$a_0; a_1^1, a_2^1; a_1^2, a_2^2; \ldots; a_1^m, a_2^m$$

(the vertices $a_1^i, a_2^i$ correspond to two distinct points of $S_i^1$, $a_0 = (e_1,\ldots,e_m)$), and $3m$ 1-simplexes

$$\{a_0, a_1^1\}, \{a_1^1, a_2^1\}, \{a_2^1, a_0\}; \ldots; \{a_0, a_1^m\}, \{a_1^m, a_2^m\}, \{a_2^m, a_0\}$$

(each group of three 1-simplexes corresponds to a circle). After this, we view the disk $D_j^2$, that corresponds to the relation $r_j$ given by a word of length $p$, as a regular $3p$-polyhedron, $j = 1,\ldots,n$. In this manner, the disk $D_j^2$ is glued to $Y$ — as we did in Theorem (VI.1.15) — by means of the closed path of geometric 1-simplexes

$$\widetilde{r}_j = (|\{a_0, a_1^{i_1}\}| \cdot |\{a_1^{i_1}, a_2^{i_1}\}| \cdot |\{a_2^{i_1}, a_0\}|)^{\varepsilon_1} \ldots (|\{a_0, a_1^{i_p}\}| \cdot |\{a_1^{i_p}, a_2^{i_p}\}| \cdot |\{a_2^{i_p}, a_0\}|)^{\varepsilon_p}.$$

On the other hand, the fundamental group of $Y$ is the free group generated by $\widetilde{g}_i$, $i = 1,\ldots,n$ where each $\widetilde{g}_i$ equals the homotopy class of the closed geometric path

$$|\{a_0, a_1^i\}| \cdot |\{a_1^i, a_2^i\}| \cdot |\{a_2^i, a_0\}|.$$

We conclude the proof by applying Theorem (VI.1.15). ∎

Theorem (VI.1.16) may be used backward for computing the fundamental group of certain two-dimensional polyhedra: actually, if a polyhedron $|K|$ is constructed as in the theorem, we have at once the generators and the relations that define the fundamental group of $|K|$. For instance, let $|K| = T^2$ be the torus; if we use the triangulation given in Sect. III.5.1 and indicate the sides $01, 12, 20$ and $03, 34, 40$, respectively, with $a$ and $b$, we see that $\pi(T^2, 0)$ is the group defined by the generators $a, b$ and the relation $aba^{-1}b^{-1}$; therefore, $\pi(T^2, 0) \cong \mathbb{Z} \times \mathbb{Z}$.

## *Exercises*

**1.** Let $X$ be the space defined by the union of the unit circle $S^1$ and the closed segment $[(4,0),(5,0)]$; take the vertices $a_0 = (1,0)$ and $a_1 = (4,0)$. Prove that the fundamental groups of $X$ based at $a_0$ and $a_1$ are not isomorphic.

**2.** Let $f\colon (Y,y_0) \to (X,x_0)$ be a homotopy equivalence. Prove that $\pi(Y,y_0) \cong \pi(X,x_0)$.

**3.** Let $(X,x_0)$ and $(Y,y_0)$ be two topological-based spaces. Prove that $\pi(X \times Y, x_0 \times y_0) \cong \pi(X,x_0) \times \pi(Y,y_0)$.

## VI.2 Fundamental Group and Homology

Let $*$ be a generic base point of either $nT^2$ or $n\mathbb{R}P^2$; by Theorem (VI.1.15),

$$\pi(nT^2,*) \cong Gp(a_1,b_1,a_2,b_2,\ldots,a_n,b_n;a_1b_1a_1^{-1}b_1^{-1}\ldots a_nb_na_n^{-1}b_n^{-1})$$

$$\pi(n\mathbb{R}P^2,*) \cong Gp(a_1,a_2,\ldots,a_n;a_1^1a_1^1\ldots a_n^1a_n^1).$$

For $n \geq 2$, these groups are not Abelian; to better understand whether they are isomorphic, we resort to a little algebraic trick: we *abelianize* them. An element of the form $ghg^{-1}h^{-1}$ of a given group $G$ is called a *commutator* of $G$; the subset of $G$

$$[G,G] = \{x \in G \,|\, x = ghg^{-1}h^{-1},\ g,h \in G\}$$

is a normal subgroup of $G$ known as the *commutator subgroup* of $G$ and is the smallest subgroup $H$ of $G$ for which $G/H$ is Abelian. The Abelian group $G/[G,G]$ is called the *abelianized* group of $G$.

**(VI.2.1) Lemma.** *A group homomorphism* $\phi : G \to H$ *induces a homomorphism*

$$\bar{\phi} : G/[G,G] \to H/[H,H]$$

*such that the following diagram commutes:*

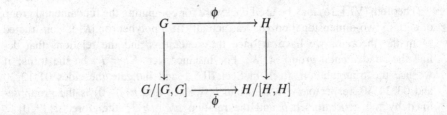

*Moreover, if* $\phi$ *is an isomorphism then also* $\bar{\phi}$ *is an isomorphism.*

*Proof.* We define the homomorphism

$$\bar{\phi}(g + [G,G]) := \phi(g) + [H,H]$$

for every $g + [G,G] \in G/[G,G]$.  ∎

Let us go back to the surfaces. The abelianized group of $G = \pi(nT^2, *)$ is the group generated by the elements

$$a_1, b_1, a_2, b_2, \ldots, a_n, b_n$$

together with the set of relations

$$R = \{a_1^1 b_1^1 a_1^{-1} b_1^{-1} \ldots a_n^1 b_n^1 a_n^{-1} b_n^{-1}; xyx^{-1}y^{-1} | x, y \in G\};$$

hence,

$$\pi(nT^2, *)/[\pi(nT^2, *), \pi(nT^2, *)] \cong \mathbb{Z}^{2n}.$$

On the other hand, the abelianized group of $H = \pi(n\mathbb{R}P^2, *)$ is the group presented as $Gp(S; R)$ where

$$S = \{a_1, a_2, \ldots, a_n\}$$
$$R = \{a_1^1 a_1^1 \ldots a_n^1 a_n^1\} \cup \{xyx^{-1}y^{-1} | x, y \in S\};$$

therefore, the abelianized group of $H$ is an Abelian group generated by the elements

$$a_1, a_2, \ldots, a_n$$

together with the relation

$$2(a_1 + a_2 + \ldots + a_n) = 1$$

and thus, by setting $h = a_1 + \ldots + a_n$, we have

$$\pi(n\mathbb{R}P^2, *)/[\pi(n\mathbb{R}P^2, *), \pi(n\mathbb{R}P^2, *)] \cong Gp(a_1, \ldots, a_{n-1}, h; 2h) \cong \mathbb{Z}^{n-1} \times \mathbb{Z}_2.$$

Since $\mathbb{Z}^{2n}$ and $\mathbb{Z}^{n-1} \times \mathbb{Z}_2$ are not isomorphic, we conclude that the two fundamental groups $\pi(nT^2, *)$ and $\pi(n\mathbb{R}P^2, *)$ are not isomorphic. By the way, this conclusion proves once more that $nT^2$ and $n\mathbb{R}P^2$ are not homeomorphic.

What is interesting to note is that the abelianized groups of the fundamental groups of $nT^2$ and $n\mathbb{R}P^2$ coincide with the corresponding homology groups $H_1(nT^2; \mathbb{Z})$ and $H_1(n\mathbb{R}P^2; \mathbb{Z})$. Hence, it is reasonable to ask whether this is merely a coincidence or it is a fact that holds true for specific types of polyhedra; the answer to this question is found in the following result:

**(VI.2.2) Theorem.** *The abelianized group of the fundamental group of a connected polyhedron $|K|$ is isomorphic to $H_1(|K|; \mathbb{Z})$.*

*Proof.* Since $K$ is connected, we may neglect writing the base point and simply refer to $\pi(|K|)$. We define the function

$$\varphi : \pi(|K|) \to H_1(|K|;\mathbb{Z})$$

as follows: let $[z]$ be a generator of $H_1(S^1;\mathbb{Z})$; then,

$$(\forall [f] \in \pi(|K|))\varphi([f]) := H_1(f;\mathbb{Z})([z]).$$

This definition is independent from the representative chosen for the homotopy class $[f]$. We now wish to make sure that $\varphi$ is a group homomorphism. In fact, for every $[f], [g] \in \pi(|K|)$,

$$\varphi([f] \overset{v}{\times} [g]) = \varphi([\sigma(f \vee g)v]) = \varphi([f]) + \varphi([g])$$

because

$$H_1(v;\mathbb{Z}) : H_1(S^1,\mathbb{Z}) \to H_1(S^1 \vee S^1;\mathbb{Z}) \cong H_1(S^1;\mathbb{Z}) \oplus H_1(S^1;\mathbb{Z}),$$

$$[z] \mapsto ([z],[z]);$$

$$H_1(f \vee g;\mathbb{Z}) : H_1(S^1;\mathbb{Z}) \oplus H_1(S^1;\mathbb{Z}) \to H_1(|K|;\mathbb{Z}) \oplus H_1(|K|;\mathbb{Z}),$$

$$H_1(f \vee g;\mathbb{Z})([z],[z]) = (H_1(f;\mathbb{Z})([z]), H_1(g;\mathbb{Z})([z]));$$

$$H_1(\sigma;\mathbb{Z}) : H_1(|K| \vee |K|;\mathbb{Z}) \to H_1(|K|;\mathbb{Z}),$$

$$(H_1(f;\mathbb{Z})([z]), H_1(g;\mathbb{Z})([z])) \mapsto H_1(f;\mathbb{Z})([z]) + H_1(g;\mathbb{Z})([z]).$$

The homomorphism $\varphi$ is surjective: in fact, let $c = \sum_i m_i\sigma_1^i$ be a 1-cycle of the one-dimensional $C_1(|K|;\mathbb{Z})$. We take a vertex $*$ of $|K|$ as the base point of $\pi(|K|)$; for each 1-simplex $\sigma_1^i$ of $c$, let $\sigma_1^i(0)$ be its first vertex and $\sigma_1^i(1)$ its second one; since $K$ is connected, for each 1-simplex $\sigma_1^i$ of $c$, there is a path of 1-simplexes $\lambda_0^i$ (respectively, $\lambda_1^i$) that joins $*$ to $\sigma_1^i(0)$ (respectively, $\sigma_1^i(1)$). The homotopy class of the loop $\lambda_0^i.\sigma_1^i.(\lambda_1^i)^{-1}$, obtained by composition, is an element $[f_i] \in \pi(|K|,*)$; hence, with a suitable triangulation of $S^1$, we may say that

$$\varphi(\overset{v}{\times}_i (f_i)^{m_i}) = \sum_i m_i\{\lambda_0^i.\sigma_1^i.(\lambda_1^i)^{-1}\}.$$

However, since $|K|$ is connected, each $\lambda_0^i.\sigma_1^i.(\lambda_1^i)^{-1}$ is homologous to $\sigma_1^i$ and so, $\varphi$ is surjective.

The surjection $\varphi : \pi(|K|) \to H_1(|K|;\mathbb{Z})$ is easily extended to a surjection

$$\overline{\varphi} : \pi(|K|)/[\pi(|K|),\pi(|K|)] \to H_1(|K|;\mathbb{Z}).$$

We now prove that $\overline{\varphi}$ is injective. Let us suppose $[f] \in \pi(|K|)$ to be such that $\varphi([f]) = 0 \in H_1(|K|;\mathbb{Z})$; then, $\varphi([f])$ is a boundary

$$\varphi([f]) = \partial_2\left(\sum_i m_i\sigma_i^2\right).$$

We now set the rule that, for every $\sigma_i^2$, $F_i^j(\sigma_i^2)$ is the $j$th-face of $\sigma_i^2$; in this manner,

$$\partial_2(\sigma_i^2) = F_i^0(\sigma_i^2) - F^1(\sigma_i^2) + F_i^2(\sigma_i^2).$$

Similarly to what we have done before for $k = 0, 1, 2$, let $\lambda_i^k(0)$ (respectively, $\lambda_i^k(1)$) be a path of 1-simplexes of $|K|$ that joins $*$ to the first (respectively, second) vertex of $F^k(\sigma_i^2)$; we now associate the loop $\mu_i$, defined by the composition of loops

$$\lambda_i^0(0).F_i^0(\sigma_i^2).(\lambda_i^0(1))^{-1} \overset{v}{\times} \lambda_i^1(0).F_i^1(\sigma_i^2).(\lambda_i^1(1))^{-1} \overset{v}{\times} \lambda_i^2(0).F_i^2(\sigma_i^2).(\lambda_i^2(1))^{-1},$$

with every 2-simplex $\sigma_i^2$ (see Sect. VI.1 on p. 195). On the other hand,

$$[f] + [\pi(|K|), \pi(|K|)] = [\overset{v}{\times}_i (\mu_i)^{m_i}] + [\pi(|K|), \pi(|K|)]$$

and, since every $\mu_i$ is homotopic to the constant loop because $|\sigma_2^i|$ is contractible, $[f] \in [\pi(|K|), \pi(|K|)]$. ∎

## VI.3 Homotopy Groups

The fundamental group $\pi(Y, y_0)$ of a based space $(Y, y_0) \in \textbf{Top}_*$, henceforth denoted by $\pi_1(Y, y_0)$, is the first of a series

$$\{\pi_n(Y, y_0) | n \geq 1\}$$

of groups associated with $(Y, y_0)$. All these groups, called *homotopy groups of Y* (with base point $y_0$) are homotopy invariants; in addition, we shall prove that, for every $n \geq 2$, all groups $\pi_n(Y, y_0)$ are Abelian, even if $\pi_1(Y, y_0)$ may not be Abelian.

As we have seen before, the principal tool in constructing the fundamental group of a based space is the comultiplication

$$v \colon S^1 \to S^1 \vee S^1, \; e^{2\pi t i} \mapsto \begin{cases} (e_0, e^{2\pi 2 t i}) & 0 \leq t \leq \frac{1}{2} \\ (e^{2\pi(2t-1)i}, e_0) & \frac{1}{2} \leq t \leq 1. \end{cases}$$

The comultiplication $v$, that we now indicate with $v_1$, has a very simple geometric interpretation: it is essentially the quotient map obtained by the identification of the points $(1, 0)$ and $(-1, 0)$ of the unit circle $S^1$. We may pursue a similar idea for defining a comultiplication in a unit sphere $S^n \subset \mathbb{R}^{n+1}$, for $n \geq 2$: let $S^{n-1}$ be the intersection of $S^n$ with the hyperplane $z_{n+1} = 0$ and let

$$q_n \colon S^n \longrightarrow S^n / S^{n-1}$$

be the quotient map; then, the comultiplication

$$v_n \colon S^n \longrightarrow S^n \vee S^n$$

is precisely the composite map of $q_n$ and the homeomorphism

$$S^n/S^{n-1} \cong S^n \vee S^n.$$

We have previously defined $v_1$ through a formula that we shall now generalize, knowing that the sphere $S^n$ is homeomorphic to the suspension of $S^{n-1}$ (see Sect. I.2); in fact, if we identify $S^n$ with $\Sigma S^{n-1}$ and then write the points of $S^n$ as $t \wedge x$, where $t \in I$ and $x \in S^{n-1}$, we have the comultiplication

$$v_n: \ S^n \longrightarrow S^n \vee S^n$$

$$v_n(t \wedge x) = \begin{cases} (\mathbf{e}_0, 2t \wedge x) & 0 \le t \le \frac{1}{2} \\ ((2t-1) \wedge x, \mathbf{e}_0) & \frac{1}{2} \le t \le 1 \end{cases}$$

(here $\mathbf{e}_0 = (1,0,\ldots,0)$ is the base point of $S^n$).

The following properties of the comultiplication $v_n$, stated here as two lemmas, are of a particular interest.

**(VI.3.1) Lemma.** *For every $n \ge 1$, the comultiplication $v_n: S^n \to S^n \vee S^n$ is associative up to homotopy, that is to say, the maps $(1_{S^n} \vee v_n)v_n$ and $(v_n \vee 1_{S^n})v_n$ are homotopic.*

*Proof.* The desired homotopy is given by the map

$$H: \ S^n \times I \longrightarrow S^n \vee S^n \vee S^n$$

such that, for every $(t \wedge x, s) \in \Sigma S^{n-1} \times I$,

$$H(t \wedge x, s) = \begin{cases} (\frac{4t}{s+1} \wedge x, \mathbf{e}_0, \mathbf{e}_0) & 0 \le t \le \frac{s+1}{4} \\ (\mathbf{e}_0, (4t-s-1) \wedge x, \mathbf{e}_0) & \frac{s+1}{4} \le t \le \frac{s+2}{4} \\ (\mathbf{e}_0, \mathbf{e}_0, \frac{4t-s-2}{2-s} \wedge x) & \frac{s+2}{4} \le t \le 1. \end{cases}$$

**(VI.3.2) Lemma.** *For every $n \ge 1$, the comultiplication $v_n: S^n \to S^n \vee S^n$ is a homotopic factor of the diagonal map $\Delta: S^n \to S^n \times S^n$; in other words, the maps $\iota v_n$ and $\Delta$ are homotopic ($\iota$ is the inclusion of $S^n \vee S^n$ in the product $S^n \times S^n$).*

*Proof.* The homotopy $\iota v_n \sim \Delta$ is given by the map

$$H: \ S^n \times I \longrightarrow S^n \times S^n$$

$$H(t \wedge x, s) = \begin{cases} (t(2-s) \wedge x, ts \wedge x) & 0 \le t \le \frac{1}{2} \\ ((st+1-s) \wedge x, (2t-1+s(1-t)) \wedge x) & \frac{1}{2} \le t \le 1. \end{cases}$$

As in the case where $n = 1$, we define the multiplication

$$\overset{v_n}{\times}: \ [S^n, Y]_* \times [S^n, Y]_* \longrightarrow [S^n, Y]_*$$

$$(\forall [f], [g] \in [S^n, Y]_*) \ [f] \overset{v_n}{\times} [g] := [\sigma(f \vee g)v_n]$$

for every $n \geq 2$ and for every $(Y, y_0) \in \mathbf{Top}_*$; also in this case, the definition does not depend on the representatives of the homotopy classes $[f]$ and $[g]$.

**(VI.3.3) Theorem.** *The set $[S^n, Y]_*$ with the multiplication $\overset{v_n}{\times}$ is a group.*

*Proof.* The associativity of $\overset{v_n}{\times}$ stems from $v_n$ being homotopy associative (see Lemma (VI.3.1)): indeed, for every $f, g, h \in \mathbf{Top}_*((S^n, e_0), (Y, y_0))$,

$$\sigma(1_Y \vee \sigma)(f \vee g \vee h)(1_{S^n} \vee v_n)v_n \sim \sigma(\sigma \vee 1_Y)(f \vee g \vee h)(v_n \vee 1_{S^n})v_n$$

and the equalities

$$\sigma(1_Y \vee \sigma)(f \vee g \vee h)(1_{S^n} \vee v_n)v_n = \sigma(f \vee \sigma(g \vee h)v_n)v_n,$$
$$\sigma(\sigma \vee 1_Y)(f \vee g \vee h)(v_n \vee 1_{S^n})v_n = \sigma(\sigma(f \vee g)v_n \vee h)v_n$$

are also valid.

The homotopy class $[c]$ of the constant map

$$c \colon S^n \to Y \, , \, t \wedge x \mapsto y_0$$

is the identity element for the multiplication defined in $\pi_n(Y, y_0)$: in fact, for every $f \in \mathbf{Top}_*((S^n, e_0), (Y, y_0))$ and by Lemma (VI.3.2), the following homotopies:

$$\sigma(f \vee c)v_n = \sigma(f \times c)\iota v_n \sim \sigma(f \times c)\Delta = f,$$
$$\sigma(c \vee f)v_n = \sigma(c \times f)\iota v_n \sim \sigma(c \times f)\Delta = f$$

hold true.

As for the inverses, we proceed in the following manner. Let $[f] \in \pi_n(Y, y_0)$ be an arbitrarily given element. We define

$$h \colon S^n \to Y \, , \, t \wedge x \mapsto f((1-t) \wedge x);$$

note that, for every $t \wedge x \in S^n$,

$$\sigma(f \vee h)v_n(t \wedge x) = \begin{cases} f(2t \wedge x) & 0 \leq t \leq \frac{1}{2} \\ h((2t-1) \wedge x) & \frac{1}{2} \leq t \leq 1. \end{cases}$$

The homotopy

$$H(t \wedge x, s) = \begin{cases} y_0 & 0 \leq t \leq \frac{s}{2} \\ f((2t-s) \wedge x) & \frac{s}{2} \leq t \leq \frac{1}{2} \\ f((2-2t-s) \wedge x) & \frac{1}{2} \leq t \leq \frac{2-s}{2} \\ y_0 & \frac{2-s}{2} \leq t \leq 1 \end{cases}$$

shows that $[h]$ is the right inverse of $[f]$; similarly, one proves that $[h]$ is also the left inverse of $[f]$. ∎

The group

$$\pi_n(Y, y_0) := [S^n, Y]*$$

is the *nth homotopy group* of the based space $(Y, y_0)$; the homotopy groups $\pi_n(Y, y_0)$ with $n \geq 2$ are also called *higher homotopy* groups of $(Y, y_0)$.

The next lemma characterizes the maps $f \colon (S^n, \mathbf{e}_0) \to (Y, y_0)$ that are homotopic to the constant map and thus characterizes the unit element of the group $\pi_n(Y, y_0)$; here the reader is asked to return to Exercise 2 of Sect. I.2 to review at least the definition of contractibility of a space. Note that the sphere $S^n$ is homeomorphic to the geometric realization of the simplicial complex $\mathring{\sigma}_{n+1}$ ($\sigma_{n+1}$ is an $(n+1)$-simplex).

**(VI.3.4) Lemma.** *A based map*

$$f \colon |\mathring{\sigma}_{n+1}| \longrightarrow Y;$$

*may be extended to* $|\overline{\sigma}_{n+1}|$ *if and only if $f$ is homotopic to the constant map.*

*Proof.* Let us suppose $f$ to be extended to a map

$$\overline{f} \colon |\overline{\sigma}_{n+1}| \longrightarrow Y.$$

Since $|\overline{\sigma}_{n+1}| \cong D^{n+1}$ is contractible, the identity map of $|\overline{\sigma}_{n+1}|$ onto itself is homotopic to the constant map $c$; let

$$H \colon |\overline{\sigma}_{n+1}| \times I \longrightarrow |\overline{\sigma}_{n+1}|$$

be the homotopy, which joins $1_{|\overline{\sigma}_{n+1}|}$ and $c$. The composite map

$$\mathring{\sigma}_{n+1} \times I \xrightarrow{\iota \times 1_I} |\overline{\sigma}_{n+1}| \times I \xrightarrow{H} |\overline{\sigma}_{n+1}| \xrightarrow{\overline{f}} Y$$

is a homotopy from $f$ to $c$.

Let us now suppose that

$$G \colon |\mathring{\sigma}_{n+1}| \times I \longrightarrow Y$$

is a homotopy from $f$ to $c$. Since the pair of polyhedra $(|\overline{\sigma}_{n+1}|, |\mathring{\sigma}_{n+1}|)$ has the Homotopy Extension Property (see Theorem (III.1.7)), there exists a map

$$H \colon |\overline{\sigma}_{n+1}| \times I \longrightarrow Y$$

such that the following diagram is commutative:

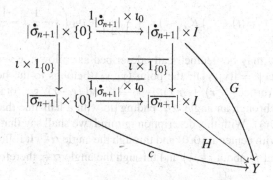

The restriction of $H$ to $|\overline{\sigma}_{n+1}| \times 1 = |\overline{\sigma}_{n+1}|$ is the desired extension of $f$.  ∎

Our next lemma is important to the proof of the two theorems that follow it.

**(VI.3.5) Lemma.** *For every $n \geq 1$, the function*

$$h \colon \Sigma(S^n \vee S^n) \longrightarrow \Sigma S^n \vee \Sigma S^n$$

*such that, for every $t \in I$ and $x \in S^n$,*

$$h(t \wedge (x, \mathbf{e}_0)) = (t \wedge x, \mathbf{e}_0), \quad h(t \wedge (\mathbf{e}_0, x)) = (\mathbf{e}_0, t \wedge x),$$

*is a homeomorphism; moreover, the maps $v_{n+1}$ and $(h\Sigma)v_n$ are homotopic.*

*Proof.* The definitions are such that, for every $t \wedge s \wedge x \in \Sigma^2 S^{n-1} \cong \Sigma S^n$,

$$h(\Sigma v_n)(t \wedge s \wedge x) = \begin{cases} (t \wedge 2s \wedge x, \mathbf{e}_0) & 0 \leq s \leq \frac{1}{2}, \\ (\mathbf{e}_0, t \wedge (2s-1) \wedge x) & \frac{1}{2} \leq s \leq 1, \end{cases}$$

$$v_{n+1}(t \wedge s \wedge x) = \begin{cases} (2t \wedge s \wedge x, \mathbf{e}_0) & 0 \leq s \leq \frac{1}{2}, \\ (\mathbf{e}_0, (2t-1) \wedge s \wedge x) & \frac{1}{2} \leq s \leq 1. \end{cases}$$

We now define the following functions:

1. $|\,|\colon \mathbb{R}^2 \longrightarrow \mathbb{R}$ , $|(y_1, y_2)| = \max(|y_1|, |y_2|)$, for every $(y_1, y_2) \in \mathbb{R}^2$
2. For every $\alpha \in [-1, 1]$,

$$\rho_\alpha \colon \mathbb{R}^2 \longrightarrow \mathbb{R}^2$$

$$(\forall (y_1, y_2) \in \mathbb{R}^2) \ \rho_\alpha(y_1, y_2) = \begin{pmatrix} \cos \alpha \frac{\pi}{2} & -\sin \alpha \frac{\pi}{2} \\ \sin \alpha \frac{\pi}{2} & \cos \alpha \frac{\pi}{2} \end{pmatrix} \begin{pmatrix} y_1 \\ y_2 \end{pmatrix}$$

3. $\widetilde{\rho}_\alpha \colon \mathbb{R}^2 \longrightarrow \mathbb{R}^2$

$$(\forall (y_1, y_2) \in \mathbb{R}^2) \ \widetilde{\rho}_\alpha(y_1, y_2) = \frac{|(y_1, y_2)|}{|\rho_\alpha(y_1, y_2)|} \rho_\alpha(y_1, y_2)$$

4. $R'_\alpha: I^2 \longrightarrow I^2$

$$(\forall(s,t) \in I^2) \, R'_\alpha(s,t) = \left(\frac{1}{2}, \frac{1}{2}\right) + \widetilde{\rho_\alpha}\left(s - \frac{1}{2}, t - \frac{1}{2}\right)$$

The function $\widetilde{\rho_\alpha}$ may be geometrically described as follows: given a vector $\vec{v} = (y_1, y_2) \in \mathbb{R}^2$, let $r = |(y_1, y_2)|$; the point $(y_1, y_2)$ belongs to the boundary of the square $Q_r$ with vertices $(r,r)$, $(-r,r)$, $(-r,-r)$, and $(r,-r)$; we rotate the half-line $(0,0)$ , $(y_1, y_2)$ through an angle $\alpha\frac{\pi}{2}$, ending up with a half-line that crosses $Q_r$ at the point $\widetilde{\rho_\alpha}(y_1, y_2)$. With this description in mind, we shall say that $\widetilde{\rho_\alpha}$ is a *square rotation* of $\mathbb{R}^2$ with center at $(0,0)$ and through the angle $\alpha\frac{\pi}{2}$. It follows that $R'_\alpha$ is a square rotation of $I^2$ about $(\frac{1}{2}, \frac{1}{2})$ and through the angle $\alpha\frac{\pi}{2}$; therefore, $R'_\alpha$ induces a map

$$R_\alpha: \Sigma\Sigma S^{n-1} \longrightarrow \Sigma\Sigma S^{n-1}$$

$$(\forall s \wedge t \wedge x \in \Sigma\Sigma S^{n-1}) \, R_\alpha(s \wedge t \wedge x) = R'_\alpha(s \wedge t) \wedge x.$$

The homotopy

$$F: \Sigma^2 S^{n-1} \times I \longrightarrow \Sigma^2 S^{n-1} \vee \Sigma^2 S^{n-1}$$

that we seek is the following composition:

$$(\forall u \in I) \, F(-, u) = (R_{-u} \vee R_{-u})\{h(\Sigma v_n)\}R_u.  \qquad \blacksquare$$

**(VI.3.6) Remark.** Notice that by iteration

$$v_{n+1} \sim (h\Sigma)^n v_1.$$

We recall that the symbol $\Omega Y$ indicates the space of paths of a based space $(Y, y_0)$ (see p. 32).

**(VI.3.7) Theorem.** *For every $n \geq 2$ and every $(Y, y_0) \in$ **Top**$_*$, the groups $\pi_n(Y, y_0)$ and $\pi_{n-1}(\Omega Y, c_{y_0})$ are isomorphic.*

*Proof.* We know that every based map

$$f: S^n \cong \Sigma S^{n-1} \to Y$$

has an adjoint map

$$\overline{f}: S^{n-1} \to \Omega Y$$

such that, for every $x \in S^{n-1}$ and every $t \in I$,

$$\{\overline{f}(x)\}(t) = f(t \wedge x).$$

In Sect. I.2, we have seen that the function

$$\Phi: M_*(\Sigma X, Y) \to M_*(X, \Omega Y) , f \mapsto \overline{f}$$

is a bijection for every pair of based spaces $X$ and $Y$; this result extends itself to the sets of homotopy classes (see Exercise 3 of Sect. I.2): the corresponding function

$$[\Phi]: [\Sigma X, Y]_* \to [X, \Omega Y]_*$$

is a bijection. We need then to prove that the bijection of sets

$$[\Phi]: \pi_n(Y, y_0) \to \pi_{n-1}(\Omega Y, c_{y_0})$$

is a group homomorphism; in other words, we must prove that

$$[\Phi]([f] \overset{v_n}{\times} [g]) = [\overline{f}] \overset{v_{n-1}}{\times} [\overline{g}]$$

for any $[f], [g] \in \pi_n(Y, y_0)$. This means that we must prove that the adjoint of $\sigma(f \vee g)v_n$, namely, $\overline{\sigma(f \vee g)v_n}$, is homotopic to $\sigma(\overline{f} \vee \overline{g})v_{n-1}$. We now have on the one hand that

$$\sigma(f \vee g)v_n \sim \sigma(f \vee g)h\Sigma v_{n-1}$$

Lemma (VI.3.5); on the other hand, we see that

$$\sigma(f \vee g)h\Sigma v_{n-1}(t \wedge s \wedge x) = \{\sigma(\overline{f} \vee \overline{g})v_{n-1}\}(s \wedge x)(t)$$

for every $t \wedge s \wedge x \in \Sigma^2 S^{n-2}$. ∎

A based space $(Y, y_0)$ is called an *H-space* if there exists a "multiplication"

$$\mu_Y: Y \times Y \longrightarrow Y$$

such that the maps

$$\mu_Y \iota, \ \sigma: Y \vee Y \longrightarrow Y$$

are homotopic. For instance, the space of paths $\Omega Y$ is an H-space. In fact, we define

$$\mu_{\Omega Y}: \Omega Y \times \Omega Y \longrightarrow \Omega Y$$

as follows:

$$(\forall (\alpha, \beta) \in \Omega Y \times \Omega Y)(\forall t \in I) \ \mu_{\Omega Y}(\alpha, \beta)(t) =
\begin{cases}
\alpha(2t) & 0 \leq t \leq \frac{1}{2} \\
\beta(2t-1) & \frac{1}{2} \leq t \leq 1.
\end{cases}$$

For proving that $\mu_{\Omega Y}$ is a multiplication, it is enough to consider the constant loop $c_{y_0}$ at the base point $y_0$ and construct the homotopy

$$H: (\Omega Y \vee \Omega Y) \times I \longrightarrow \Omega Y,$$

defined by

$$H((\alpha, c_{y_0}), s)(t) =
\begin{cases}
\alpha(\frac{2t}{s+1}) & 0 \leq t \leq \frac{s+1}{2} \\
y_0 & \frac{s+1}{2} \leq t \leq 1,
\end{cases}$$

and

$$H((c_{y_0},\beta),s)(t) = \begin{cases} y_0 & 0 \le t \le \frac{1-s}{2} \\ \beta(\frac{2t+s-1}{s+1}) & \frac{1-s}{2} \le t \le 1. \end{cases}$$

**(VI.3.8) Theorem.** *Let* $(Y,y_0)$ *be a given H-space; then, for every* $n \ge 1$, *the homotopy group* $\pi_n(Y,y_0)$ *is Abelian.*

*Proof.* Due to the definitions of multiplication in $Y$ and comultiplication in $S^n$, and also to $[c_{y_0}]$ being the identity element of $\pi_n(Y,y_0)$, the homotopies

$$f \sim \sigma(f \vee c_{y_0})v_n \sim \mu_Y(f \times c_{y_0})\Delta = f'$$
$$g \sim \sigma(c_{y_0} \vee g)v_n \sim \mu_Y(c_{y_0} \times g)\Delta = g'$$

hold true for every $f,g \in \mathbf{Top}_*(S^n,Y)$; therefore,

$$\sigma(f \vee g)v_n \sim \sigma(f' \vee g')v_n \sim \mu_Y(f' \times g')\Delta$$
$$\sigma(g \vee f)v_n \sim \sigma(g' \vee f')v_n \sim \mu_Y(g' \times f')\Delta.$$

However, for every $x \in S^n$, either $f'(x)$ or $g'(x)$ must equal $y_0$ and so, $(f' \times g')\Delta$ is a map from $S^n$ to $Y \vee Y$; consequently, we have the homotopies

$$\mu_Y(f' \times g')\Delta \sim \sigma(f' \times g')\Delta$$
$$\mu_Y(g' \times f')\Delta \sim \sigma(g' \times f')\Delta.$$

We end this proof by observing that

$$\sigma(f' \times g')\Delta = \sigma(g' \times f')\Delta. \qquad \blacksquare$$

**(VI.3.9) Theorem.** *For every* $n \ge 2$ *and every* $(Y,y_0) \in \mathbf{Top}_*$, $\pi_n(Y,y_0)$ *is Abelian.*

*Proof.* It is a consequence of Theorems (VI.3.7), (VI.3.8), and of the fact that $\Omega Y$ is an H-space. $\qquad \blacksquare$

**(VI.3.10) Theorem.** *For every* $n \ge 2$,

$$\pi_n : \mathbf{Top}_* \longrightarrow \mathbf{Ab}$$

*is a covariant functor.*

*Proof.* A based map $k : (Y,y_0) \longrightarrow (X,x_0)$ induces a group homomorphism

$$\pi_n(k) : \pi_n(Y,y_0) \longrightarrow \pi_n(X,x_0)$$

as follows: for every $[f] \in \pi_n(Y,y_0)$,

$$\pi_n(k)([f]) := [kf].$$

We must use the fact that $\sigma(k \vee k) = k\sigma$. $\qquad \blacksquare$

## VI.3.1 *Action of the Fundamental Group on the Higher Homotopy Groups*

Like the fundamental group, the higher homotopy groups depend on the choice of a base point. In fact, in Sect. VI.1 we have proved that a path $\lambda: I \to Y$ between two points $y_0$ and $y_1$ of $Y$ defines an isomorphism

$$\phi_\lambda: \pi_1(Y, y_0) \longrightarrow \pi_1(Y, y_1)$$

as follows: for every $[f] \in \pi_1(Y, y_0)$,

$$\phi_\lambda([f]) := [\lambda^{-1} * (f * \lambda)].$$

When $\lambda$ is a closed path at $y_0$, that is to say, a loop of $Y$ with base at $y_0$, we have an isomorphism from $\pi_1(Y, y_0)$ onto itself and this defines an action

$$\phi: \pi_1(Y, y_0) \times \pi_1(Y, y_0) \longrightarrow \pi_1(Y, y_0)$$

given by the conjugation:

$$(\forall [f], [g] \in \pi_1(Y, y_0)) \; \phi([f], [g]) = [f]^{-1} \overset{v}{\times} [g] \overset{v}{\times} [f].$$

For the higher homotopy groups, we avail ourselves of the Homotopy Extension Property for polyhedra. Let $\lambda$ be a path from $y_0$ to $y_1$ in $Y$. We identify $S^n$ with the polyhedron $|\dot{\sigma}_{n+1}|$ and suppose that the base point $e_0$ of $S^n$ is identified with a vertex of $|\dot{\sigma}_{n+1}|$. By the Homotopy Extension Property, applied to the pair of polyhedra $(|\dot{\sigma}_{n+1}|, e_0)$, the function $f$ and the path $\lambda$ give rise to a homotopy (not necessarily unique)

$$F: |\dot{\sigma}_{n+1}| \times I \to Y$$

through the diagram

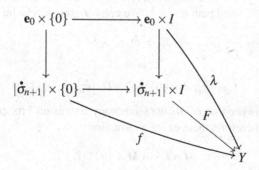

Let $\overline{f} := F(-,1)$. Note that the maps $f$ and $\overline{f}$ are homotopic to each other by means of a *free* homotopy (actually, the restriction of $F$ to $\mathbf{e}_0 \times I$ coincides with the path $\lambda$); moreover, $\overline{f}(\mathbf{e}_0) = y_1$. We thus define the relation

$$\phi_\lambda^n : \pi_n(Y,y_0) \longrightarrow \pi_n(Y,y_1)$$

with the condition $\phi_\lambda([f]) = [\overline{f}]$.

**(VI.3.11) Theorem.** *The relation $\phi_\lambda^n$ defined by the path $\lambda$ is a group isomorphism such that:*

1. *If $\mu : I \to Y$ is a path from $y_0$ to $y_1$ homotopic rel $\partial I$ to the path $\lambda$, then $\phi_\mu^n = \phi_\lambda^n$.*
2. *The constant path $c_{y_0}$ at $y_0$ induces the identity isomorphism*

$$\phi_{c_{y_0}}^n : \pi_n(Y,y_0) \to \pi_n(Y,y_0).$$

3. *If $\eta : I \to Y$ is a path from $y_1$ to $y_2 \in Y$, then*

$$\phi_{\lambda*\eta}^n = \phi_\eta^n \phi_\lambda^n.$$

*Proof.* We have not yet established whether $\phi_\lambda^n$ is well defined; this will follow from the proof of 1. Let

$$G : |\dot{\sigma}_{n+1}| \times I \to Y$$

be a homotopy induced by $f$ and $\mu$, in other words, such that

$$G(-,0) = f \text{ and } G(\mathbf{e}_0,t) = \mu(t).$$

We define $\overline{g} = G(-,1)$ and note that the composite homotopy

$$H = F^{-1} * G : |\dot{\sigma}_{n+1}| \times I \longrightarrow Y$$

is a free homotopy from $\overline{f}$ to $\overline{g}$, that may be transformed into a based homotopy, as follows. From the condition $\mu \sim \lambda$ rel $\partial I$, we obtain a based homotopy

$$\overline{K} : I \times I \longrightarrow Y$$

such that $\overline{K}(-,0) = \mu^{-1} * \lambda$ and $\overline{K}(-,1) = c_{y_1}$. Note that the restriction $H|\mathbf{e}_0 \times I$ coincides with the closed path $\mu^{-1} * \lambda$. We consider the polyhedra

$$M = |\dot{\sigma}_{n+1}| \times I$$

and

$$L = |\dot{\sigma}_{n+1}| \times \{0\} \cup \mathbf{e}_0 \times I \cup |\dot{\sigma}_{n+1}| \times \{1\};$$

since $L$ is a subpolyhedron of $M$, the Homotopy Extension Property holds for the pair $(M,L)$ and, therefore, there exists a retraction

$$r : M \times I \longrightarrow M \times \{0\} \cup L \times I.$$

We now construct the following maps:

$$\overline{F}: (|\dot{\sigma}_{n+1}| \times \{0\}) \times I \to Y$$

and

$$\overline{G}: (|\dot{\sigma}_{n+1}| \times \{1\}) \times I \to Y$$

such that, for every $x \in |\dot{\sigma}_{n+1}|$ and every $t \in I$,

$$\overline{F}(x,0,t) = \overline{f}(x) \text{ and } \overline{G}(x,1,t) = \overline{g}(x).$$

Note that

$$\overline{F}(\mathbf{e}_0,0,t) = \overline{G}(\mathbf{e}_0,1,t) = y_1.$$

Let

$$\overline{H}: M \times \{0\} \cup L \times I \longrightarrow Y$$

be the map defined by the union of the maps $\overline{H} \cup (\overline{F} \cup \overline{K} \cup \overline{G})$ (note that the restriction of $H$ to $L \times \{0\}$ coincides with $\overline{f} \cup \mu^{-1} * \lambda \cup \overline{g}$); we now consider the homotopy

$$\theta: M \times I \xrightarrow{\quad r \quad} M \times \{0\} \cup L \times I \xrightarrow{\quad \overline{H} \quad} Y$$

in other words, the function defined by the commutative diagram

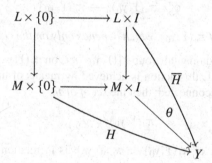

The homotopy

$$\widetilde{H} = \theta(-,-,1): |\dot{\sigma}_{n+1}| \times I \to Y$$

is a free homotopy from $\overline{f}$ to $\overline{g}$. Hence, $[\overline{f}] = [\overline{g}] \in \pi_n(Y,y_1)$ and thus, by setting $\mu = \lambda$, we see that $\phi_\lambda^n([f])$ does not depend on the choices of $\overline{f}$, the representative of the class $[f]$, or the homotopy $F$ with the required properties. We have then proved that $\phi_\lambda^n$ is a (well defined) function that satisfies property 1 stated in the theorem.

We leave the proof of properties 2 and 3 to the reader (anyway, the results follow easily from the definitions). A consequence of properties 1, 2, and 3 is that $\phi_\lambda^n$ is injective and surjective.

We now prove that $\phi_\lambda^n$ is a group homomorphism. Let $[f], [g] \in \pi_n(Y,y_0)$ be given arbitrarily and let

$$F,G: S^n \times I \to Y$$

be two homotopies such that $F(-,0) = f$ , $G(-,0) = g$ and, for every $t \in I$, the equality $F(e_0,t) = G(e_0,t) = \lambda(t)$ holds; besides, let $\overline{f} = F(-,1)$ and $\overline{g} = G(-,1)$. We must prove that

$$\phi_\lambda^n([f] \overset{v_n}{\times} [g]) = \phi_\lambda^n([f]) \overset{v_n}{\times} \phi_\lambda^n([g])$$

that is to say

$$[\overline{\sigma(f \vee g)v_n}] = [\sigma(\overline{f} \vee \overline{g})v_n]$$

with $\overline{\sigma(f \vee g)v_n} = K(-,1)$; here

$$K\colon S^n \times I \longrightarrow Y$$

is any homotopy such that $K(-,0) = \sigma(f \vee g)v_n$ and $K(e_0,t) = \lambda(t)$, for every $t \in I$. We choose

$$K := \sigma(F \vee G)(v_n \times 1_I)\colon S^n \times I \longrightarrow Y;$$

in this case, a simple calculation shows that

$$(\forall t \wedge x \in S^n)\ K(t \wedge x, 1) = \sigma(\overline{f} \vee \overline{g})v_n(t \wedge x)$$

and so $\phi_\lambda^n$ is a homomorphism.                                                                            ∎

The following result is a direct consequence of the previous theorem.

**(VI.3.12) Corollary.** *For every closed loop at $y_0$, the function*

$$\phi_\lambda^n\colon \pi_n(Y,y_0) \longrightarrow \pi_n(Y,y_0)$$

*is an automorphism of $\pi_n(Y,y_0)$, which depends only on the class of $\lambda$.*

We say that the fundamental group $\pi_1(Y,y_0)$ *acts on* $\pi_n(Y,y_0)$; as we have already remarked, when $n = 1$, this action is achieved by means of inner automorphisms. Suppose $Y$ to be path-connected; then the *set of orbits*

$$\pi_n(Y,y_0)/_{\pi_1(Y,y_0)},$$

induced by the action of $\pi_1(Y,y_0)$ on $\pi_n(Y,y_0)$, is in relation with the set of free homotopy classes $[S^n, Y]$; actually, we have the following

**(VI.3.13) Theorem.** *Let $(Y,y_0)$ be a path-connected based space. Then there exists a bijection*

$$\phi\colon \pi_n(Y,y_0)/_{\pi_1(Y,y_0)} \longrightarrow [S^n,Y].$$

We do not prove this theorem here; the reader, who wishes to read a proof of this result, is asked to seek Corollary 7.1.3 in [26].

Let us suppose $Y$ to be *simply connected*, that is to say, $\pi_1(Y,y_0) = 0$; any loop of $Y$ with base at $y_0$ is homotopic to the constant loop $c_{y_0}$; thus, by part 2 of Theorem (VI.3.11), $\phi_\lambda^n$ is the identity automorphism of $\pi_n(Y,y_0)$ and consequently

$$\pi_n(Y,y_0) \equiv [S^n,Y].$$

The following definition is important to Sect. VI.4.

**(VI.3.14) Definition.** A path-connected space $Y$ is *n-simple* if, for a given $y_0 \in Y$, the action of $\pi_1(Y,y_0)$ on $\pi_n(Y,y_0)$ is trivial; in other words, if $\phi_\lambda^n$ is the identity isomorphism, for every loop with base at $y_0$. This happens, for instance, if $Y$ is *simply connected*, that is to say, if $\pi_1(Y,y_0) \cong 0$. Therefore, the $n$-sphere $S^n$ is $n$-simple. In other words, $Y$ is $n$-simple if, for every pair of points $y_0$ and $y_1$ of $Y$, $\pi_n(Y,y_0) \cong \pi_n(Y,y_1)$, and this isomorphism does not depend on the choice of the path from $y_0$ to $y_1$.

## VI.3.2 On the Homotopy Groups of Spheres

We know that spheres have a very simple triangulation; in fact, the $n$-dimensional sphere is homeomorphic to the polyhedron $|\overset{\bullet}{\sigma}_{n+1}|$. This gives us an easy way to prove that the fundamental group of $S^1$ is isomorphic to $\mathbb{Z}$ and that $\pi_1(S^2, \mathbf{e}_0) \cong 0$ (see Sect. VI.1). We now prove that the "lower groups" of the spheres are trivial, that is to say,

**(VI.3.15) Theorem.** *For every $n \geq 2$ and every $r$ with $1 \leq r \leq n-1$,*

$$\pi_r(S^n, \mathbf{e}_0) \cong 0.$$

*Proof.* Let us choose $[f] \in \pi_r(S^n, \mathbf{e}_0)$ arbitrarily; we must prove that the map $f$ is homotopic to the constant map $c_{\mathbf{e}_0}$. Let $K = (X, \Phi)$ and $L = (Y, \Psi)$ be two simplicial complexes such that $|K| \cong S^r$ and $|L| \cong S^n$. By the Simplicial Approximation Theorem, there exists a simplicial function $g \colon K^{(t)} \to L$, of a suitable barycentric subdivision of $K$ such that $|g| \sim fF$, where $F \colon |K^{(t)}| \to |K|$ is the homeomorphism defined in Sect. III.1. Let us suppose that $|K|$ and $|K^{(t)}|$ are identified with $S^r$; in addition, let us identify $L$ with $S^n$ and let us assume that $|g|$ and $f$ are maps from $S^r$ to $S^n$; hence, notwithstanding the homeomorphisms, we have that $|g| \sim f$. Yet, since $\dim K^{(t)} = r$ and $r < n$, the simplicial function $g$ cannot be surjective and so

$$|g| \colon S^r \longrightarrow S^n$$

is not surjective. Let $p$ be a point of $S^n$ that does not belong to the image of $|g|$; hence,

$$|g| \colon S^r \longrightarrow S^n \setminus \{p\} \cong \mathbb{R}^n$$

(the homeomorphism $\phi \colon S^n \setminus \{p\} \to \mathbb{R}^n$ is a stereographic projection). Now, let $c \colon S^r \to \mathbb{R}^n$ be the constant map at the point $\phi(\mathbf{e}_0)$; from the identification $S^n \setminus \{p\} \equiv \mathbb{R}^n$, we conclude that $|g|$ and $c$ are homotopic, with the homotopy given by the map

$$H \colon S^r \times I \longrightarrow \mathbb{R}^n$$

$$(\forall (x,t) \in S^r \times I) \ H(x,t) = (1-t)|g|(x) + tc.$$

We may then say that $f$ is homotopic to a constant function. ∎

**(VI.3.16) Remark.** Let $S^0$ be the 0-*dimensional sphere*, that is to say, the pair $\{-1,1\}$ of points of $\mathbb{R}$ with the discrete topology and the base point $(1)$. For any based space $(Y, y_0)$, let $\pi_0(Y)$ be the set $[S^0, Y]_*$; it is easily seen that, if $Y$ is path-connected, then $\pi_0(Y) = 0$: in fact, two based maps $f, g\colon S^0 \to Y$ are always homotopic, the homotopy being given by a path that joins $f(1)$ and $g(1)$; thus, there exists only one homotopy class in $[S^0, Y]_*$, that of the constant map.

In particular, since for each $n \geq 1$, $S^n$ is path-connected, $\pi_0(S^n) = 0$.

We know that $\pi_1(S^1, e_0) \cong \mathbb{Z}$; what can we tell about the groups $\pi_n(S^n, e_0)$? We have the following result:

**(VI.3.17) Theorem.** *For every* $n \geq 1$, $\pi_n(S^n, e_0) \cong \mathbb{Z}$.

Before proving the theorem, we recall that the *degree* of a map $f\colon S^n \to S^n$ (see p. 121) is homotopy invariant and so, it induces a function

$$d\colon \pi_n(S^n) \longrightarrow \mathbb{Z}.$$

**(VI.3.18) Lemma.** *For every* $n \geq 1$, *the degree function*

$$d\colon \pi_n(S^n, e_0) \longrightarrow \mathbb{Z}$$

*is a group isomorphism.*

*Proof.* Let $[f], [g] \in \pi_n(S^n, e_0)$ be given arbitrarily; we wish to determine the degree of the function

$$\sigma(f \vee g)v_n\colon S^n \longrightarrow S^n.$$

First of all we note that, by Theorem (III.4.3),

$$H_n(S^n \vee S^n; \mathbb{Z}) \cong H_n(S^n; \mathbb{Z}) \oplus H_n(S^n; \mathbb{Z});$$

we now identify the components $\{e_0\} \times S^n$ and $S^n \times \{e_0\}$ of $S^n \vee S^n$ with $S^n$ (in other words, we interpret $S^n \vee S^n$ as a "union" of two spheres $S^n$) and consider the "projections"

$$p_1\colon S^n \vee S^n \to S^n \text{ and } p_2\colon S^n \vee S^n \to S^n;$$

it is easy to verify that $p_1 v_n \sim 1_{S^n}$ and $p_2 v_n \sim 1_{S^n}$; hence,

$$H_n(v_n)\colon H_n(S^n, \mathbb{Z}) \longrightarrow H_n(S^n; \mathbb{Z}) \oplus H_n(S^n; \mathbb{Z})$$

is such that $H_n(v_n)(\{z\}) = \{z\} \oplus \{z\}$. Moreover,

$$H_n(\sigma)\colon H_n(S^n; \mathbb{Z}) \oplus H_n(S^n; \mathbb{Z}) \longrightarrow H_n(S^n; \mathbb{Z}), \ x \oplus y \mapsto x + y$$

for every $x \oplus y \in H_n(S^n; \mathbb{Z}) \oplus H_n(S^n; \mathbb{Z})$. Therefore,

$$H_n(\sigma(f \vee g)v_n)(\{z\}) = H_n(\sigma)(H_n(f)(\{z\}) \oplus H_n(g)(\{z\}))$$
$$= H_n(f)(\{z\}) + H_n(g)(\{z\}) = d(f) + d(g).$$

Since the map $1_{S^n}: S^n \to S^n$ has obviously degree 1, the degree function that we have defined is surjective.

It must be proved that the degree function is injective; we do not prove it here, but the reader can find the proof in the work of H. Whitney [34]. However, Theorem (VI.3.17) may also be proved in different ways (see, for instance, Sect. V.3 of [26]).

∎

Lemma (VI.3.18) proves Theorem (VI.3.17).

So far, not all homotopy groups of the spheres are known; yet, we have two important results due to Jean–Pierre Serre [31] and based on difficult techniques of homological algebra:

**(VI.3.19) Theorem.** *If $n$ is odd and $m \neq n$, then $\pi_m(S^n, e_0)$ is a finite group.*

**(VI.3.20) Theorem.** *If $n$ is even, then*

1. $\pi_m(S^n, e_0)$ *is finite if $m \neq n$ and $m \neq 2n - 1$.*
2. $\pi_{2n-1}(S^n, e_0)$ *is the direct sum of an infinite cyclic group and a finite group (eventually trivial).*

## VI.3.3 Another Approach to Homotopy Groups

In the literature, the homotopy groups are also described in a different way but equivalent to ours; sometimes it is convenient to study the homotopy groups under this other point of view. We begin our work with a lemma that is very important to the development of our theme.

**(VI.3.21) Lemma.** *Given two maps $f, g \in \mathbf{CTop}((X,A),(Y,B))$ such that $f|A = g|A$, let*

$$H: (X \times I, A \times I) \longrightarrow (Y, B)$$

*be a homotopy relative to A from f to g. Then, there exists a based homotopy*

$$\overline{H}: X/A \times I \longrightarrow Y/B$$

*between the maps*

$$\overline{f}, \overline{g}: X/A \longrightarrow Y/B$$

*induced by f and g.*

*Proof.* The quotient space $X/A$ is given by the following pushout

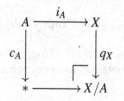

where $i_A$ is the inclusion map and $q_X \colon X \to X/A$ is a quotient map. By Corollary
(I.1.40), also the following diagram is a pushout:

$$
\begin{array}{ccc}
A \times I & \xrightarrow{\ i_A \times 1_I\ } & X \times I \\
\downarrow & & \downarrow{\scriptstyle q_X \times 1_I} \\
* \times I & \xrightarrow{\qquad} & X/A \times I
\end{array}
$$

Let $q_Y \colon Y \to Y/B$ be the quotient map and $c \colon * \times I \to Y/B$ be the constant map at
the base point of $Y/B$. Since $(q_Y H)(i_A \times 1_I) = c(c_A \times 1_I)$, by the Universal Property
of Pushouts, there exists a homotopy

$$\overline{H} \colon X/A \times I \longrightarrow Y/B$$

with the required properties.                                                                    ∎

We now choose a special pair: for every $n \geq 1$, $(I^n, \partial I^n)$ is the pair defined by the
$n$-dimensional hypercube and its boundary $\partial I^n$. Given any map

$$f \in C\mathbf{Top}((I^n, \partial I^n), (Y, y_0)),$$

we consider the following pushout diagram

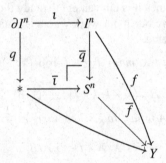

where $\iota$ is the inclusion map and $\overline{q}$ is a quotient map $q_{I^n}$ followed by the homeomorphism $I^n/\partial I^n \cong S^n$. By the preceding lemma, if

$$g \in C\mathbf{Top}((I^n, \partial I^n), (Y, y_0))$$

is homotopic to $f$ rel$\partial I^n$, then $f$ and $g$ are homotopic through a based homotopy. Let

$$[(I^n, \partial I^n), (Y, y_0)]_{\mathrm{rel}\partial I^n} := C\mathbf{Top}((I^n, \partial I^n), (Y, y_0))/\mathrm{rel}\partial I^n$$

be the set of the homotopy classes rel$\partial I^n$ of maps of pairs from $(I^n, \partial I^n)$ to $(Y, y_0)$.
The following result is easily obtained from our preceding remarks.

**(VI.3.22) Theorem.** *The function of sets*

$$\psi\colon [(I^n,\partial I^n),(Y,y_0)]_{\mathrm{rel}\,\partial I^m} \longrightarrow \pi_n(Y,y_0)$$

$$[f]_{\mathrm{rel}\,\partial I^m} \mapsto [f]$$

*is a bijection.*

We now define the operation

$$\overset{\mathrm{rel}\,\partial I^m}{\times}\colon [(I^n,\partial I^n),(Y,y_0)]_{\mathrm{rel}\,\partial I^m} \times [(I^n,\partial I^n),(Y,y_0)]_{\mathrm{rel}\,\partial I^m} \longrightarrow$$

$$\longrightarrow [(I^n,\partial I^n),(Y,y_0)]_{\mathrm{rel}\,\partial I^m}$$

as follows: let $[f]_{\mathrm{rel}\,\partial I^m}$ and $[g]_{\mathrm{rel}\,\partial I^m}$ be any two elements in $[(I^n,\partial I^n),(Y,y_0)]_{\mathrm{rel}\,\partial I^m}$; let us suppose that these two classes are represented, respectively, by the functions $f$ and $g$; we now consider the function

$$f * g\colon (I^n,\partial I^n) \to (Y,y_0)$$

such that

$$(f*g)(x_1,\ldots,x_n) = \begin{cases} f(2x_1,\ldots,x_{n-1},x_n) & 0 \le x_1 \le \frac{1}{2} \\ g(2x_1-1,\ldots,x_{n-1},x_n) & \frac{1}{2} \le x_1 \le 1. \end{cases}$$

Note that

$$(\forall (x_1,\ldots,x_n) \in \partial I^n)\ (f*g)(x_1,\ldots,x_n) = y_0.$$

Thus, by definition,

$$[f]_{\mathrm{rel}\,\partial I^m} \overset{\mathrm{rel}\,\partial I^m}{\times} [g]_{\mathrm{rel}\,\partial I^m} := [f*g]_{\mathrm{rel}\,\partial I^m}.$$

This operation is well defined, in other words, it does not depend on the element representing the class.

**(VI.3.23) Theorem.** *The set* $[(I^n,\partial I^n),(Y,y_0)]_{\mathrm{rel}\,\partial I^m}$ *with the operation* $\overset{\mathrm{rel}\,\partial I^m}{\times}$ *is a group isomorphic to* $\pi_n(Y,y_0)$.

*Proof.* By Theorem (VI.3.22), it is sufficient to prove that the function

$$\psi\colon [(I^n,\partial I^n),(Y,y_0)]_{\mathrm{rel}\,\partial I^m} \longrightarrow \pi_n(Y,y_0)$$

keeps the operations. Let

$$f,g\colon (I^n,\partial I^n) \longrightarrow (Y,y_0)$$

be two given functions; let us break them down into functions through $S^n$, that is to say,

$$f\colon I^n \xrightarrow{\ \overline{q}\ } S^n \xrightarrow{\ \overline{f}\ } Y$$

$$g\colon I^n \xrightarrow{\ \overline{q}\ } S^n \xrightarrow{\ \overline{g}\ } Y.$$

On the other hand, we define the map $\theta\colon I^n \to I^n \vee I^n$ such that, for every

$$(x_1,\ldots,x_n) \in I^n ,$$

we have

$$\theta(x_1,\ldots,x_n) = \begin{cases} (*,(2x_1,\ldots,x_n)) & 0 \le x_1 \le \tfrac{1}{2} \\ ((2x_1-1,\ldots,x_n),*) & \tfrac{1}{2} \le x_1 \le 1; \end{cases}$$

we then notice that the following diagram is commutative:

$$
\begin{array}{ccc}
I^n & \xrightarrow{\ \overline{q}\ } & S^n \equiv \Sigma S^{n-1} \\
\theta \downarrow & & \downarrow v_n \\
I^n \vee I^n & \xrightarrow[\overline{q}\vee\overline{q}]{} & S^n \vee S^n.
\end{array}
$$

By directly applying the definitions, we have

$$(\sigma(\overline{f}\vee\overline{g})v_n)\overline{q} = \sigma(\overline{f}\vee\overline{g})(\overline{q}\vee\overline{q})\theta =$$
$$= \sigma(f\vee g)\theta = f*g .\ \blacksquare$$

## Exercises

**1.** Let $X$ be any based space. Prove that the suspension $\Sigma X$ of $X$ is a space with an associative comultiplication.

**2.** Prove that for every $(X,x_0),(Y,y_0) \in \textbf{Top}_*$, the set of based homotopy classes $[\Sigma X,\Omega Y]_*$ is an Abelian group.

**3.** Let $f\colon A \to B$ and $g\colon Y \to B$ be two given maps; take the space

$$X = \{(a,y) \in A \times Y \,|\, f(a) = g(y)\}$$

with the projections $\mathrm{pr}_1\colon X \to A$ and $\mathrm{pr}_2\colon X \to B$. Prove that $f\mathrm{pr}_1 = g\mathrm{pr}_2$. Furthermore, prove that, for every topological space $Z$ and any maps $h\colon Z \to Y$ and $k\colon Z \to A$ such that $fk = gh$, there exists a unique map $\ell\colon Z \to X$ such that $\mathrm{pr}_1\ell = \mathrm{pr}_2\ell$. This is an example of *pullback*, the pushout dual in **Top**, which was defined in Sect. I.2. This situation is depicted by the following commutative diagram:

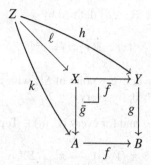

**4.** Let $f \in \mathbf{Top}_*((A, a_0), (B, b_0))$ be a given map; construct the space of based functions

$$PB = \{\lambda : I \to B \mid \lambda(0) = b_0\}$$

(space of paths beginning at $b_0$) and the map

$$g : PB \longrightarrow B, \ \lambda \mapsto \lambda(1).$$

Then, construct the pullback diagram determined by $f$ and $g$ to obtain the space

$$C_f = \{(a, \lambda) \in A \times PB \mid f(a) = g(\lambda)\}$$

with the maps $\bar{f}$ and $\bar{g}$. Prove that for every $n$ the sequence of homotopy groups

$$\pi_n(C_f, *) \xrightarrow{\ \pi_n(\bar{g})\ } \pi_n(A, a_0) \xrightarrow{\ \pi_n(f)\ } \pi_n(B, b_0)$$

is exact in $\pi_n(A, a_0)$.

**5.** A map $f : A \to B$ is a *fibration* if,

$$(\forall X \in \mathbf{Top})(\forall g \in \mathbf{Top}(X, A))(\forall H \in \mathbf{Top}(X \times I, B)) \mid H \, i_0 = fg,$$

there exists $G : X \times I \to A$ such that $G(-, 0) = g$ and $fG = H$. Prove that a projection map $f : X \times Y \to X$ is a fibration. Moreover, prove that the map $p : PB \to B$ of Exercise 4 above is a fibration.

**6.** Prove that, if the map $f : (A, a_0) \to (B, b_0)$ of Exercise 4 above is a fibration, then the space $C_f$ is of the same homotopy type as the fiber $f^{-1}(b_0)$ over $b_0$.

**7.** Prove that, if $f : (A, a_0) \to (B, b_0)$ is a fibration, there exists an (left) infinite exact sequence of homotopy groups

$$\cdots \longrightarrow \pi_n(f^{-1}(b_0), a_0) \longrightarrow \pi_n(A, a_0) \longrightarrow \pi_n(B, b_0)$$

$$\longrightarrow \pi_{n-1}(f^{-1}(b_0), a_0) \longrightarrow \pi_{n-1}(A, a_0) \longrightarrow \cdots$$

**8.** Prove that the function $p \colon \mathbb{R} \to S^1$ defined by

$$(\forall t \in \mathbb{R})\ p(t) = e^{2\pi i t}$$

is a fibration with fiber $\mathbb{Z}$ over every point of $S^1$; with this result and the previous exercise, prove that $\pi_n(S^1) \cong 0$, for every $n \geq 2$.

**9.** Prove that, for every $n \geq 1$ and for every $(Y, y_0) \in \mathbf{Top}_*$, the function

$$\Sigma_* \colon \pi_n(Y, y_0) \longrightarrow \pi_{n+1}(\Sigma Y, [y_0])$$

defined by

$$(\forall [f] \in \pi_n(Y, y_0))\ \Sigma_*([f]) = [\Sigma f]$$

is a group homomorphism.

# VI.4 Obstruction Theory

In this last section, we put together the homotopy groups and the cohomology with coefficients in a homotopy group to study the map extension problem. More precisely, let $|K|$ be a polyhedron, $|L|$ a subpolyhedron of $|K|$, and $W$ a topological space. We intend to study under what conditions a map $f \colon |L| \to W$ can be *extended* to a map $g \colon |K| \to W$, in other words, when it is possible to find $\overline{f} \colon |K| \to W$ such that the diagram

commutes. Here, $\iota$ is the inclusion map. We answer this question in the case where $W$ is $n$-simple, with $1 \leq n \leq \dim K - 1$ (see Definition (VI.3.14)). Note that the homotopy groups of $W$ do not depend on the choice of a base point; then, we forgo the base point and just write $\pi_n(W)$ for such groups.

Let us suppose that $K = (X, \Phi)$ and $L = (Y, \Psi)$; since $L$ is a subcomplex of $K$, it follows that $Y \subset X$ and $\Psi \subset \Phi$. We start by giving an orientation to $K$ (and consequently also to $L$) so that we are able to compute their homology and cohomology; let $K^n$ be the $n$-dimensional subcomplex of $K$ (in other words, the union of all simplexes whose dimension is less than or equal to $n$). Suppose that we have extended $f$ to a map $f^n \colon |K^n \cup L| \to W$. We note that, for every $(n+1)$-simplex $\sigma$ of $K$, the simplicial complex $\dot{\sigma}$ is a subcomplex of $K^n \cup L$. Let $f^n_\sigma$ be the restriction of $f^n$ to $|\dot{\sigma}|$; since $|\dot{\sigma}| \cong S^n$, we may regard $f^n_\sigma$ as a map from $S^n$ to $W$; then this map defines an element $[f^n_\sigma] \in \pi_n(W)$. By linearity, we define the homomorphism

$$c_f^{n+1}: C_{n+1}(K;\mathbb{Z}) \longrightarrow \pi_n(W)$$

that takes each $\sum_i m_i \sigma^i \in C_{n+1}(K;\mathbb{Z})$ into $\sum_i m_i[f_{\sigma^i}^n]$. We notice that if $\sigma \in \Psi$, then by Lemma (VI.3.4), $[f_\sigma^n] = 0$: in fact, $f$ is defined in $|L|$ and $f^n$ can be therefore extended to $|\overline{\sigma_{n+1}}|$. This allows us to conclude that $c_f^{n+1} \in C^{n+1}(K, L; \pi_n(W))$, that is to say, $c_f^{n+1}$ is a cochain.

The next lemma is useful for proving that $c_f^{n+1}$ is a cocycle.

**(VI.4.1) Lemma.** *Let $W$ be an $n$-simple space and $S^{n+1}$ be the sphere viewed as an $(n+1)$-manifold, which is triangulated by a simplicial complex $K = (X, \Phi)$, as in Theorem (V.1.5); let $\sigma^i$, with $i = 1, 2, \ldots, s$, be the $(n+1)$-simplexes of $K$; finally, let $K^n = (X, \Phi^n)$ be the simplicial $n$-dimensional subcomplex of $K$, where*

$$\Phi^n = \{\sigma \in \Phi \mid \dim \sigma \le n\}.$$

*Then, for any map $f: |K^n| \to W$,*

$$\sum_{i=1}^s [f_\sigma^i] = 0.$$

*Proof.* By Theorem (V.1.5), every $n$-simplex of $K$ is a face of exactly two $(n+1)$-simplexes; besides, by Definition (V.3.1), each $n$-simplex inherits opposite orientations from its two adjacent $(n+1)$-simplexes. The spaces $|\mathring{\sigma}^i|$ are homeomorphic to the sphere $S^n$ whose elements may be considered as $t \wedge x$, with $x \in S^{n-1}$; this way of viewing the elements of $S^n$ gives us an idea of the orientation of the sphere; in other words, if we take the elements in the format $t \wedge x$, we travel the sphere with a "positive" orientation but, if we take them in the format $(1-t) \wedge x$, we travel $S^n$ with the opposite orientation, that is to say, we give $S^n$ a "negative" orientation. With this in mind, we observe that the function $f$ is applied twice on each $|\mathring{\sigma}^i|$: once, viewed as the function $f(t \wedge x)$ and once, as the function $f((1-t) \wedge x)$; on the other hand, the product by $\overset{v_n}{\times}$ of the homotopy classes of these functions is the trivial class of $\pi_n(W)$ (see Theorem (VI.3.3)).

We note that the base point of each homotopy class is irrelevant because $W$ is $n$-simple. ∎

**(VI.4.2) Theorem.** *The cochain $c_f^{n+1}$ is a cocycle.*

*Proof.* It is necessary to prove that, for every $(n+2)$-simplex $\sigma_{n+2}$ of $K^n \cup L$,

$$d^{n+1}(c_f^{n+1})(\sigma_{n+2}) = c_f^{n+1}(d_{n+2}(\sigma_{n+2})) = 0.$$

This derives directly from the preceding lemma when we interpret $S^{n+1}$ as $|\mathring{\sigma}_{n+2}|$ and, writing as usual $\sigma_{n+2} = \{x_0, \ldots, x_{n+1}\}$, if we consider the orientation of its $(n+1)$-simplexes given by

$$(-1)^i \{x_0, \ldots, \widehat{x_i}, \ldots, x_{n+1}\}$$

(see the beginning of Sect. II.2.3).                                                ■

The cocycles $c_f^{n+1}$ are of particular interest, as we may realize from what follows.

**(VI.4.3) Theorem.** *An extension $f^n$: $|K^n \cup L| \to W$ of $f$: $|L| \to W$ can be extended to $|K^{n+1} \cup L|$ if and only if $c_f^{n+1}$: $C_{n+1}(K; \mathbb{Z}) \to \pi_n(W)$ is the trivial homomorphism.*

*Proof.* The map $f^n$ can be extended to $|K^{n+1} \cup L|$ if and only if $f^n$ can be extended to $|\overline{\sigma_{n+1}}|$, for every $\sigma_{n+1}$ of $K^{n+1} \cup L$ (see Lemma (VI.3.4)); therefore, $f^n$ can be extended to $|K^{n+1} \cup L|$ if and only if $c_f^{n+1} = 0$.                              ■

Somehow, $c_f^{n+1}$ indicates whether there are obstructions to the extension of $f^n$; this is why $c_f^{n+1}$ is known as *obstruction cocycle* . We now go over some examples of possible applications of Theorem (VI.4.3).

1. Let a polyhedron $|K|$ with dimension $n \geq 2$ be given and let $|L|$ be a subpolyhedron; if, for every $0 \leq i \leq n - 1$, $\pi_i(Y) \cong 0$, then every map $f$: $|L| \to Y$ can be extended to a map $\overline{f}$: $|K| \to Y$. For instance, if $Y = S^2$, $|K|$ is the torus $T^2$ with the triangulation shown in Sect. III.5.1 and $|L|$ is the geometric realization of a generating 1-cycle of the homology of $T^2$ (for example, $L$ is the simplicial complex with vertices $\{0\}$, $\{3\}$, $\{4\}$ and 1-simplexes $\{0,3\}$, $\{0,4\}$, $\{3,4\}$), then every map $f$ : $|L| \to S^2$ can be extended to a map $\overline{f}$: $T^2 \to S^2$. The construction of $\overline{f}$ is easy: we choose a point of $y_i \in S^2$ for each vertex of $K$ distinct from $\{0\}$, $\{3\}$, $\{4\}$ (for these, we have $y_0 = f(\{0\})$, $y_3 = f(\{3\})$, and $y_4 = f(\{4\})$); then, since $S^2$ is path-connected, we choose a path of $S^2$ for each 1-simplex of $K$; in this way, we extend $f$ to $f^1$: $|K^1 \cup L| \to S^2$; finally, we apply Theorem (VI.4.3) to extend $f^1$ to $|K|$.

2. **(VI.4.4) Theorem.** *Let $K$ be a simplicial complex of dimension $n \geq 2$ and $Y$ a space such that $\pi_i(Y) \cong 0$, for every $0 \leq i \leq n$. Then, any two maps $f, g$: $|K| \to Y$ are homotopic.*

*Proof.* Let $f, g$: $|K| \to Y$ be two maps given arbitrarily. The product $|K| \times I$ is an $(n+1)$-dimensional polyhedron; let $|L| = |K| \times \partial I$ and let $h$: $|L| \to Y$ be the map such that

$$h| \, |K| \times \{0\} = f \text{ and } h| \, |K| \times \{1\} = g.$$

For each vertex $\{x\} \in K$ we choose a path $h_x$: $\{x\} \times I \to Y$ (this is possible because $Y$ is path-connected) and, in doing so, we obtain an extension $h^1$ of $h$ to $|K^1 \cup L|$. By Theorem (VI.4.3), we have an extension of $h^1$ to $|K^2 \cup L|$, and so on, arriving to a homotopy

$$H: |K| \times I \to Y$$

from $f$ to $g$.                                                                    ■

3. **(VI.4.5) Theorem.** *Let $K$ be a simplicial complex of dimension $n \geq 2$. Then, $|K|$ is contractible if and only if $\pi_i(|K|) \cong 0$, for every $0 \leq i \leq n$.*

*Proof.* If $|K|$ is contractible, the statement is evident. If $|K|$ is $n$-simple and $\pi_i(|K|) \cong 0$ for every $0 \leq i \leq n$, then, by the preceding result, the identity map $1_{|K|}$ and any constant map from $|K|$ onto itself are homotopic. ∎

The condition $c_f^{n+1} = 0$ is unfortunately too strong for the more general cases, even if it works well in cases as the ones previously mentioned. The results obtained when the obstruction cocycles are cohomologous to 0 are much more interesting. Let us consider these cases. For the next definition (when necessary), we consider the homotopy group $\pi_n(W)$ with the structure given by Theorem (VI.3.23). Let $f, g \colon |K^n \cup L| \to W$ be two maps whose restrictions to $|K^{n-1} \cup L|$ coincide; in addition, let $f_i$ and $g_i$ be the restrictions of $f$ and $g$ to $|\overline{\sigma_n^i}|$, the geometric realization of the simplicial complex generated by an $n$-simplex $\sigma_n^i$ of $K^n \cup L$. We identify the space $|\overline{\sigma_n^i}|$ with the hypercube $I^n$ and interpret $f_i$ and $g_i$ as maps $I^n \to W$. Since $f_i$ and $g_i$ coincide at $|\dot{\sigma}_n^i| \equiv \partial I^n$, the restriction of the map

$$f_i * g_i^{-1}(x_1, \ldots, x_n) = \begin{cases} f_i(2x_1, \ldots, x_n) & 0 \leq x_1 \leq \tfrac{1}{2} \\ g_i((2 - 2x_1), \ldots, x_n) & \tfrac{1}{2} \leq x_1 \leq 1 \end{cases}$$

to $\partial I^n$ is homotopic to a constant map, and so

$$f_i * g_i^{-1} \colon (I^n, \partial I^n) \longrightarrow (W, w_0)$$

for a suitable $w_0$. We define

$$\delta^n(f, g)(\sigma_n^i) := [f_i * g_i^{-1}]_{\mathrm{rel}\,\partial I^n} \in \pi_n(W).$$

We point out to the reader that had $\sigma_n^i$ been in $L$, then

$$\delta^n(f, g)(\sigma_n^i) = 0.$$

We have thus defined a homomorphism

$$\delta^n(f, g) \colon C_n(K; \mathbb{Z}) \longrightarrow \pi_n(W)$$

$$\delta^n(f, g)\{\Sigma_i m_i \sigma_n^i\} := \Sigma_i m_i [f_i * g_i^{-1}]_{\mathrm{rel}\,\partial I^n};$$

$\delta^n(f, g) \in C^n(K; \pi_n(W))$ is the *difference $n$-cochain* of $f$ and $g$.

**(VI.4.6) Theorem.** *If the maps $f, g, h \colon |K^n \cup L| \to W$ coincide in $|K^{n-1} \cup L|$, then*

1. $\delta^n(g, f) = -\delta^n(f, g)$
2. $\delta^n(f, f) = 0$
3. $\delta^n(f, g) + \delta^n(g, h) = \delta^n(f, h)$
4. $d^n(\delta^n(f, g)) = c_f^{n+1} - c_g^{n+1}$

*Proof.* The first three assertions are direct consequences of the definition of difference cochain. For proving the 4th one, we take any $(n+1)$-simplex

$$\sigma = \{x_0, x_1, \ldots, x_{n+1}\}$$

of $K$ and let $f_\sigma$ (respectively, $g_\sigma$) be the restriction of $f$ (respectively, $g$) to $|\dot{\sigma}|$; we
intend to prove that

$$\{(c_f^{n+1} - c_g^{n+1}) - d^n(\delta^n(f,g))\}(\sigma_{n+1}) = 0$$

that is to say

$$[f_\sigma] - [g_\sigma] - (\Sigma_{i=0}^{n+1}(-1)^i \delta^n(f,g)(\sigma^i)) = 0$$

where $\sigma^i = \{x_0, x_1, \ldots, \widehat{x_i}, \ldots, x_{n+1}\}$. To obtain this result, we identify $S^{n+1}$ with the
boundary of $I^{n+2} \cong |\overline{\sigma}| \times I$, that is to say,

$$S^{n+1} = \partial(|\overline{\sigma}| \times I) = |\overline{\sigma}| \times \{0\} \cup |\overline{\sigma}| \times \{1\} \cup (\cup_{i=0}^{n+1} |\overline{\sigma^i}| \times I);$$

after this, we define the map of

$$F: |\dot{\sigma}| \times \{0\} \cup |\dot{\sigma}| \times \{1\} \cup (\cup_i |\dot{\sigma^i}| \times I) \to W$$

given by the union of maps

$$f: |\dot{\sigma}| \times \{0\} \to W,$$

$$g: |\dot{\sigma}| \times \{1\} \to W,$$

$$\cup_{i=0}^{n+1} f_i * g_i^{-1}: (\cup_i |\overline{\sigma^i}| \times I) \to W$$

and finally, we apply Lemma (VI.4.1) to $F$.                                                  ∎

**(VI.4.7) Remark.** If $g$ is a constant map, we conclude that

$$[f| \, |\dot{\sigma}|] = \Sigma_{i=0}^{n+1}[f| \, |\overline{\sigma^i}|].$$

This is the so-called *Homotopy Addition Theorem*; it states that the (based) homotopy class of a map $f: S^n \to Y$ is the sum of the homotopy classes of the restrictions
of $f$ to the geometric $n$-simplexes of the triangulation of the sphere. It is interesting to note that the Homotopy Addition Theorem appeared (without proof) in the
literature for the first time in [35]; its first formal proof was written by S-J. Hu [20].

Part 4 of Theorem (VI.4.6) shows that two extensions to $|K^n \cup L|$ of a
map $f: |K^{n-1} \cup L| \to W$ have cohomologous $(n+1)$-obstruction cocycles and,
therefore, these cocycles produce the same element of the cohomology group
$H^{n+1}(K, L; \pi_n(W))$ (provided that $W$ be $n$-simple). We now prove the converse of
this result.

**(VI.4.8) Theorem.** *Let $W$ be an $n$-simple space and $f^n: |K^n \cup L| \to W$ an extension
of $f$ whose obstruction $(n+1)$-cocycle $c_f^{n+1}$ is cohomologous to a cocycle $z^{n+1} \in
C^{n+1}(K, L; \pi_n(W))$. Then, there is an extension $g: |K^{n+1} \cup L| \to W$ of the restriction
of $f^n$ to $|K^{n-1} \cup L|$ such that $c_g^{n+1} = z^{n+1}$.*

*Proof.* The hypothesis $c_f^{n+1} \sim z^{n+1}$ points to the existence of an $n$-cochain $c^n \in C^n(K,L;\pi_n(W))$ such that

$$c_f^{n+1} - z^{n+1} = d^n(c^n).$$

We construct $g$ over each $n$-simplex $\sigma^i$ of $K^n \cup L$ (we may suppose that $\sigma^i$ is not in $L$, as indicated by Remark (II.4.8)) as follows. We suppose that $c^n(\sigma^i) = [h_i]_{\mathrm{rel}\,\partial I^n}$; since

$$(|\overline{\sigma^i}|, |\overset{\bullet}{\sigma^i}|) \cong (I^n, \partial I^n),$$

we may view $h_i$ as a map

$$h_i \colon (|\overline{\sigma^i}|, |\overset{\bullet}{\sigma^i}|) \to (W, w_0),$$

for a suitable $w_0$. On the other hand, the restriction of $f^n$ to $|K^{n-1} \cup L|$ is homotopic to a constant map, and we note that

$$[f^n * (h_i * f^n)^{-1}]_{\mathrm{rel}\,\partial I^n} = [h_i]_{\mathrm{rel}\,\partial I^n} = c^n(\sigma^i).$$

In this way, we have constructed a map

$$g_i = h_i * f^n \colon (|\overline{\sigma^i}|, |\overset{\bullet}{\sigma^i}|) \to (W, w_0)$$

whose homotopy class rel $\partial I^n$ coincides with $c^n(\sigma^i)$. By proceeding like this for each $n$-simplex of $K^n \cup L$, we obtain an extension $g \colon |K^{n+1} \cup L| \to W$ of the restriction of $f^n$ to $|K^{n-1} \cup L|$ such that $c_g^{n+1} = z^{n+1}$. ∎

We finally prove the most important theorem of this section.

**(VI.4.9) Theorem.** *Let an $n$-simple space $W$, a polyhedron $|K|$ with one of its sub-polyhedra $|L|$, and a map $f \colon |L| \to W$ be given; in addition, let $f^{n-1} \colon |K^{n-1} \cup L| \to W$ be an extension of $f$ and suppose that $f^{n-1}$ extends to $|K^n \cup L|$. Then, $f^{n-1}$ can be extended to $|K^{n+1} \cup L|$ if and only if $c_f^{n+1}$ is cohomologous to zero.*

*Proof.* If $f^{n-1}$ can be extended to $|K^{n+1} \cup L|$, then $f^{n-1}$ has an extension $g \colon |K^n \cup L| \to W$ that can be extended to $|K^{n+1} \cup L|$; therefore, by Theorem (VI.4.3), we have $c_g^{n+1} = 0$. However, $c_f^{n+1} \sim c_g^{n+1}$ (see Theorem (VI.4.6)) and so $c_f^{n+1} \sim 0$.

Reciprocally, if $c_f^{n+1} \sim 0$, we conclude that there exists an extension $g \colon |K^n \cup L| \to W$ of $f^{n-1}$ such that $c_g^{n+1} = 0$ (see Theorem (VI.4.8)). Hence, by Theorem (VI.4.3), it is possible to extend $g$ over the entire polyhedron $K^{n+1} \cup L|$. ∎

# References

1. J.W. Alexander – *A proof of the invariance of certain constants of analysis situs*, Trans. Am. Math. Soc. 16 (1915), 148–154.
2. J.W. Alexander – *On the chains of a complex and their duals*, Proc. Natl. Acad. Sci. USA 21 (1935), 509–511.
3. P.S. Aleksandrov – *Poincaré and Topology*, In: Proceedings of Symposia in Pure Mathematics, vol. 39, Part 2, American Mathematical Society, Providence (1983).
4. M. Barr – *Acyclic Models*, American Mathematical Society, Providence 2002.
5. E. Betti – *Sopra gli spazi di un numero qualunque di dimensioni*, Ann. Mat. Pura Appl. 4 (1871), 140–158.
6. E. Čech – *Höherdimensionale Homotopiegruppen*, In: Proceedings of the International Congress of Mathematicians. vol. 3, Zürich (1932).
7. A. Dold – *Lectures on Algebraic Topology*, Springer, New York 1972.
8. P.H. Doyle and D.A. Moran – *A Short Proof that Compact 2-Manifolds Can Be Triangulated*, Inventiones Math. 5 (1968), 160–168.
9. B. Eckmann and P. J. Hilton – *Groupes d'homotopie et dualité*, C. R. Acad. Sci. Paris Sér. A-B. 246 (1958), 2444–2447.
10. S. Eilenberg – *Cohomology and Continuous Mappings*, Ann. Math. 41 (1940), 231–251.
11. S. Eilenberg and S. MacLane – *General Theory of Natural Equivalences*, Trans. Am. Math. Soc. 58 (1945), 231–294.
12. S. Eilenberg and S. MacLane – *Acyclic Models*, Am. J. Math. 75 (1953), 189–199.
13. S. Eilenberg and N. Steenrod – *Axiomatic Approach to Homology Theory*, Proc. Natl. Acad. Sci. USA 31 (1945),117–120.
14. R. Fritsch and R.A. Piccinini – *Cellular Structures in Topology*, Cambridge University Press, Cambridge 1990.
15. K. Gruenberg – *Introduzione all'Algebra Omologica*, Pitagora Editrice, Bologna 2002.
16. P.J. Hilton and U. Stammbach – *A Course in Homological Algebra*, GTM 4, Springer, New York 1971.
17. P. Hilton and S. Wylie – *Homology Theory*, Cambridge University Press, Cambridge 1960.
18. H. Hopf – *Über die Abbildungen der dreidinsionalen Sphäre auf die Kugelfläche*, Math. Ann. 104 (1931), 637–665.
19. W. Hurewicz – *Beiträge zur Topologie und Deformationen*, Proc. Akad. Wetensch. Amsterdam 38 (1935), 112–119.
20. Sze-Tsen Hu – *The Homotopy Addition Theorem*, Ann. Math. 58 (1953), 108–122.
21. H. Künneth – *Über die Bettische Zahlen einer Produktmannigfeltigkeit*, Math. Ann. 91 (1923), 65–85.
22. S. Lefschetz – *Algebraic Topology*, Colloquium Publications, vol. XXVII, American Mathematical Society, New York 1942.
23. J.B. Listing – *Vorstudiem zur Topologie*, Gott. Stud. 1 (1847), 811–875.

24.  C.R.F. Maunder – *Algebraic Topology*, Cambridge University Press, Cambridge 1970.
25.  W.S. Massey – *Algebraic Topology: an introduction*, Harcourt, Brace & World, New York 1967.
26.  R.A. Piccinini – *Lectures on Homotopy Theory*, Elsevier, North-Holland 1992.
27.  H. Poincaré – *Analysis Situs*, J. École Polytechnique 1 (1895), 1–121.
28.  H. Poincaré – *Analyse de ses travaux scientifiques*, Summary written in 1901 by H. Poincaré at the request of Mittag-Leffler, published in Acta Math. 38 (1921), 36–135.
29.  T. Radó – *Über den Begriff der Riemannschen Fläche*, Acta Litt. Sci. Szeged 2 (1925), 101–121.
30.  H. Seifert and W. Threlfall – *Lehrbuch der Topologie*, Teubner Verlag, Stuttgart 1934.
31.  J.-P. Serre – *Groupes d'homotopie et classes de groupes abéliens*, Ann. Math. 58 (1953), 258–294.
32.  E. Spanier – *Algebraic Topology*, Springer, New York 1966.
33.  G.W. Whitehead – *Elements of Homotopy Theory*, Springer, New York 1978.
34.  H. Whitney – *On Maps of An n-Sphere into Another n-Sphere*, Duke Math. J. 3 (1937), 51–55.
35.  H. Whitney – *On Products in a Complex*, Ann. Math. 39 (1938), 397–432.

# Index